17. COLLOQUIUM DER
GESELLSCHAFT FÜR PHYSIOLOGISCHE CHEMIE
AM 21./23. APRIL 1966 IN MOSBACH/BADEN

MOLEKULARE BIOLOGIE DES MALIGNEN WACHSTUMS

BEARBEITET VON H. HOLZER
UND A. W. HOLLDORF

MIT 137 ABBILDUNGEN

SPRINGER-VERLAG BERLIN HEIDELBERG GMBH
1966

ISBN 978-3-540-03481-0   ISBN 978-3-642-87539-7 (eBook)
DOI 10.1007/978-3-642-87539-7

Alle Rechte, insbesondere das der Übersetzung in fremde Sprachen, vorbehalten

Ohne ausdrückliche Genehmigung des Verlages ist es auch nicht gestattet, dieses Buch oder Teile daraus auf photomechanischem Wege (Photokopie, Mikrokopie) oder auf andere Art zu vervielfältigen

© by Springer-Verlag Berlin Heidelberg 1966
Ursprünglich erschienen bei Springer-Verlag Berlin Heidelberg New York 1966

Library of Congress Catalog Card Number 52-3250

Die Wiedergabe von Gebrauchsnamen, Handelsnamen, Warenbezeichnungen usw. in diesem Werk berechtigt auch ohne besondere Kennzeichnung nicht zu der Annahme, daß solche Namen im Sinn der Warenzeichen- und Namenschutz-Gesetzgebung als frei zu betrachten wären und daher von jedermann benutzt werden dürfen

Titel Nr. 4345

Vorwort

Der vorliegende Band enthält die Vorträge und Diskussionen des 17. Mosbacher Colloquiums der Gesellschaft für Physiologische Chemie, das vom 21.—23. April 1966 unter dem Titel „Molekulare Biologie des malignen Wachstums" stattfand. Es war nicht die Absicht der Veranstalter, alle Aspekte dieses Problemkreises in dem Colloquium erschöpfend zu behandeln. Vielmehr wurde versucht, einige besonders aktuelle Probleme hervorzuheben und diese von verschiedenen Standpunkten aus zu behandeln. Es sind während des Colloquiums deshalb Biochemiker, Morphologen, Immunologen und Virologen zu Wort gekommen. Die große Teilnehmerzahl am Colloquium mag als Anzeichen für das breite Interesse an den behandelten Problemen gewertet werden.

Der Dank der Veranstalter gilt allen Rednern und Diskussionsteilnehmern, die zum Gelingen des Colloquiums beigetragen haben, insbesondere den Gästen aus Israel, den Vereinigten Staaten, Großbritannien und den benachbarten europäischen Ländern. Mit besonderer Freude wurde von allen Teilnehmern eine Abordnung von 15 Mitgliedern der „Arbeitsgemeinschaft Biochemie in der Gesellschaft für experimentelle Medizin der DDR" begrüßt.

Die Organisation der Veranstaltung lag weitgehend in den Händen von Herrn Dr. AUHAGEN, dem besonderer Dank gebührt. Die Tonbandaufnahmen der Diskussionen wurden in hervorragender Weise von Herrn Prof. Dr. HEINKEL u. Mitarb. (Wiesbaden) durchgeführt. Ihnen und den Herren der Stadtverwaltung und des Verkehrsamtes von Mosbach, die wesentlich zum Gelingen der Tagung beigetragen haben, sei an dieser Stelle gedankt.

Fast alle Diskussionsbeiträge haben vor der Drucklegung noch einmal den Rednern zur Durchsicht vorgelegen. Kürzungen an den Beiträgen gegenüber der gesprochenen Diskussion wurden nur von den Rednern selbst vorgenommen. Die Redaktion der Diskussion lag in den Händen von Herrn Dr. A. W. HOLLDORF (Freiburg i. Br.).

Freiburg i. Br., im Sommer 1966   HELMUT HOLZER

Inhalt

Vorwort III

Über die Ursache des Krebses (O. WARBURG, Berlin) 1

Nuclear RNA, histones and differentiation (H. BUSCH, K. HIGASHI, S. T. JACOB, T. NAKAMURA, S. M. SCHWARTZ, and S. J. SMITH, Houston, Texas) 17

The regulation of enzyme synthesis and it's role in the neoplastic process (H. C. PITOT, Madison, Wisconsin) 43

Zur Frage tumorspezifischer Histone (H. LETTRÉ, Heidelberg) 58

Allosterie-Effekte an Enzymen aus normalen und leukämischen Leukocyten (A. W. HOLLDORF, Freiburg) 59

Minderung der Euchromatinanteile in den Zellkernen von Tumorzellen (E. HARBERS, Göttingen) 61

Regulation der Thymidinkinase im Zellteilungscyclus (W. SACHSENMAIER, Heidelberg) 63

Stoffwechsel der Ribonucleinsäure in der Leber während der Applikation von N-Nitroso-morpholin (H. KRÖGER, Freiburg) 67

Untersuchungen über den Nucleinsäurestoffwechsel in Zellkernen und Mitochondrien von Morris-Hepatomen (D. NEUBERT, Berlin) 69

Diskussion (Leitung: H. HOLZER, Freiburg)

Alkylation of nucleic acids and carcinogenesis (P. N. MAGEE, Carshalton, Surrey) 79

Zur Frage der Wechselwirkung zwischen Nucleinsäuren und aromatischen Kohlenwasserstoffen und Aminen (H. DANNENBERG, München) 96

Die cocarcinogene Wirkung der Phorbolester (E. HECKER, Heidelberg) 105

Neue Reaktionen der carcinogenen Kohlenwasserstoffe 3,4-Benzpyren, 9,10-Dimethyl-1,2-benzanthracen und 20-Methylcholanthren (M. WILK, W. BEZ und J. ROCHLITZ, Frankfurt a. M.) 117

Diskussion (Leitung: F. BERGEL, London)

Mechanism of action of alkylating agents: comparisons with other cytotoxic, mutagenic and carcinogenic agents (P. D. LAWLEY, London) 126

DPNase-Induktion, RNase-Erhöhung und Hemmung der DNS-Synthese durch cytostatische Agenzien (H. HILZ, Hamburg) 142

Einfluß alkylierender Cytostatika auf biochemische Parameter und die Transplantierbarkeit von Tumorzellen (E. LISS, Berlin) 145

Untersuchungen über die Wirkung von Trenimon (2,3,5-Trisäthyleniminobenzochinon-1,4) auf die DNS von Ehrlich-Ascites-Tumorzellen (H. GRUNICKE, Freiburg) 147

Some new points of view concerning the mode of action of biological alkylating agents (I. P. HORVÁTH, and L. INSTITÓRIS, Budapest) 149

Diskussion (Leitung: TH. BÜCHER, München)

Fluorinated pyrimidines, biochemically and clinically useful antimetabolites (CH. HEIDELBERGER, Madison, Wisconsin) 156

Thymidylat-Synthese in Leukämiezellen unter der Einwirkung von Folsäure-Antagonisten (W. WILMANNS, Tübingen) 177

Hemmstoffe des Abbaus von Fluordesoxyuridin und Joddesoxyuridin (P. LANGEN, Berlin-Buch) 179

Diskussion (Leitung: TH. BÜCHER, München)

Onkogene Viren (A. GRAFFI, Berlin-Buch) 184

An analysis of the mechanism of carcinogenesis by Polyoma Virus, hydrocarbons, and X-irradiation (L. SACHS, Rehovoth) 242

Struktur der Nucleinsäuren onkogener Viren (H. BIELKA, Berlin-Buch) 256

Untersuchungen über die cytocide und onkogene Wirkung des Simian-Virus (SV-)40 (R. HAAS, Freiburg) 267

Untersuchungen am einstufigen Vermehrungscyclus von SV-40 mit Hilfe von Aktinomycin D (K. MUNK, H. FISCHER und W. BRÜMMER, Heidelberg) 280

Diskussion (Leitung: W. SCHÄFER, Tübingen)

Zusammenfassung und Ausblick (F. BERGEL, London) 285

Über die Ursache des Krebses

Von O. WARBURG

Max-Planck-Institut für Zellphysiologie, Berlin

Mit 9 Abbildungen

Wenn man von der Ursache einer Krankheit spricht, muß man vorerst sagen, was man damit meint. Es gibt entfernte und letzte Ursachen von Krankheiten. Zum Beispiel ist die letzte Ursache der Pest der Pestbacillus, aber entfernte Ursachen der Pest sind der Schmutz und die Ratten und die Flöhe, die den Pestbacillus von den Ratten auf den Menschen übertragen.

Krebs zeichnet sich vor allen anderen Krankheiten dadurch aus, daß es unzählig viele Krebsursachen gibt. Fast alles kann Krebs erzeugen, wofür ich im folgenden eine thermodynamische Erklärung geben werde; aber auch beim Krebs gibt es nur eine einzige letzte Ursache, auf die alle entfernten Ursachen hinauslaufen.

In wenigen Worten zusammengefaßt ist die letzte Ursache des Krebses die Umwandlung der obligat aeroben normalen Körperzellen in die fakultativ anaeroben gärenden Krebszellen, deren bestes Analogon die gärenden Kulturhefen sind. Vom Standpunkt der Physik und der Chemie des Lebens ist dieser Unterschied zwischen normalen Zellen und Krebszellen so groß, daß man ihn sich nicht größer vorstellen kann. Der Sauerstoff, der Spender der Energie alles höheren Lebens, ist in den Krebszellen entthront und ersetzt durch die energieliefernde Reaktion der niedersten Lebewesen.

Obwohl die ersten Anzeichen der Anaerobiose der Krebszellen schon 1923 im Kaiser-Wilhelm-Institut in Dahlem entdeckt worden waren, hat es über 40 Jahre gedauert, bis die Anaerobiose der Krebszellen bewiesen worden ist. Inzwischen hat man unablässig versucht, andere letzte Ursachen des Krebses zu finden, doch ist diesen Bemühungen kein Erfolg beschieden gewesen. Heute ist das Suchen nach anderen letzten Krebsursachen so sinnlos, wie nach der Entdeckung des Pestbacillus, das Suchen nach einer anderen letzten Ursache der Pest wäre.

In diesem Vortrag werde ich einige entscheidende Experimente der letzten Jahre beschreiben, die zum Teil von DEAN BURK, aus dem National Cancer Institute in Bethesda, stammen, zum Teil von meinen Mitarbeitern GAWEHN, GEISSLER und LORENZ ausgeführt worden sind[1—5]. DEAN BURK ist Abteilungsleiter in Bethesda.

1. Über den Sauerstoffdruck beim Wachstum der Krebszellen

Man kann heute die Asciteskrebszellen der Maus leicht in vitro züchten und dabei den niedersten Sauerstoffdruck bestimmen, bei dem die Krebszellen noch wachsen können. Wir fanden[6], daß der niedrigste Sauerstoffdruck bei etwa 20 mm Wasser liegt (Abb. 1).

$$5\% \ CO_2\text{-Luft}$$

Q_{O_2}	$Q_M^{O_2}$	$Q_M^{N_2}$
—7	+35	+70

$$5\% \ CO_2\text{-}0{,}2\% \ O_2\text{-Ar}$$

Q_{O_2}	$Q_M^{O_2}$	$Q_M^{N_2}$
—0,2	+70	+70

Abb. 1. Stoffwechselquotienten der Asciteskrebszellen der Maus bei Sauerstoffdrucken von 20 oder 2000 mm Wasser

In Bild 1 sind die Stoffwechselquotienten verzeichnet, die bei diesem niedrigen Druck von 20 mm sowie bei einem 100mal größeren Sauerstoffdruck, in Luft, gefunden worden sind.

Man kann aus diesen Zahlen berechnen, daß bei dem niedrigen Sauerstoffdruck von 20 mm Wasser, bei dem die Asciteskrebszellen wachsen, 99% der Energie aus der Gärung und nur 1% aus der Atmung stammt. (Die Summe von Atmungs- und Gärungsenergie ist bei dem niedrigsten Sauerstoffdruck und in Luft ungefähr gleich.) Deshalb nennen wir das Wachstum anaerobes Wachstum, obwohl wir wissen, daß die Krebszellen, wie die meisten fakultativen Anaerobier, *ganz* ohne Sauerstoff nicht wachsen können.

Daß die Asciteskrebszellen auch in vivo bei sehr niedrigen Sauerstoffdrucken wachsen, hat Dr. KAYSER in meinem Institut gezeigt, indem er in den Ascites von Krebsmäusen Sporen von Chlostridium butyricum injizierte, die nur bei Sauerstoffdrucken unter 50 mm Wasser auskeimen. Diese Sporen keimten in dem Asciteskrebs aus und bewiesen damit den niedrigen Sauerstoffdruck beim Wachstum in vivo.

2. Beweis, daß das anaerobe Wachstum spezifisch für die Tumoren ist

Injiziert man Sporen von Tetanusbacillen, die zu den Chlostridien gehören, in das Blut gesunder Mäuse, so erkranken sie nicht an Tetanus, weil die Sporen nirgends im Körper, auch nicht in den Embryonen, Orte finden, an denen der Sauerstoffdruck kleiner ist als 50 mm Wasser. Injiziert man aber die Tetanussporen in das Blut tumortragender Mäuse, so erkranken die Mäuse an Tetanus, weil der Sauerstoffdruck in den Tumoren so niedrig ist, daß die Sporen auskeimen können, also niedriger als 50 mm Wasser ist. Diese schöne Demonstration unserer Ergebnisse stammt von den Amerikanern MALMGREN und FLANEGAN [7] und beweist, daß das anaerobe Leben und Wachstum im Körper streng spezifisch für die Tumoren ist.

3. Beweis, daß das normale Wachstum ein rein aerober Vorgang ist [8]

Auf eine andere Weise hat DEAN BURK gezeigt, daß das normale Wachstum im Körper ein rein aerober Vorgang ist. Amputiert man Ratten einen Teil ihrer Lebern, so wächst die Leber schneller nach, als die meisten Tumoren wachsen. BURK bestimmte den Stoffwechsel dieses schnellen normalen Wachstums und fand einen rein aeroben Stoffwechsel. Es war damit auf sehr direkte Weise die Behauptung widerlegt, daß alles Wachstum, auch das normale Wachstum, ein anaerober Vorgang sei. Tatsächlich ist, nach den Erfahrungen von 100 Jahren, nichts empfindlicher gegen Erniedrigung des Sauerstoffdrucks, als das normale Wachstum im Körper.

4. Zusammenhang von Gärung und Malignität

Eine in-vitro-Kultur von Krebszellen, die EARLE und SANFORD aus einer einzigen normalen Körperzelle gezüchtet hatten, spaltete sich in zwei Linien sehr verschiedener Malignität. DEAN BURK [9] hat Atmung und Gärung der beiden Linien gemessen und fand, daß die Gärung der Linie geringer Malignität klein ist und daß die Gärung der Linie großer Malignität groß ist — ein schöner Beweis des Zusammenhangs von Gärung und Malignität.

5. Stoffwechsel der Morris-Tumoren

H. P. MORRIS erzeugte bei Ratten mit verschiedenen Carcinogenen Lebertumoren sehr verschiedener Malignität, die transplantierbar waren und ihre verschiedene Malignität nach der Transplantation beibehielten. DEAN BURK[10] hat den Stoffwechsel in Serien dieser Tumoren gemessen und gefunden, daß die Malignität um so größer ist, je größer die Gärung ist (Abb. 2). In der Figur sind die Gärungsgrößen Q gegen die Wachstumsgeschwindigkeiten α aufgetragen. $\alpha = 1$ bedeutet, daß sich das Tumorgewicht in 0,7 Monaten verdoppelt. $\alpha = 10$ bedeutet, daß sich das Tumorgewicht in 0,07 Monaten verdoppelt. Wie man sieht, nimmt die Gärungsgröße so erheblich mit der Wachstumsgeschwindigkeit zu, daß an einem Zusammenhang kein Zweifel bestehen kann. Auch diejenigen Tumoren, die keine Morris-Tumoren sind, passen in die Kurve hinein, z. B. der Ascetiskrebs der Maus, dessen $\alpha = 21$ und dessen Gärung = 70 ist. So scheint es, daß die Kurve von BURK allgemein den Zusammenhang zwischen Wachstumsgeschwindigkeit und Gärungsgröße wiedergibt.

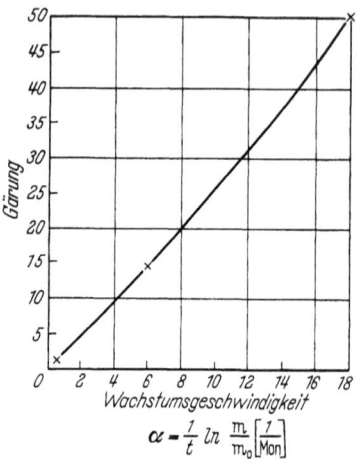

Abb. 2. Gärung als Funktion der Wachstumsgeschwindigkeit (Kurve nach Versuchen von DEAN BURK und MARK WOODS)

Zur Frage, wie sich BURKS und WOODS Kurve in der Nähe des Nullpunkts verhält, war in den USA ein Streit ausgebrochen. Nach den Experimenten von DEAN BURK und MARK WOODS [10, 12] ist in der Nähe des Nullpunkts die Tumorgärung in vitro immer größer, als die Lebergärung desselben Versuchstieres in vitro; während seine Gegner[11] behaupteten, daß zwischen den langsamsten Morristumoren und den normalen Lebern in vitro *kein* Unterschied der Gärung bestehe; was bedeuten würde, daß in den langsamsten Morristumoren Tumoren gefunden worden seien, die nicht gären; daß also die Anaerobiose *nicht* die letzte Ursache des Krebses sein könne[13].

PIETRO GULLINO[12, 13] hat diese Streitfrage in Bethesda mit einer neuen Methode in vivo geprüft. GULLINO transplantierte Morris-Tumoren in die Ovarien von Ratten und erhielt Tumoren (Abb. 3), in die durch einen Stiel nur *eine* Arterie eintrat und nur *eine* Vene austrat, so daß durch Punktionen in vivo der Stoffwechsel wochenlang gemessen werden konnte. Er bestätigte mit dieser Anordnung alles, was wir früher für Rattentumoren in vivo gefunden hatten:

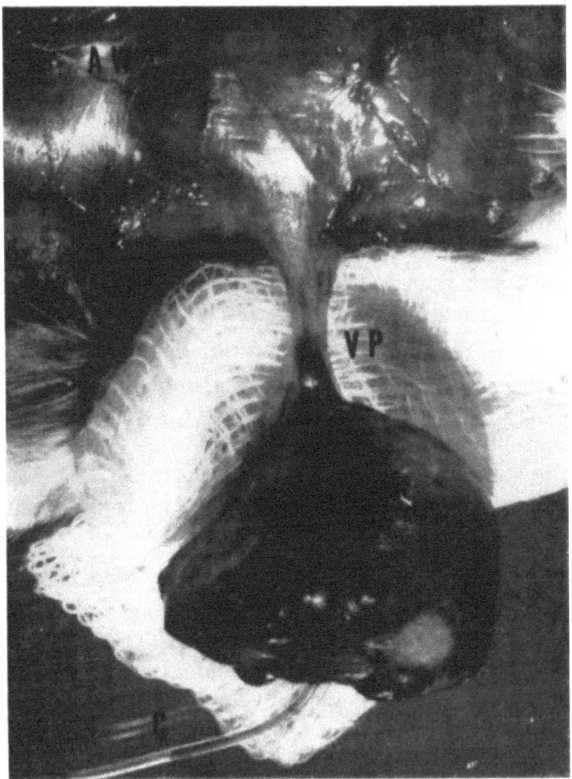

Abb. 3. In das Rattenovar transplantierter Morris-Tumor

daß der Zucker im Blutstrom durch die Tumoren abnimmt, daß das Bikarbonat dabei abnimmt, und die Milchsäure in entsprechender Weise zunimmt. Auch der langsamste Morristumor, nahe am Nullpunkt von BURKS und WOODS Kurve, machte keine Ausnahme. Bedenkt man, daß die Leber in vivo zwecks Glykogenbildung immer Milchsäure aus dem Blut herausnimmt, wie schon CLAUDE BERNARD vor 100 Jahren und GULLINO neuerdings noch überzeugender gezeigt hat, so ist also der Unterschied zwischen den langsamsten Morristumoren bezüglich der Milchsäure wie $+$ gegen $-$, also sogar unendlich groß.

Ich möchte nunmehr, ehe ich mit der Beschreibung der Versuche fortfahre, einige thermodynamische und chemische Betrachtungen einschalten.

6. Zur Thermodynamik

Wie die Gesteinskunde lehrt, gab es schon Lebewesen auf der Erde, als die Erdatmosphäre noch keinen freien Sauerstoff enthielt. Diese ohne Sauerstoff lebenden Organismen waren gärende, wenig differenzierte Einzeller. Erst als — vor etwa 800 Millionen Jahren — der freie Sauerstoff in der Erdatmosphäre auftrat[14], setzte fast plötzlich die Höherentwicklung des Lebens ein und es entstanden die Königreiche der Pflanzen und Tiere. Die „Evolution Creatrice" der Philosophen des Lebens ist also das Werk des Sauerstoffs oder genauer ausgedrückt, das Werk der Sauerstoffatmung.

Der umgekehrte Vorgang, die Entdifferenzierung der hochentwickelten Körperzellen erfolgt heute vor unseren Augen in größtem Ausmaß bei der Krebsentstehung. Bei der Krebsentstehung sinkt die Sauerstoffatmung, es erscheint die Gärung und es entstehen wieder die wenig differenzierten Anaerobier, die keine Funktion mehr im Körper haben, und die sich deshalb im Körper ungehemmt und unheilbringend vermehren. —

Die Wirkung der Sauerstoffatmung auf die Differenzierung zeigt, daß die Differenzierung der Zufuhr von Arbeit bedarf oder, was dasselbe ist, daß die Entdifferenzierung ein von selbst verlaufender Vorgang ist. In der Tat, wie die Erzeugung von Temperaturdifferenzen in einem gleichtemperierten Gas, ist die Differenzierung ein unwahrscheinlicher Vorgang, vielleicht der unwahrscheinlichste Vorgang der Welt; während die Entdifferenzierung ein wahrscheinlicher Vorgang ist, wie der Ausgleich von Temperaturdifferenzen in einem ungleich temperierten Gas. Die Boltzmannsche Theorie der Thermodynamik, auf das Krebsproblem angewendet, erklärt so aus dem Wahrscheinlichkeitsprinzip, warum alles, sogar die Zeit selbst, Krebs erzeugt. Eine Kugel auf einer schiefen Ebene kann durch jede Erschütterung, gleichgültig wie sie erzeugt wird, ins Rollen gebracht werden.

Die Thermodynamik erklärt also eines der größten Rätsel des Krebses, aber sie erklärt nicht, warum nur die Atmung, aber nicht die Gärung Differenzierung schaffen und erhalten kann. Würde es diese Minderwertigkeit der Gärungsenergie nicht geben, so würde es keinen Krebs geben. Denn dann würde, beim Ersatz der Atmung durch die Gärung, die Gärung der Kreator der Differenzierung werden. Offenbar ist diese Minderwertigkeit der Gärungsenergie ein Problem, das nicht die Thermodynamik, sondern nur die Chemie lösen kann. Der Biochemiker denkt dabei an den Unterschied zwischen der Atmungs- und der Gärungs-Phophorylierung.

7. Zur Chemie*

Was kann die Chemie allgemein zur Erklärung der Krebsentstehung beitragen?

In Abb. 4 sind die beiden Wege des Kohlenhydratabbaus, der Weg der Atmung und der Gärung, eingetragen. Wie man sieht, sind die

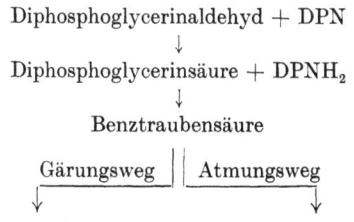

Abb. 4. Die Wege von Atmung und Gärung

Wege der Atmung und Gärung bis zur Brenztraubensäure gemeinsam. Auf dem gemeinsamen Weg wird 1 Molekül Nicotinsäureamid zur Oxydation des Diphosphoglycerinaldehyds verbraucht.

Von der Brenztraubensäure an trennen sich die beiden Wege. Die Gärung erfordert bis zu ihrem Endprodukt nur noch eine einzige Reaktion, nämlich die Hydrierung der Brenztraubensäure zu Milchsäure durch Dihydro-Nicotinsäureamid; während die Verbrennung der Brenztraubensäure zu Kohlensäure und Wasser nach Sir HANS KREBS noch rund 30 weitere Reaktionen erfordert: 5 weitere Moleküle Nicotinsäureamid müssen noch dehydriert werden, 5 weitere Moleküle Dihydro-Nicotinsäureamid müssen über je 5 Glieder der Atmungskette durch Sauerstoff reoxydiert werden.

---

* O. WARBURG: Wasserstoffübertragende Fermente. Berlin 1948

Rechnet man noch 3 Decarboxylierungen hinzu sowie je eine Anlagerung und Abspaltung von Wasser, so hat man mindestens 30 Reaktionen auf dem Atmungsweg gegenüber einer einzigen Reaktion auf dem Gärungsweg. Der Zweck dieser Aufzählung war es, klar zu machen, daß es ungeheuer viel wahrscheinlicher ist, daß im Leben die Atmung einen Schaden erleidet, als daß die Fähigkeit zur Gärung geschädigt wird; daß also Chemie und Thermodynamik darin übereinstimmen, daß die Entdifferenzierung der wahrscheinliche Vorgang ist.

Es ist aber keineswegs so, daß immer, wenn die Atmung sinkt, Krebszellen entstehen. Die Regel ist vielmehr, daß bei einem Ausfall von Atmung in allen Körperzellen Milchsäurebildung auftritt, mit der sie dann aus Energiemangel zugrunde gehen, wie die kernlosen roten Blutzellen im Kreislauf oder wie die herausgeschnittene Netzhaut in vitro. Krebszellen entstehen in der Folge von Atmungssenkungen nur dann, wenn zwei weitere Bedingungen erfüllt sind: erstens daß die anaerobe Gärung so erheblich ansteigt, daß sie den Ausfall von Atmung energetisch kompensiert; und zweitens, daß diese angestiegene Gärung zur energieliefernden Reaktion des Wachstums wird. Man muß also unterscheiden zwischen *Glykolyse*, die Absterben durch Milchsäurebildung bedeutet und *Anaerobiose*, die Wachstum durch Milchsäurebildung bedeutet.

8. Verwandlung von Körperzellen in vitro in Anaerobier [1—5]

Wenn ich nunmehr zur Beschreibung von Experimenten zurückkehre, so werde ich zeigen, daß es eine Methode gibt, um in-vitro-Körperzellen in 48 Stunden quantitativ in unbegrenzt wachsende fakultative Anaerobier zu verwandeln.

Unser Versuchsmaterial waren mit Trypsin zerteilte embryonale Zellen des Huhns oder der Maus, von denen man leicht soviel gewinnen kann, daß man wie mit Hefezellen in vitro biochemisch experimentieren kann. Bringt man solche Zellen in geeigneten Medien in Petrischalen und stellt sie in Exsiccatoren, die mit geeigneten Gasmischungen gefüllt sind (Abb. 5), so wachsen die embryonalen Zellen am Boden der Petrischalen an. Sie verdoppeln dabei bei 38° in 48 Std ihr Gewicht, wobei sie sich einmal teilen. Vorbedingung ist, daß die Schalen nicht bewegt werden. Werden sie bewegt, so wachsen die Zellen nicht. So hat man die Möglichkeit, indem man die Schalen bewegt oder nicht bewegt, unter sonst gleichen äußeren Bedingungen immer wachsende Zellen und nicht wachsende Zellen zur Verfügung zu haben.

Nur wenn die Zellen wachsen, ereignet sich etwas, was wir seit vielen Jahren gesucht haben. Der bei der Einsaat rein aerobe Stoffwechsel schlägt in 48 Std quantitativ um in Krebsstoffwechsel, mit dem nunmehr die Zellen in vitro in Subkulturen unbegrenzt weiterwachsen.

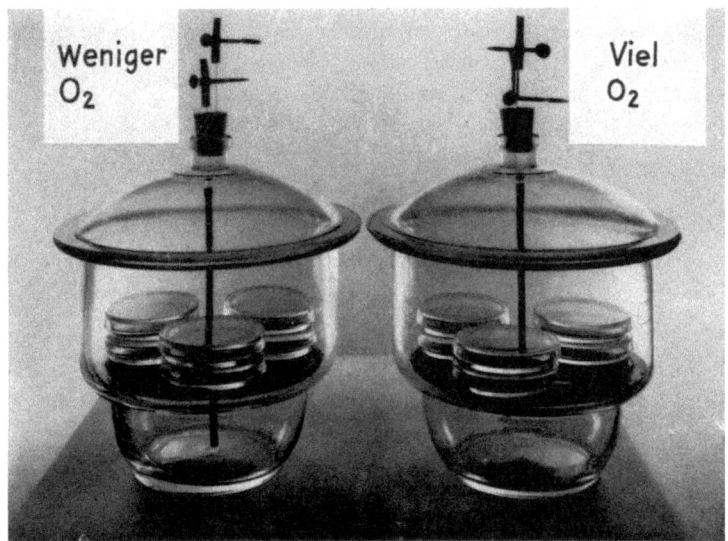

Abb. 5. Wachstum der embryonalen Zellen in Petrischalen bei verschiedenen Sauerstoffdrucken

Was ist die Ursache dieses Umschlags des embryonalen Stoffwechsels in den Krebsstoffwechsel? Die nähere Untersuchung hat gezeigt, daß der Umschlag nur dann erfolgt, wenn der Sauerstoffdruck in der Wachstumsschicht am Boden der Kulturschalen so niedrig ist, daß die Atmung gehemmt ist. Die Ursache des Stoffwechselumschlags ist also die Hemmung der Atmung, durch den verminderten Sauerstoffdruck, während des Wachstums.

Die Erzeugung und die Messung des kritischen Sauerstoffdrucks beim Umschlag bedarf einer Erläuterung, da man, wie erwähnt, während des Wachstums die Schalen nicht bewegen darf und also der Sauerstoff nur durch Diffusion in die Wachstumsschicht am Boden der Kulturschalen gelangen kann. Die Hydrodiffusion aber ist ein sehr langsamer Vorgang. Stand zum Beispiel Luft über den ruhenden Schalen, so war der Sauerstoffdruck in der Wachstumsschicht am Boden der Schalen etwa 20 mal kleiner und betrug nur

	Q_{O_2}	$Q_M^{O_2}$	$Q_M^{N_2}$
vor dem Umschlag	−11	0	+13
nach dem Umschlag ...	− 4	+29	+45

Abb. 6. Stoffwechselumschlag der embryonalen Zellen durch Wachsen bei vermindertem Sauerstoffdruck

etwa 100 mm Wasser, der mit der Sauerstoffmikroelektrode der Firma Beckman gemessen wurde. Dies ist ein Druck, der so niedrig ist, daß die Atmung dabei schon merklich gehemmt ist. Stand andererseits über den ruhenden Schaalen statt Luft Sauerstoff vom

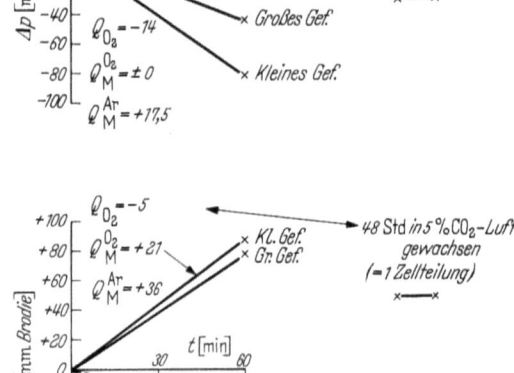

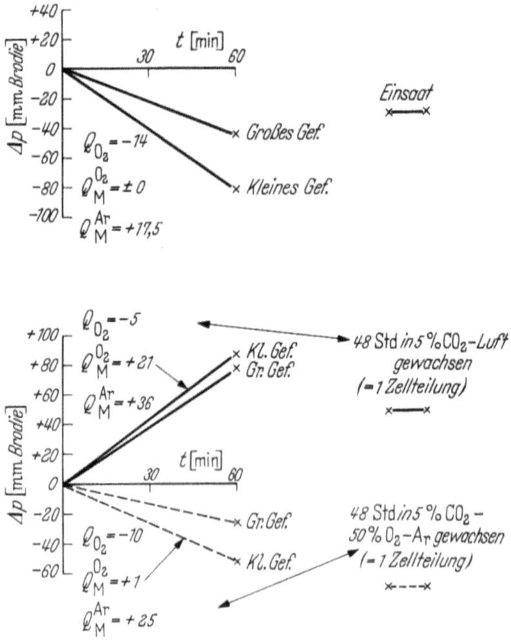

Abb. 7. Manometrische Messung des Stoffwechsels embryonaler Hühnerzellen vor und nach dem Umschlag des Stoffwechsels

Druck einer halben Atmosphäre, so war der Sauerstoffdruck in der Wachstumsschicht am Boden der Schalen so groß, daß die Atmung nicht gehemmt wurde. Dann schlug auch der Stoffwechsel nicht um.

Abb. 6 zeigt die Stoffwechselquotienten vor dem Umschlag und nach dem Umschlag bei einem Sauerstoffdruck von etwa 100 mm Wasser in den Wachstumsschichten.

Damit Sie beurteilen können, aus wie großen manometrischen Ausschlägen diese Quotienten berechnet sind, zeige ich zwei graphische Darstellungen der Messungen (Abb. 7). Die Bilder sind so zu verstehen, daß Proben der Zellen während des Wachstums aus

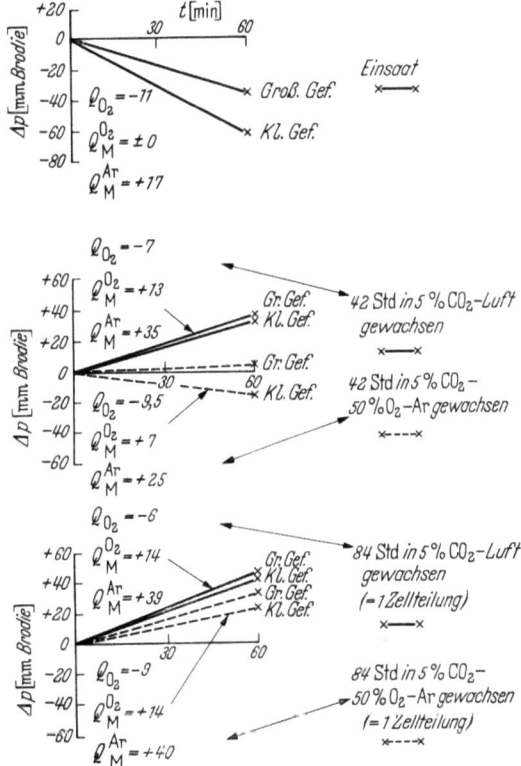

Abb. 8. Manometrische Messung des Stoffwechsels embryonaler Mäusezellen vor und nach dem Umschlag des Stoffwechsels

den Kulturschalen herausgenommen würden und daß der Stoffwechsel unter Schütteln der Manometriegräße, also ohne Wachstum, mit der üblichen Zweigefäßmethode gemessen wurde. Sie sehen, daß die Einsaat, weil der embryonale Stoffwechsel rein aerob ist, negative Drucke gibt und daß die umgeschlagenen Zellen, wegen der Gärung, positive Drucke geben; und Sie sehen ferner, daß die gemessenen Drucke von der Größenordnung 50 bis 100 mm Wasser pro Std waren. Dies ist ein Versuch mit Hühnerzellen, das nächste Bild zeigt einen Versuch mit Mäusezellen.

Wir haben schließlich eine Methode ausgearbeitet, um die Atmungshemmung während des Stoffwechselumschlags manometrisch zu messen. Wir züchteten zu dem Zweck in Manometriegefäßen (Abb. 9), die während des Wachstums 24 oder 48 Std ruhig standen

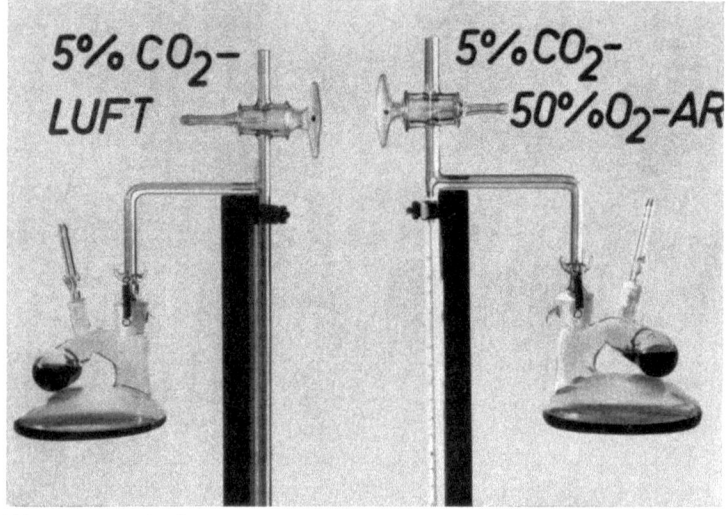

Abb. 9. Manometriegefäße zur Messung des Stoffwechsels während des Stoffwechselumschlags

und nur vor den Ablesungen je 10 min geschüttelt wurden. Wenn Luft über den ruhenden Schalen stand und der Stoffwechsel in Krebsstoffwechsel umschlug, fanden wir eine Atmungshemmung von 35%. Bei den höheren Sauerstoffdrucken, bei denen die Atmung nicht umschlug, fanden wir keine Atmungshemmung. Dieses Ergebnis ist auch medizinisch wichtig. Denn Sauerstoffdrucke, die die Atmung zu 35% hemmen, können im Körper am venösen Ende der Blutcapillaren durchaus vorkommen, so daß nunmehr Carcinogenese im Körper durch verminderten Sauerstoffdruck durchaus möglich erscheint.

Jedenfalls wird eine bisher unverständliche Carcinogenese durch unsere Versuche nunmehr erklärt. Impft man Ratten dünne Schnitte von festen Körpern unter die Haut, so werden sie bald mit einer Gewebekapsel eingehüllt, die durch einen beweglichen Stiel mit Blutgefäßen versorgt wird. Nicht selten, sondern sehr häufig entstehen in diesen Kapseln Sarkome, wobei es gleichgültig ist, ob die eingeimpften Scheiben aus Kunststoffen oder aus Gold oder aus Elfenbein bestehen. Was hier Krebs erzeugt, ist nicht die chemische Natur der Scheiben, sondern die bei der Bewegung der Ratten wechselnde und unzureichende Versorgung mit Sauerstoff durch den Stiel der Kapseln.

Im übrigen ist durch unsere Versuche mehr als wahrscheinlich geworden, daß die Ursache der Carcinogenese bei der in-vitro-Kultur der Körperzellen oft, wenn nicht immer, unzureichende Sauerstoffzufuhr gewesen ist — unbewußt, weil man sich der Langsamkeit der Hydrodiffusion nicht bewußt war.

9. Ist der Stoffwechselumschlag reversibel?

Wir haben weiterhin untersucht, ob der Umschlag des embryonalen Stoffwechsels in den Krebsstoffwechsel reversibel ist. Züchtet man embryonale Hühnerzellen, deren Stoffwechsel bei niedrigem Sauerstoffdruck in Krebsstoffwechsel umgeschlagen ist, in Subkulturen bei höheren Sauerstoffdrucken weiter, so verschwindet der Krebsstoffwechsel und es erscheint wieder der rein aerobe Stoffwechsel der aeroben Zellen. Der Stoffwechselumschlag in den Hühnerzellen ist sofort beim Wechsel des Sauerstoffdrucks vollständig reversibel.

Da aber die Carcinogenese ein irreversibler Vorgang ist, haben wir nach irreversiblen Stoffwechselumschlägen gesucht. Wir fanden, daß der Umschlag zum Krebsstoffwechsel, der in embryonalen *Mäusezellen* durch niedrige Sauerstoffdrucke bewirkt wird, vollständig irreversibel ist[15]. Wurde der Umschlag zunächst bei niederen Sauerstoffdrucken erzeugt und wurde dann in Subkulturen bei höheren Sauerstoffdrucken weitergezüchtet, so blieb der Krebsstoffwechsel quantitativ erhalten. Dies war ein wichtiger weiterer Schritt in Richtung der Carcinogenese in vitro.

10. Wachstum in vivo

Obwohl die Mäusezellen mit ihrer irreversiblen Gärung in vitro unbeschränkt und bei beliebigen Sauerstoffdrucken weiterwachsen, wachsen sie *nicht* weiter in vivo, wenn man sie auf Mäuse transplantiert, auch nicht, wenn man sie auf den gleichen Inzuchtstamm transplantiert, von dem die gärenden Mäusezellen stammen. Zwar weiß man, daß alle in vitro gezüchteten Körperzellen nach Monaten oder Jahren zu transplantierbaren Krebszellen werden, aber man weiß nicht, was in dieser Latenzzeit in vitro vor sich geht.

Wir haben deshalb untersucht, ob man in vitro einen chemischen Unterschied findet zwischen den transplantierbaren und den nicht transplantierbaren gärenden Mäusezellen. Wir fanden, daß die transplantierbaren Asciteskrebszellen in vitro bei einem Sauerstoffdruck von 20 mm Wasser noch wachsen können, daß aber für die nicht transplantierbaren gärenden Mäusezellen ein Sauerstoffdruck von mindestens 100 mm Wasser zum Wachstum notwendig ist. Man kann daraufhin sagen, daß trotz völlig gleicher Stoffwechselquotienten die Fähigkeit zur Anaerobiose in den transplantierbaren gärenden Zellen rund fünfmal größer ist, als in den nicht transplantierbaren Zellen. Wahrscheinlich wird sich zeigen, daß auch die Transplantierbarkeit mit derjenigen Substanz zusammenhängt, die bisher der Schlüssel zu allen Problemen des Krebses gewesen ist. Ich meine den Sauerstoff.

11. Anwendungen

Was nutzen uns diese Ergebnisse?

In Nordskandinavien kommt ein Krebs des Rachens und der Speiseröhre vor, dessen Vorbote das sog. Plummer-Vinsonsche Syndrom ist, das man mit Hilfe von Röntgenstrahlen leicht diagnostizieren kann. Diese Präcancerose kann geheilt werden, wenn man der Nahrung die Wirkungsgruppen der Atmungsfermente zusetzt: Nicotinsäureamid, Flavin, Thiamin, Panthotensäure und Eisensalze. Da man die Vorboten heilen kann, kann man den Krebs verhüten. Nach ERNEST WYNDER[16] vom Sloane-Kettering-Institut in New York ist man zur Zeit dabei, in Skandinavien diesen Krebs des Rachens und der Speiseröhre auszurotten.

Ich möchte daraufhin raten, größere Mengen der Wirkungsgruppen der Atmungsfermente allgemein unserer Nahrung zuzusetzen. Da es keine Überdosierung der Wirkungsgruppen gibt, riskiert man mit den Zusätzen nichts, während man die Chance hat, daß auch andere Krebsarten verhütet werden, da die Atmung *aller* Arten von Krebszellen gestört ist.

Ich würde noch weiter gehen und versuchen, nach Operationen Metastasen durch große Dosen der Wirkungsgruppen der Atmungsfermente zu hemmen. Eine Redifferenzierung der Metastasen infolge eines Wiederanstiegs des Anstiegs der Atmung darf man dabei nicht erwarten, da in der kurzen Lebensdauer eines Menschen die Wahrscheinlichkeit einer Zurückdifferenzierung gleich

Null ist. Aber man kann erwarten, daß die Atmung steigt und dadurch die Gärung sinkt und daß dann die entdifferenzierten Zellen so ungefährlich werden, wie etwa die langsamsten Morris-Tumoren, die die Lebenszeit einer Ratte benötigen, um heranzuwachsen; oder wie die Krebszellen in der Prostata der alten Männer, die man „schlafende Krebszellen" genannt hat und die das Wohlbefinden der Träger nicht stören. Man wird zu derartigen Vorschlägen ermutigt[17] durch die große und längst nicht genügend gewürdigte Entdeckung, daß Nicotinsäureamid, die allmächtige Wirkungsgruppe der Atmungsfermente, die Tuberkulose heilen kann.

Ein anderes Beispiel soll von einem Versuch berichten, die Gärung der Krebszellen für die Therapie auszunutzen. Der Physiker MANFRED V. ARDENNE hat den Entschluß gefaßt, für den wir alle dankbar sind, seine großen technischen Fähigkeiten für die Therapie des Krebses einzusetzen. v. ARDENNE geht davon aus, daß die gärenden Krebszellen saurer sind, als die nicht gärenden normalen Zellen; und daß sie deshalb gegen Überhitzung empfindlicher sind, als normale Zellen. Auf der Basis dieser von ihm genau untersuchten Tatsachen erhitzt er Krebspatienten, nach der chirurgischen Entfernung des primären Tumors, auf 43 °C in der Hoffnung, daß die Metastasen abgeschwächt oder getötet werden. Es ist noch nicht entschieden, ob dieser Idee ein praktischer Erfolg beschieden sein wird. Aber schon die bisher vorliegenden Arbeiten von v. ARDENNE über die Gärung der Krebszellen im Körper sind meines Erachtens von größter Bedeutung auf einem Gebiet, auf dem bisher alle Hoffnungen der Chemo-Therapie fehlgeschlagen sind.

Schließlich noch einige Worte über ein drittes Anwendungsgebiet. Jeder weiß, daß sich der Krebs im Zeitalter der Technik zu einer Seuche entwickelt hat. Von den 25 Millionen Männern, die heute in der Bundesrepublik leben, werden nach einer Schätzung von K. H. BAUER[18] mindestens eine Million an Krebs der Luftwege sterben, ein mehrfaches davon wird an anderen Krebsarten sterben. Bedenkt man, daß der Krebs, im Gegensatz zur Pest, eine permanente Seuche ist, so ist der Krebs heute eine der gefährlichsten Seuchen in der Geschichte der Medizin geworden.

Unter diesen Umständen muß man sich fragen, warum nichts zur Verhütung des Krebses getan wird, obwohl sich die Experten darüber einig sind, daß die Mehrzahl der Krebsvorkommen verhütet werden könnte, wenn man die wohlbekannten Carcinogene von den

Menschen fernhalten würde. Die Verhütung des Krebses würde dabei keine größeren Kosten verursachen, als die Verhütung der Pest oder der Malaria oder des gelben Fiebers und besonders wären vorbereitende Forschungen nicht mehr erforderlich.

Die Antwort auf die gestellte Frage ist meines Erachtens, daß von einer Schar von Krebsforschern das Dogma verbreitet wird, daß man nicht weiß, was der Krebs ist. Wie aber sollte man etwas verhüten, von dem man nicht weiß, was es ist ?

Wenn man aber nunmehr weiß, was der Krebs ist, genauer als man weiß, was die Pest ist, so muß das Dogma des Agnosticismus fallen und man hat keine Ausrede mehr, die Verhütung des Krebses hinauszuschieben. Der Nutzen der vorgetragenen Versuche also könnte sein, daß man mit der Verhütung beginnt.

Literatur

[1] WARBURG, O., K. GAWEHN, A. W. GEISSLER und S. LORENZ: Hoppe-Seylers Z. physiol. Chem. **321**, 252 (1960).
[2] —, K. GAWEHN und T. TERRANOVA: In Weiterentwicklung der zellphysiologischen Methoden. Stuttgart und New York 1962, S. 562.
[3] WARBURG, O., A. W. GEISSLER und S. LORENZ: Z. Naturforsch. **17 b**, 758, 772 (1962).
[4] — — — Klin. Wschr. **43**, 289 (1965).
[5] — — — Z. Naturforsch. **20 b**, 1070 (1965).
[6] — — — Z. Naturforsch. **17 b**, 758 (1962).
[7] MALMGREN, R. M., and C. C. FLANEGAN: Cancer Res. **15**, 473 (1955).
[8] BURK, D.: Symposium on Respiratory Enzymes, University of Wisconsin Press 1942, p. 235.
[9] — Science **123**, 313 (1956).
[10] —, and M. WOODS: Proc. Amer. Ass. Cancer Res. **6**, 9, 69 (1965).
[11] VAN POTTER: Prisma **63**, 1963 (Firma Boehringer u. Sohn, Ingelheim/Rh.).
[12] BURK, D., M. WOODS, and J. HUNTER: J. nat. Cancer Instit. 1966, in Press.
[13] GULLINO, P.: Proc. Amer. Ass. Cancer Res. **7**, 27 (1966).
[14] THODE, H. G., J. MACNAMARA und W. H. FLEMING: Geochimica Acta **3**, 235 (1953).
[15] Z. Naturforsch., Juli-Heft 1966 (Im Druck).
[16] WYNDER, E. L., S. HULTBERG, F. JACOBSON and I. J. BROK: Cancer, **10**, 470 (1957); **18**, 167 (1965).
[17] Vital Chorin, C. R. Séance du 15. janvier 1945, FURT, B., und A. STUDER, Schweiz. Z., allg. Path. **14**, 523 (1951).
[18] BAUER, K. H.: Das Krebsproblem, 2. Aufl. Berlin-Göttingen-Heidelberg: Springer 1963.

Nuclear RNA, Histones and Differentiation*

By H. BUSCH, K. HIGASHI, S. T. JACOB, T. NAKAMURA, S. M. SCHWARTZ, and S. J. SMITH

Baylor University College of Medicine, Department of Pharmacology, Houston, Texas, USA

With 18 Figures

The problem of neoplasia is one that has been in the forefront of medicine for almost 100 years. Studies made then by VIRCHOW and his colleagues established many of the characteristics of neoplastic cells. As summed up in a series of observations which have formed the basis for the thoughts of pathologists, neoplastic cells have marked destructive characteristics, which result from the metastasis, invasiveness and pressures by neoplasms. Among the significant changes found in neoplastic cells are aberrations in the cytoplasm and the constituents of the cytoplasm of the tumor cells. These variations may be generalized under the heading of "Phenotypic Repression". The extent of phenotypic repression of neoplasms varies markedly from neoplasm to neoplasm of the same type and also from tumor to tumor in an individual. In man and in animals these variations have been a source of much concern. In recent years, MORRIS and his associates at the National Institutes of Health have developed a series of hepatomas in which phenotypic repression varies such that some tumors are recognizable only as highly anaplastic and malignant lesions and others are neoplasms in which virtually all of the enzymes of the liver are present and yet the cells are neoplastic cells. To some, these data are very confusing with respect to neoplasia. To others, they indicate that since enormous variations may occur in the cytoplasm among cells that are neoplastic, the cytoplasm has little if any relationship to the neoplastic process. Such negative evidence suggests but does not establish that in fact, neoplasia is a nuclear disease.

---

* These studies were supported in part by grants from the American Cancer Society, The Jane Coffin Childs Fund, the National Science Foundation, and USPHS Grant CA-08182.

Of course, the variations in cytoplasmic elements found by MORRIS and his associates are really not unique to animal neoplasms but in fact, similar types of aberrations have been known for human neoplasms for many years. Some tumors such as pancreatic adenomas may be, phenotypically, virtually identical to the normal β-cells of the pancreas and in addition, it should be clear that the neoplasms of many tissues begin as phenotypically differentiated tumors. However, in metastasis and in animal neoplasms many tumors "tend to converge to a common cell type" (GREENSTEIN, 1954) and this has led to the broad concept that there are fundamental characteristics of neoplastic tissues. Although in one sense the importance of the "convergence theory" of GREENSTEIN (1947, 1954) has been somewhat diminished with the finding of various remnants of differentiated cytoplasmic functions in neoplasms, these remnants may only mean that some of the phenotypically differentiated templates are still functional and have a role in protein synthesis in these cells.

Interestingly, WEBER and MORRIS (1963) have documented the fact that there is a relationship between growth rate of neoplasms and the loss of phenotypically active enzymes, i.e., the fewer the phenotypically characteristic enzymes, the greater the growth rate.

From the point of view of the biochemist, the problem of neoplastic cell still has not been satisfactorily resolved even into constituent parts that might provide a basis for the ultimate goal of demonstration of "cancer-specific" lesions. The fact that in biochemical studies in neoplasia the biochemist is confronted with a wide variety of aberrations that may occur along with ,,cancer-specific" changes has accordingly resulted in much confusion in literature. However, the hereditary characteristics of the neoplastic lesion and its transmissibility to daughter cells have provided the present basis for increasing studies on the molecular biology of cancer (BUSCH, 1962).

The "positive hypothesis of biochemistry in oncology" is that genetically transmissable neoplastic aberrations are expressed through formation of cancer-specific messenger RNAs which in turn serve as templates for production of a few cancer-specific proteins (BUSCH and STARBUCK, 1964). It is possible that one to ten templates are either altered or derepressed in the oncogenic process, but it is an unfortunate fact that accompanying such changes or derepressions is a series of unrelated changes that are not necessarily specific to the process. The phenotypic variance in neoplasia in any one tissue and in a whole group of tissues has been the object of fruitless study by pathologists and biochemists for many years. However, the only common histological features of neoplastic cells are nuclear changes, including increased stainability, pleomorphic nucleoli, dense chromatin and aberrant mitosis.

The nucleus

Regardless of the degree of phenotypic differentiation of tumor cells, the fact remains that neoplasia manifests itself cytologically by nuclear aberrations. These aberrant mitoses, excessive chromatin

Table 1. *Nuclear Constituents*

Chromatin
  DNA ⎫
  Histones ⎬ in chromosomes
  RNA: AU-rich, GC rich ⎫
  Acidic proteins ⎬ perichromosomal

Nuclear ribonucleoprotein network

  Nucleolus
    perinucleolar DNA
    intranucleolar DNA
    AU-rich RNA
    GC-rich RNA, 45 S ———→ 35 S ———→ 28 S RNA +
    acidic proteins ———→ nascent 55—60 S particle, histones
    enzymes — RNA polymerase, ,,Convertase", RNase, DPN
    pyrophosphorylase, ATPases, lipoproteins

  Network
    acidic proteins ⎫
    18 S RNA ⎬ nascent 30 S particle
    lipoproteins ⎭

Soluble nuclear fraction

  Nuclear ribosomal precursors
  Lipoproteins
  RNA
  Enzymes
  RNase; inhibitors

Nuclear membrane

  Outer layer — related to endoplasmic reticulum
  Inner layer — related to porous lipoprotein shell

in the nucleus, aberrations in chromosomes and particularily pleomorphism in the nucleoli (Fig. 1) are part of a complex which

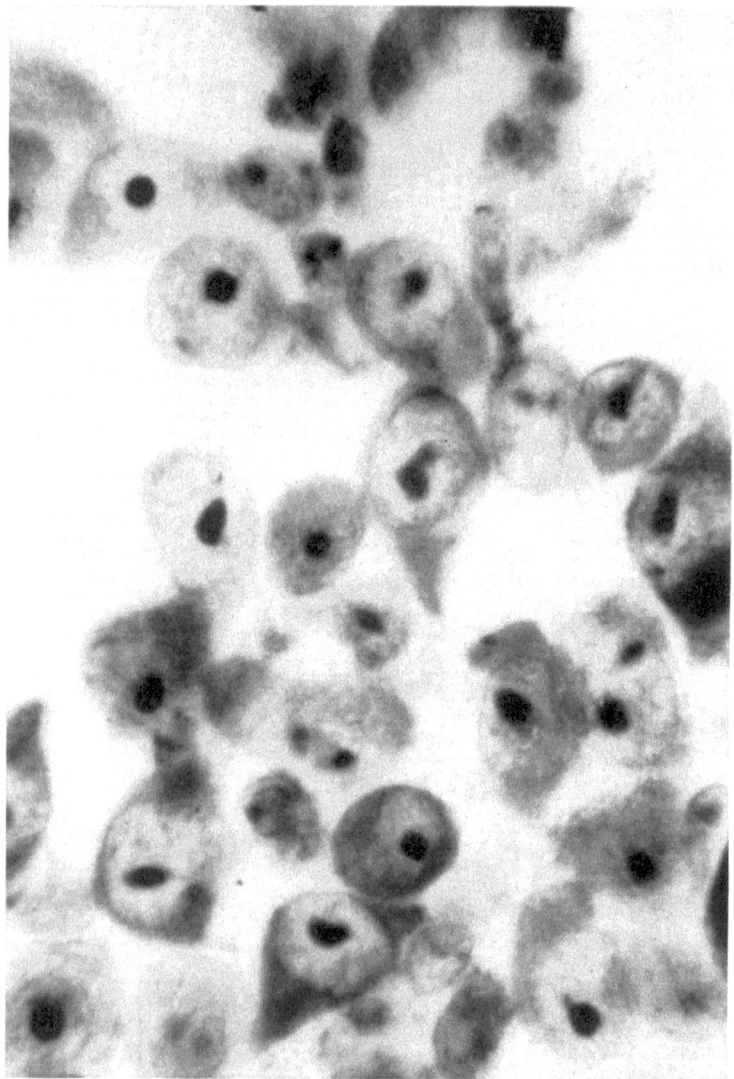

Fig. 1. Preparation of Walker tumor stained with Azure C to demonstrate nucleolar pleomorphism

appears to be the result of one or more genes grouped under the heading of the cancer polyoperon or polyoperon V (BUSCH et al., 1964).

If the nucleus is indeed the cellular segment that is primarily involved in neoplastic disease and even if one can not state this definitively at present, it behooves one to make a systematic study of the nucleus of the cancer cell and its components. The nucleus contains a large number of specialized elements some of which are found primarily or only in nuclei and are not present or are present in different types or degrees in non-nuclear elements. Among the constituents that are found primarily in nuclei are those listed in Tab. 1.

Nuclear constituents in cell division: There is an ordered pattern of biosynthetic sequences in cells about to replicate (BUSCH, 1965). If one views interphase as a relatively steady state, one or more stimuli shown in Fig. 2 may cause the cell to undertake the reactions necessary for new cell synthesis. The first of these seems to be a rapid synthesis of GC-rich RNA. In regenerating liver, it was found in this laboratory (MURAMATSU and BUSCH, 1965) that one of the very earliest events is that of the synthesis of GC-rich nucleolar RNA; this synthesis is initiated within three hours after the cell has begun to respond to the insult of the hepatectomy.

Other workers including HOLBROOK et al. (1962) have noted that the synthesis of RNA occurs very early and subsequently, synthesis of histones seems to occur. Almost at the same time as histone synthesis is initiated, there is a rise in the cell content of DNA polymerase which may or may not occur in correlated fashion with the synthesis of the enzyme series previously reported to be responsible for the production in increased amounts of TTP. As a result of the activities of these enzymes and the availability of their appropriate substrates, DNA synthesis occurs which must be initiated at the end of a number of chains. The synthesis of DNA chains takes place in a defined order and a dual compliment of DNA is developed.

According to the experiments of GURLEY et al. (1964) at the same time as the histones are synthesized, and DNA polymerase and appropriate enzymes for DNA production are being made, a separate pathway is already involved in the biosynthesis of inhibitors of DNA synthesis. This pathway is one which has been

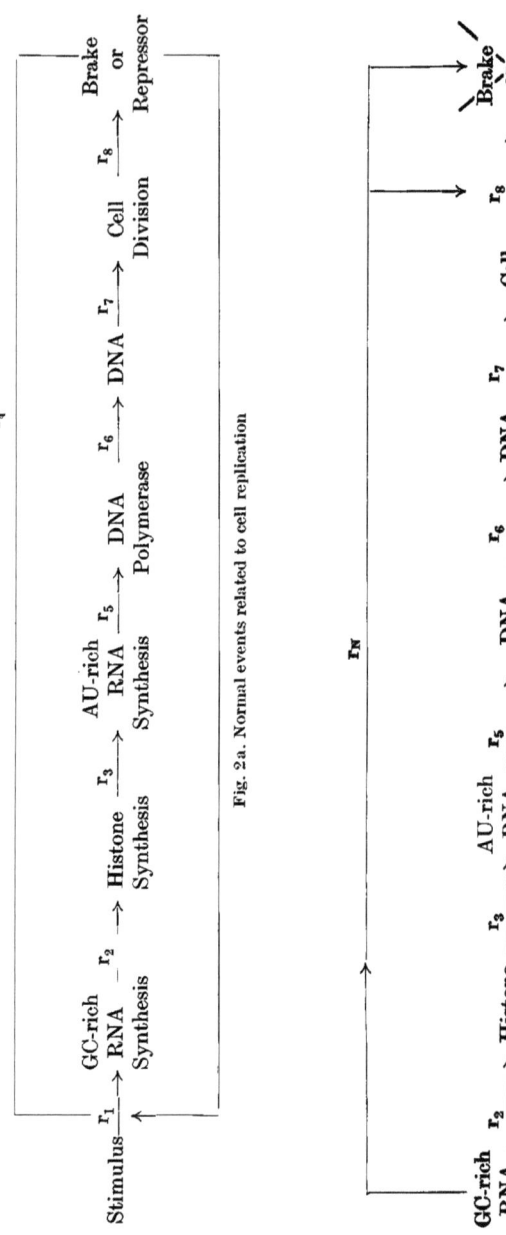

Fig. 2a. Normal events related to cell replication

Fig. 2b. Events in neoplasia related to cell replication. r_N is reaction induced by „cancer genes" which either inhibit the brake, inhibit the activation of the brake (r_8) or inhibit the product of the brake

related to synthesis of the histones although other proteins may be very important in this process. When the synthesis of DNA is

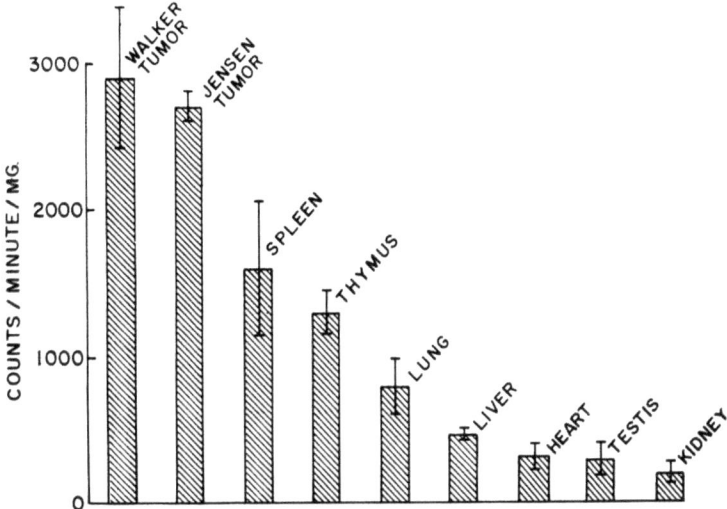

Fig. 3. Specific activities of the chromosomal proteins of tissues of tumor-bearing rats one hour after injection of 10 μcuries of L-lysine-U-^{14}C. The standard errors are shown

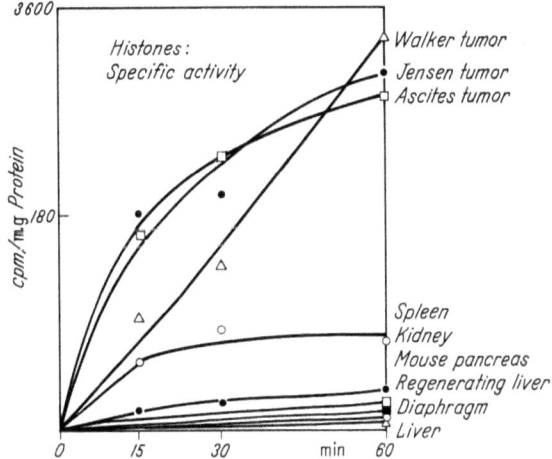

Fig. 4. Specific activities of histones of tumors and other tissues at various times after incubation of tissue slices with labeled lysine or arginine

complete, the cell proceeds to the mysterious "dance of the chromosomes" involved in mitosis, two nuclear membranes reform and

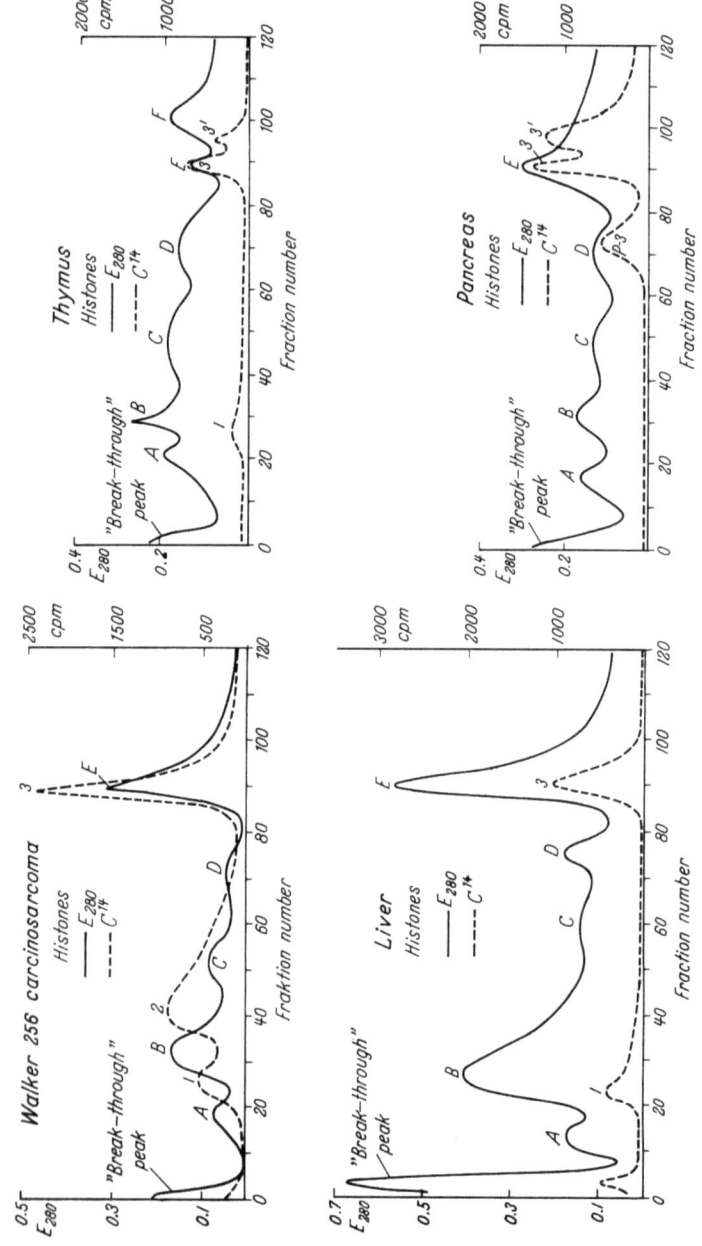

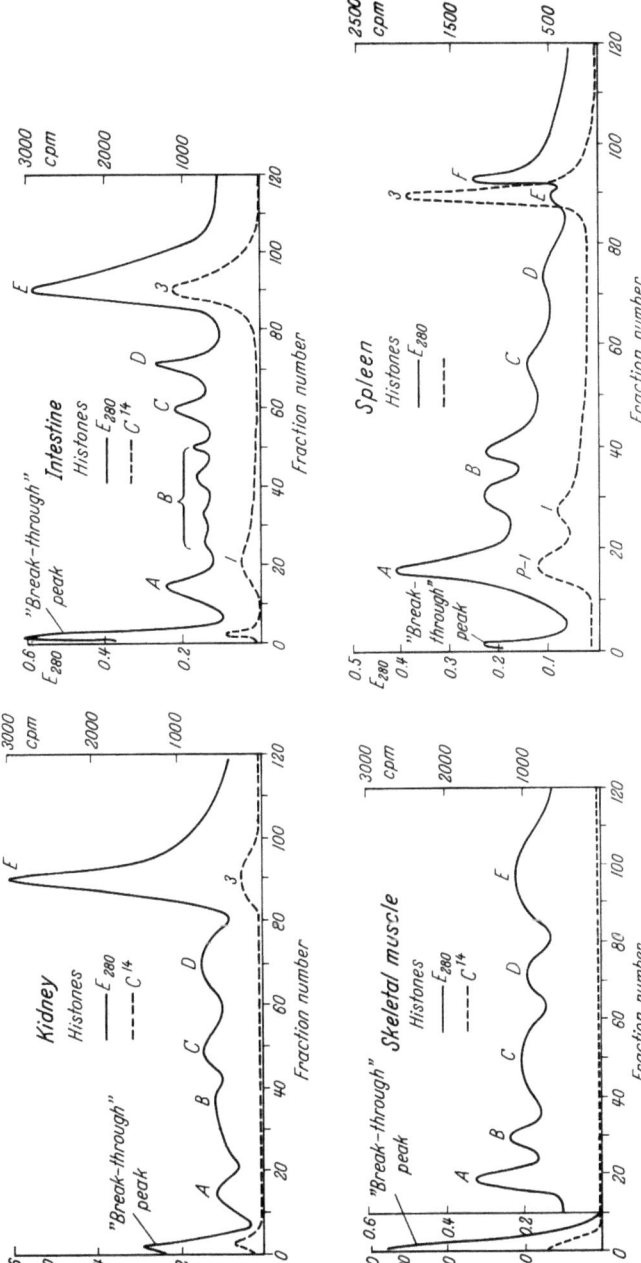

Fig. 5. Chromatography of acid extractable nuclear proteins of the Walker tumor and other tissue on carboxymethylcellulose (DAVIS and BUSCH, 1959)

simultaneously the cytoplasmic membrane or cell membrane makes appropriate constrictions and two perfectly formed daughter cells result.

The events set in motion very early may cause the formation of sufficient amounts of inhibitor of DNA synthesis or production of a protein such that now the brake or a repressor is activated which renders the cell refractory to the stimulus and/or a block in production of DNA synthesis now ensues. This is the normal situation.

What is the relationship of this scheme to neoplastic disease? In neoplastic disease, a short-circuiting of a series of pathways of one of a number of reactions may be occurring as shown in Fig. 2 b. For example, the production of GC-rich RNA proceeds at a rapid pace as exemplified by the very high rates of synthesis of nucleolar RNA (MURAMATSU and BUSCH, 1964). All the events involved in production of new cells occur but apparently the brake is not activated. The brake may be present, and its products may be inhibited by products of cancer genes. In neoplasia, the cancer genes may produce elements that stimulate GC-rich RNA synthesis, inhibit the receptor for the brake, inhibit the brake or inhibit the product of the brake. The relationship of these possibilities to studies in our laboratory is the primary topic of this report.

Nuclear proteins and neoplasia

The work in our laboratory on the nucleus began with observations that the uptake of amino acids into nuclear proteins was disproportionately high in neoplastic cells (DAVIS and BUSCH, 1959, 1960). Later studies established that one of the reasons for this very high labeling of nuclear proteins in tumors was the rapid uptake of isotope into chromosomal proteins, particularly the histones (Fig. 3, 4). This high uptake of amino acids into histones was associated in tumors with a very rapid labeling of all histone fractions (BUSCH, 1965). Interestingly, extraction of nuclear proteins with acid was performed and differences were found in the chromatographic behavior of the acid extractable proteins of tumors and of tissues. In the samples from tumors a radioactive peak appeared in a region referred to as RP2-L (radioactive peak 2-lysine, since lysine was used as a precursor). This peak was not found in onn-tumor tissues (Fig. 5) in our studies although later there were reports that such a peak or similar peaks were present in bone marrow.

This RP2-L peak contained a complex mixture of histones and some acidic proteins (BUSCH et al., 1962). The possibility that one

```
                 ↗ DNP COMPLEXES
    HISTONES
                 ↘ ACIDIC PROTEIN COMPLEXES
```

Fig. 6. Alternate pathways for histones. Excessive amounts of acidic proteins may release gene segments for transcription (BUSCH et al., 1963)

of the acidic proteins might represent one of the groups present either in the soluble or deoxyribonucleoprotein fractions was considered and efforts were made to characterize these acidic proteins.

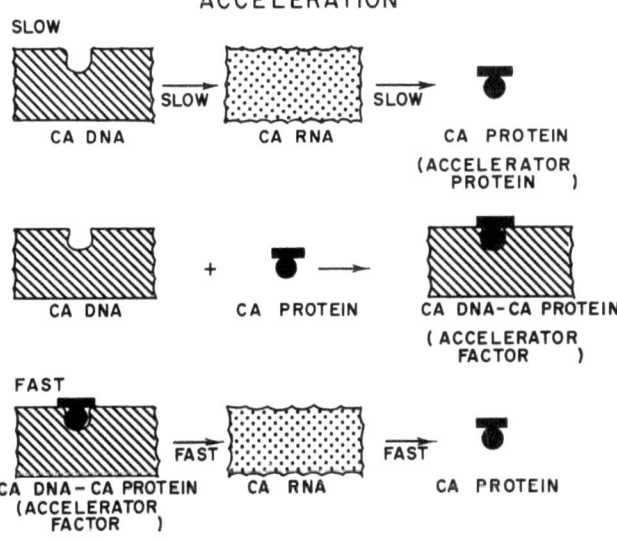

Fig. 7. Stage of acceleration in neoplasia characterized by production of cancer RNA and cancer protein which may act to accelerate macromolecular readouts from the genome

Their presence might produce the removal of histones from DNA surfaces or DNP complexes and a resulting release of the genome involved in neoplastic transformation (Fig. 6). It is an unfortunate fact that these proteins have not been well characterized even to the present largely because of the limited solubility which they have and the fact that they are not readily amenable to fractionation on any of the usually employed systems.

The thought that tumors might have some aberrant protein synthesis in the nucleus and that this aberrant synthesis might reflect the production of cancer-specific RNA (Fig. 7) led to further experiments designed to determine what types of abnormalities might exist either in templates forming nuclear proteins or in the products that were produced.

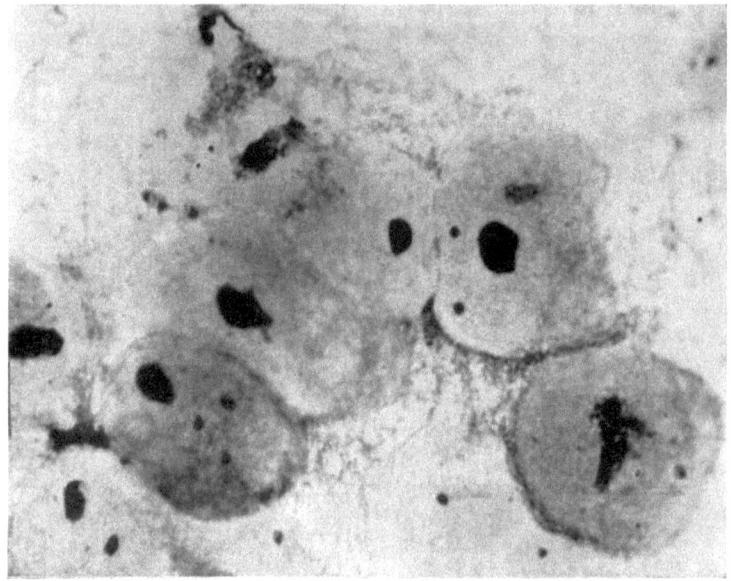

Fig. 8. Nuclear preparation of the Walker tumor stained with toluidine blue

Isolation of nucleoli: The nucleoli represent the largest single aggregate of RNA in most cells and indeed have been an object of interest among biochemists for many years. However, methods for isolation of nucleoli for mammalian cells were not practical until 1963 when a number of workers reinvestigated the sonication technique of Dounce and the compression decompression procedure employing the French cell. It was found that the critical requirement for maintenance of the integrity of nucleoli was the presence of high concentrations of calcium ion in the medium (BUSCH et al., 1963). Under these conditions, satisfactorily nucleoli could be isolated with the aid of various techniques designed to destroy nucleolar integrity (MURAMATSU et al., 1963, DESJARDINS et al., 1965, Fig. 9 to 11).

Nuclear RNA, Histones and Differentiation 29

Even before nucleoli were isolated, autoradiographic studies had been made by many workers which provided evidence that nucleoli

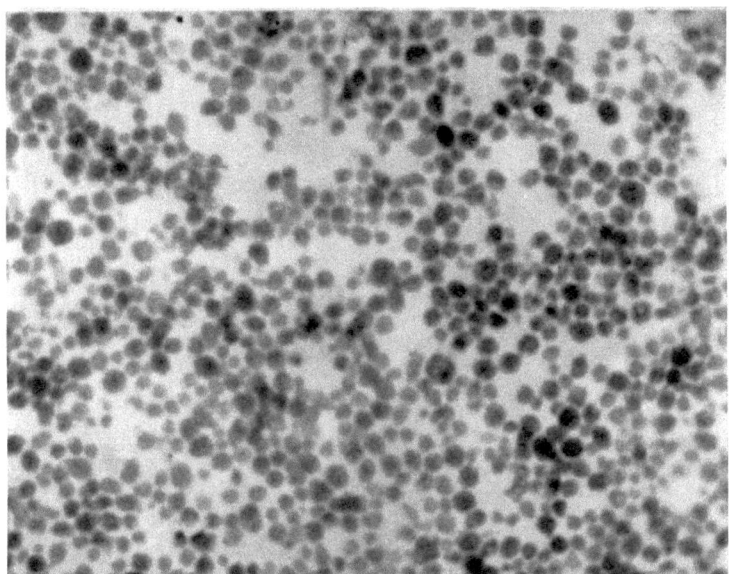

Fig. 9a. Isolated liver nucleoli

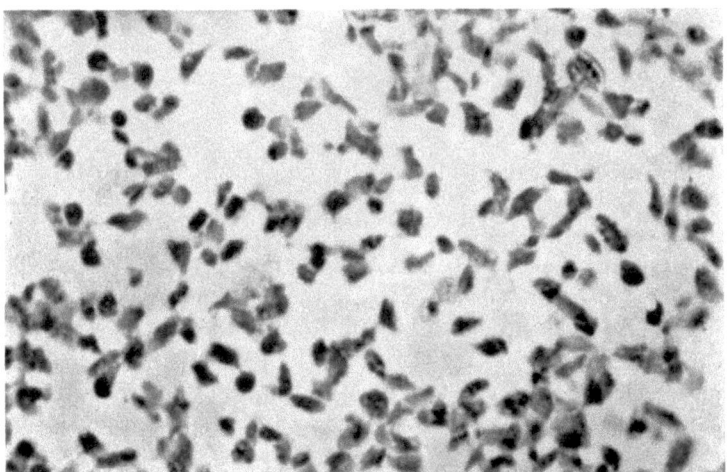

Fig. 9b. Isolated nucleoli of Walker tumor

were in fact very active sites of RNA synthesis (BUSCH et al., 1963). Among the earliest studies made in this laboratory on isolated nucleoli were those in which the nucleoli were analyzed both with respect to the base composition and distribution of isotope into

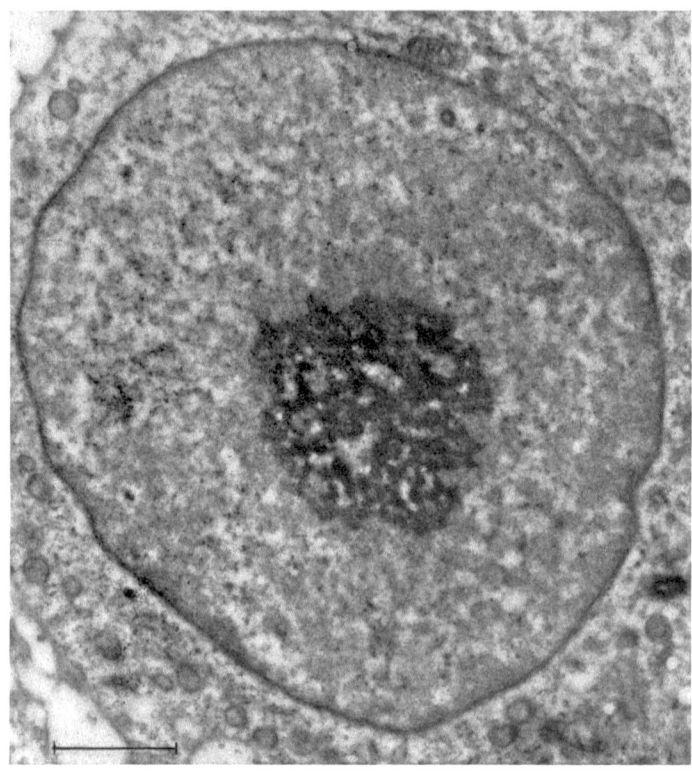

Fig. 10a. Electronmicrograph of the nucleus of the Walker tumor cell showing nucleolar substructure

RNA. It was found that in the Walker tumor almost 70% of the total ^{32}P incorporated into RNA in the nucleus was incorporated into RNA in the nucleolus. The comparable value was only 15% for the liver. These results indicated that the distribution of precursors into nucleolar RNA was very great in tumors by comparison with the liver. Of great interest was the fact that the base composition of the RNA of the Walker tumor was markedly different from that of the normal liver. In the base composition of newly synthesized

nucleolar RNA of the tumor, the value for adenylic acid was low
i.e., 13% by comparison with a value of 20% for the liver. A

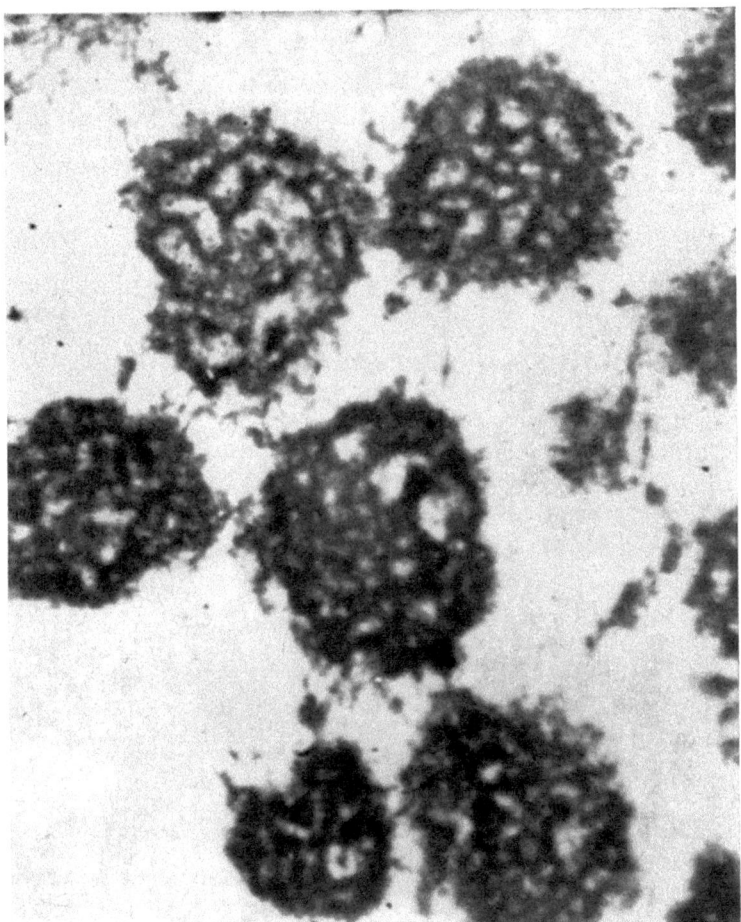

Fig. 10b. Electronmicrograph of isolated nucleoli

somewhat higher value for U was found in the nucleolar RNA of
the tumor as compared with that of the liver (Tab. 2).

In studies on biosynthesis of RNA of tumors and other tissues
significant differences have been found between newly synthesized
nucleolar RNA of the Walker tumor and normal liver (MURAMATSU

and BUSCH, 1964). With a 15 min. pulse of ^{32}P-orthophosphate as a label, the A, U, G and C contents of newly synthesized nucleolar RNA of the Walker tumor were 13, 21, 35 and 31% of total

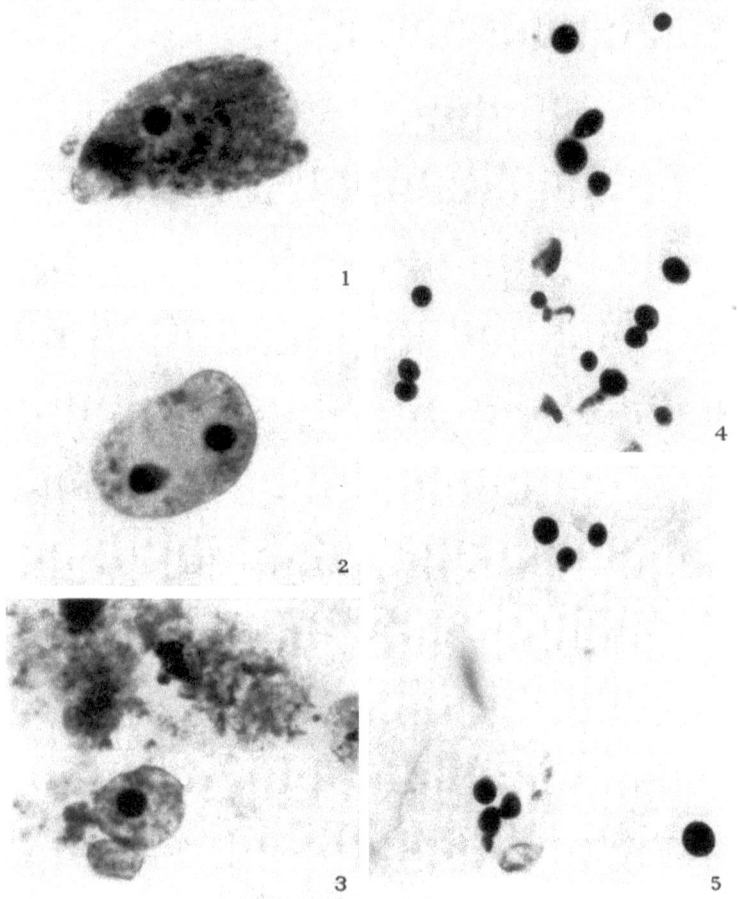

Fig. 11. Stages of isolation of nucleoli of a human cancer of the lung

nucleotides, respectively. In normal liver, the corresponding values for A, U, G and C were 21, 17, 37 and 24, and in regenerating liver, the corresponding values were 20, 17, 38 and 26, respectively. Improved procedures for isolating nucleoli and separation of RNA of varying sedimentation constants permit determination of the

^{32}P base composition of RNA of individual sedimentation classes. In normal liver, the ^{32}P base composition of the nucleolar 45 S RNA was A-20, U-17, G-38 and C-25. For Walker tumor, the corresponding values were 13, 21, 34 and 31. The UV base composition of

Table 2. *Base composition of nucleolar RNA — ^{32}p*

	A	U	G	C
Tumors				
Ehrlich ascites	12.8	21.2	34.3	31.6
Novikoff hepatoma	12.0	22.1	32.9	32.9
Walker carcinosarcoma	13.0	21.1	34.7	31.3
Liver				
Normal	21.0	17.2	37.3	24.5
Regenerating (6 hrs)	20.6	16.1	36.9	26.2
Regenerating (18 hrs)	19.7	16.8	37.7	25.6

nucleolar 45 S RNA is A-15, U-21, G-34 and C-30. It appears that markedly different nucleolar 45 S RNAs are formed rapidly in the Walker tumor and liver. The similarity in the ^{32}P base composition of nucleolar 45 S RNA in regenerating and normal liver indicates that the differences found between the Walker tumor and liver are not due to growth alone. Since the half-life time for 45 S RNA in the regenerating liver is less than that of the Walker tumor (4.5 vs 6.8 min), the differences in newly synthesized RNA are not due to rates of synthesis. Further studies are in progress on other tumors and non-tumor tissues to determine whether the differences in newly synthesized RNA are due to neoplastic transformation.

Fractionation of nucleolar RNA

It was recognized very early in studies on RNA that many species of molecules are present. Gradient sedimentation analysis of nuclear and nucleolar RNA indicated that nucleolar RNA contains much more rapidly sedimenting RNA than whole nuclear RNA and contains very small amounts if any 18 S RNA (Fig. 12). Since most of the radioactivity following a pulse label was present in the rapidly sedimenting 45 S RNA, the nucleolar 45 S RNA was purified and subjected to further analytical procedures (Fig. 13). The technique employed for purification of subfractions is that of countercurrent distribution from which a number of fractions were obtained in our most recent studies. These fractions consist of subunits varying in composition according to the predicted distributions on the basis of Kirby's papers (KIRBY et al., KIDSON and KIRBY). It is not yet possible to specify whether any one of

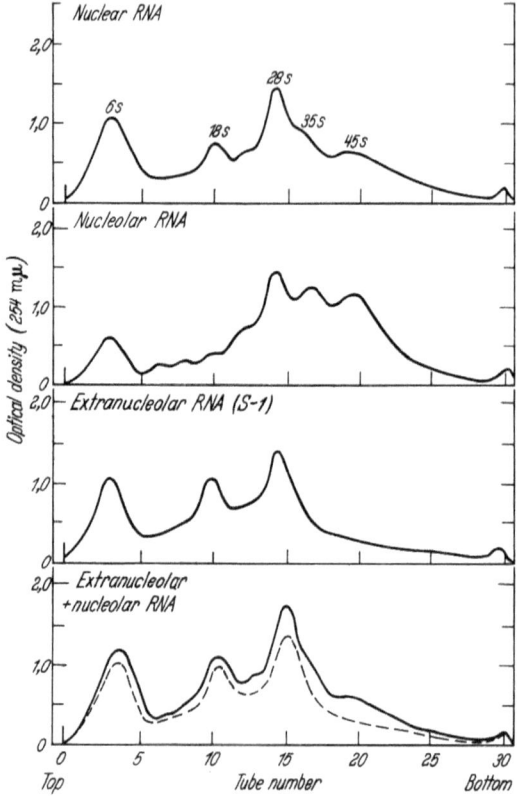

Fig. 12. Sucrose density gradients of RNA isolated from nuclear fractions

these fractions contains a tumor specific component (Fig. 14). Some have much higher content of adenylic acid than others and the lowest content of adenylic acid that has been found thus far in these fractions is 5% which is present in a rather large amount of the RNA of the Walker tumor (Fig. 14 to 16).

Turnover of nucleolar RNA: Studies with actinomycin D have shown that at various time intervals after a pulse of actinomycin D a rather marked shift of the distribution of nucleolar RNA occurs such there is a relatively rapid loss of 45 and 55 S RNA, a slower loss of 35 S RNA and finally, a 28 S RNA peak remains in the nucleoli, as a residue of the rapidly sedimenting RNA. It has thus become possible to determine the half-lives or turnover times of the rapidly sedimenting RNA of the Walker tumor and other

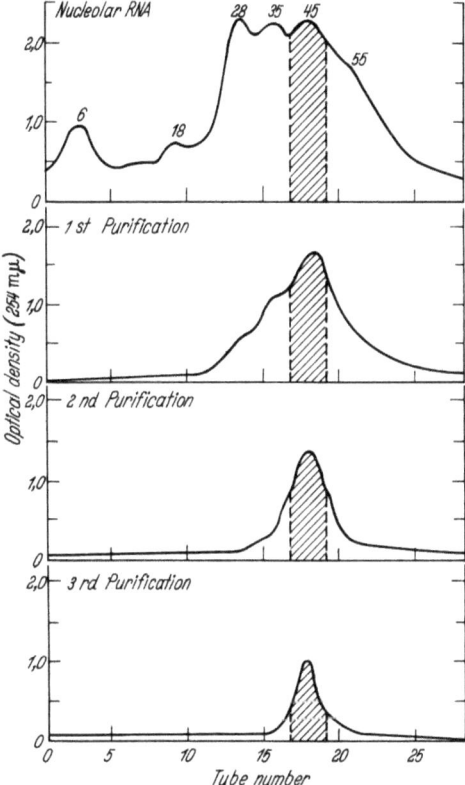

Fig. 13. Purification of 45 S nucleolar RNA of Walker tumor

tissues. In the Walker tumor, the half lifetime of 45 S RNA was only 6.8 min; that of regenerating liver was 4 to 5 min. In normal liver nucleoli, the half-life of 45 S RNA was 8 to 10 min. Since there is so much more 45 S RNA in nucleoli of the Walker tumor than there is in the nucleoli of the regenerating liver, it was of interest to determine the total turnover.

On the basis of the previously determined amounts of RNA and the half lives found by analysis of the actinomycin D treated

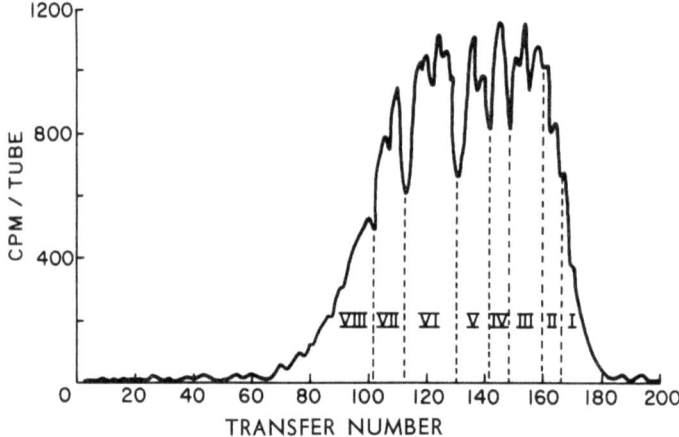

Fig. 14. Countercurrent distribution pattern for nucleolar 45 S RNA of the Walker tumor

samples, the rate of synthesis of nucleolar 45 S RNA in the Walker tumor was approximately 32 femtograms of RNA per min per

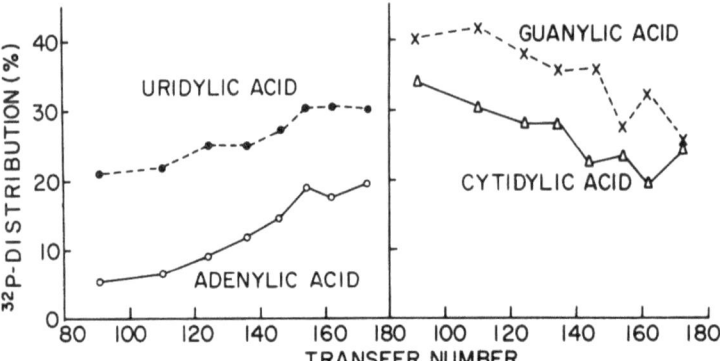

Fig. 15. Per cent of total bases for each individual base in various fractions of countercurrent distribution

nucleolus by comparison with approximately 16 in the regenerating liver. In normal liver nucleoli, the rate was much slower, i.e., 4 femtograms of RNA per min.

The role of the rapidly sedimenting nucleolar RNA is apparently to serve as a precursor for the ribosomal RNA although it is possible that other functions exist. The series of reactions established is 45 → 35 → 28 S RNA.

From our present data, it seems that the rapidly sedimenting RNA represents either hydrogen bonded precursor or larger molecular weight precursor which must be subjected either to changes

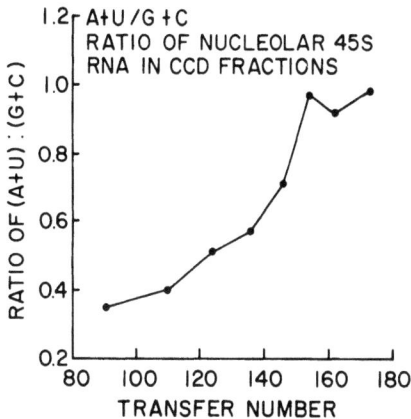

Fig. 16. A + U/G + C ratio for various transfer fractions in the countercurrent distribution isolation

in tertiary structure or to cleavage in order to yield the 28 S RNA of the ribosomes (MURAMATSU and BUSCH, 1966).

Nuclear RNA

As one gropes ones way into the functions of the nucleus, one wonders about the effects of the changes in nucleolar RNA synthesis on the metabolism of the whole nucleus. As pointed out recently (OKAMURA and BUSCH, 1965), marked differences exist in the base compositions of the nuclear RNA fractions of the Walker tumor and the liver, particulary in the rapidly sedimenting RNA fractions. A much higher content of adenylic acid is found in the rapidly sedimenting RNA of the nuclei of normal liver as compared to that of the Walker tumor (Fig. 17). Interestingly, there is relatively little change in the composition of nuclear RNA of regenerating liver by comparison to normal liver (Tab. 3).

The present studies have served to point a direction of difference between neoplastic cells and other cells, in that there appear to be substantial differences in the nucleolar product and possibly also in the whole nuclear product of the tumors. It should be noted that several tumors have now been studied and surprisingly the

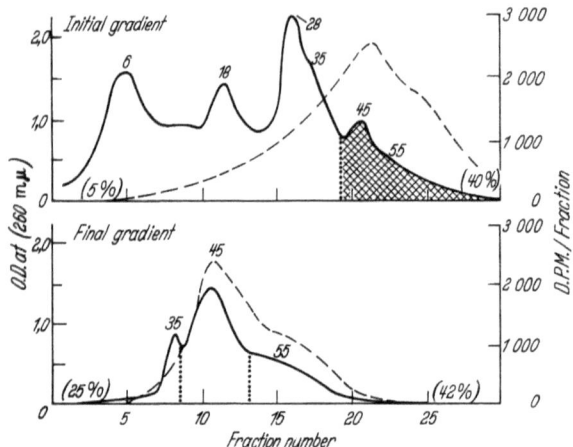

Fig. 17. Density gradient fractionation of nuclear RNA and stage of purification

Table 3. *Base composition of high molecular weight nuclear RNA determined by ultraviolet absorption*

	Liver		Walker Tumor	
	45 S	55 S	45 S	55 S
Adenine (A)	19.1	20.8	16.5	17.5
Uracil (U)	25.7	28.4	20.5	23.6
Guanine (G)	31.0	30.7	35.6	34.1
Cytosine (C)	24.2	20.1	27.3	24.8
A + U/G + C	0.81	0.97	0.59	0.70

Distribution of ^{32}P in nucleotides of early-labeled nuclear RNA

	Liver		Walker Tumor	
	45 S	55 S	45 S	55 S
Adenylic acid (A)	24.9	26.1	15.3	17.6
Uridylic acid (U)	21.2	22.6	22.7	24.0
Guanylic acid (G)	29.5	26.5	32.3	30.0
Cytidylic acid (C)	24.4	24.8	29.7	28.5
A + U/G + C	0.86	0.95	0.61	0.71

composition of the newly synthesized nucleolar RNA is essentially identical to that of the Walker tumor and is markedly different from that of the normal or regenerating liver (Tab. 2).

Similar studies have been carried out on whole nuclei of kidney and the results have been very much the same, that is to say, a very high A content is present by comparison with the A content of the normal liver. Both tissues have RNA with much higher contents of adenylic acid than the Walker tumor in their nuclei.

Discussion

Can we understand from this data that there are really the distinctions that we have hoped to find between the nuclei of tumors and other tissues on the basis of the positive hypothesis of biochemistry in oncology? Do the changes in RP 2-L, the rapid rate of synthesis of nuclear proteins, the pleomorphic nucleoli with their different newly synthesized RNA and the extremely rapid rates of synthesis of nucleolar RNA reflect differences between tumor cells and other cells in terms of the cancer-specific genes or do they reflect the excessively active processes of growth which are required in order to produce new cells at rapid rates? The only evidence that the latter is not the case emerges from studies on regenerating liver which can best be described as probing studies rather than definitive ones because regenerating liver is a complex mixture of functional and dividing cells. One can be cautiously optimistic now that systems are being developed for fractionation of RNA such that individual molecular species of RNA may be isolated. One can only hope that cancer-specific RNAs will be found by ourselves or our successors, and further that use can be made of either such nucleic acids or their protein products for chemotherapy.

For the pharmacologist, the crucial question is whether the events in the neoplastic cell are biochemically normal and that, in fact, all the products are normal, in which case there would be nothing to attack except excessive rates of synthesis. Most workers agree that such success as exists in cancer chemotherapy at present is based upon these differential growth phenomena.

Conversely, it is possible that one or more elements produced in neoplastic cells may be abnormal and the products of a cancer gene. Such a cancer gene might simply render the brake inactive.

Through production of a cancer-specific RNA and/or its protein such a gene might function to remove the normal protein and its blocking effect as was postulated some time ago for RP 2-L. Such a gene might simply function to produce a large amount of a

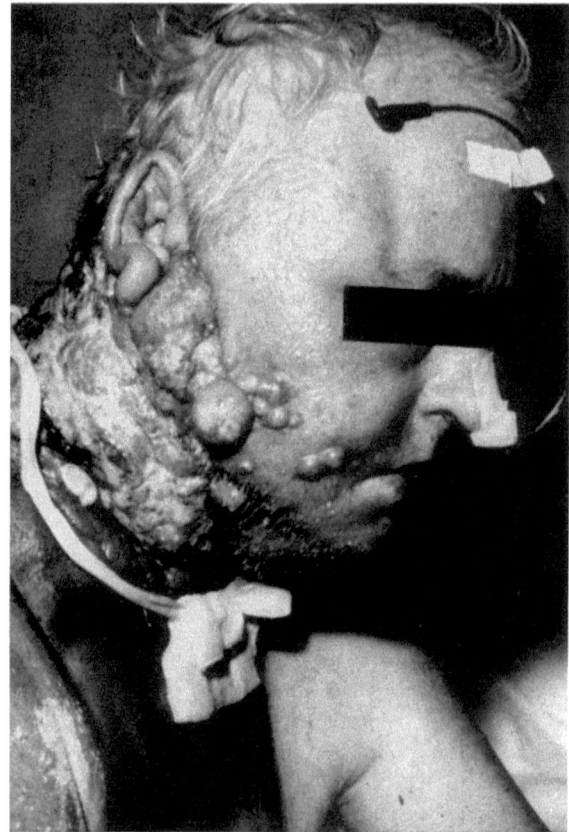

Fig. 18. Patient with epidermoid carcinoma showing local invasiveness and metastasis

given molecule which blocks normal destructive reactions. ROTH has made the suggestion that in neoplastic cells the rates of production of RNase inhibitors is very high. Indeed, some histochemists find no RNase activity in tumor nuclei.

The unanswered question which is always posed is, what good will come of identification of one or several cancer-specific lesions?

In what way can drugs be developed which will suppress the growth of cancer cells or destroy them completely? Of course, it is not possible to give an answer to this type of question at the present time, although one can readily imagine almost 75 years ago someone asking the early microbiologists, what good will come of identifying the causative organisms of disease, and asking the second question, even if you can find them, what will you then do for therapy? In 1966, happily, that question is answered, but solution to chemotherapy of cancer needs to be attained.

At present, the biochemist interested in oncology still must adhere to the positive hypothesis that one or several aberrations occur in neoplastic cells such that abnormal products are produced. Certainly no evidence has arisen to make this hypothesis untenable. It is the hope of the workers in our laboratory that through these experiments on nuclear structure and components, some leads, both as to research areas and possible aberrant substances, are now being obtained which may ultimately be meaningful in the lives of the patients afflicted with cancer (Fig. 18).

References

BUSCH. H.: Tex. Rep. Biol. Med. **19**, 1 (1961).
— An introduction to the Biochemistry of the cancer cell. New York: Academic Press 1962.
— The histones and other nuclear proteins. New York: Academic Press 1965.
—, P. BYVOET, and K. SMETANA: Cancer Res. **23**, 313 (1963).
—, L. S. HNILICA, S. C. CHIEN, J. R. DAVIS, and C. TAYLOR: Cancer Res. **22**, 637 (1962).
—, M. MURAMATSU, H. R. ADAMS, K. SMETANA, W. J. STEELE, and M. C. LIAU: Exp. Cell Res. Suppl. **9**, 150 (1963).
—, W. C. STARBUCK. T. S. RO, and E. SINGH: Chromosomal proteins. In The role of chromosomes in development, M. LOCKE, Ed. 23rd Symp. Soc. Development and Growth, p. 51. New York: Academic Press 1964.
— — Ann. Rev. Biochem. **33**, 519 (1964).
—, W. J. STEELE, L. S. HNILICA, C. W. TAYLOR, and H. MAVIOGLU: J. cell. comp. Physiol. **62**, Suppl. 1, 95 (1963).
DAVIS, J. R., and H. BUSCH: Cancer Res. **19**, 1157 (1959).
— — Cancer Res. **20**, 1208 (1960).
DESJARDINS, R.. K. SMETANA, W. J. STEELE, and H. BUSCH: Cancer Res. **23**, 1819 (1963).
— —, and H. BUSCH: Exp. Cell. Res. **40**, 353 (1965).
— —, D. GROGAN, K. HIGASHI, and H. BUSCH. Cancer Res. **26**, 97 (1966).
GREENSTEIN, J. P.: Biochemistry of cancer. New York: Academic Press 1954.

GURLEY, L. R., J. L. IRVIN, and D. J. HOLBROOK: Biochem. biophys. Res. Commun. **14**, 527 (1964).
HOLBROOK, D. J., Jr., J. H. EVANS, and J. L. IRVIN: Exp. Cell Res. **28**, 120 (1962).
KIDSON, C., and K. S. KIRBY: Fractionation of DNA. Biochim. biophys. Acta (Amst.) **76**, 624 (1963).
— — J. molec. Biol. **10**, 187 (1964).
— — Nature (Lond.) **203**, 599 (1964).
KIRBY, K. S.: Biochim. biophys. Acta (Amst.) **61**, 506 (1962).
—, J. R. B. HASTINGS, and M. A. O'SULLIVAN: Biochim. biophys. Acta (Amst.) **61**, 978 (1962).
MORRIS, H. P.: Progr. exp. Tumor Res. (Basel) **3**, 370 (1963).
— Advanc. Cancer Res. **9**, 227 (1965).
MURAMATSU, M., K. SMETANA, and H. BUSCH: Cancer Res. **23**, 510 (1963).
—, and H. BUSCH: Cancer Res. **24**, 1028 (1964).
— — J. biol. Chem. **240**, 3960 (1965).
—, J. L. HODNETT, and H. BUSCH: J. biol. Chem. **241**, 1544 (1966).
OKAMURA, N., and H. BUSCH: Cancer Res. **25**, 693 (1965).
ROTH, J.: Ribonucleases in cancer cells. In Methods in cancer research, HARRIS BUSCH, Ed. New York: Academic Press (In press).
SWEENEY, M. J., J. ASHMORE, H. P. MORRIS, and G. WEBER: Cancer Res. **23**, 995 (1963).
WAGLE, S. R., H. P. MORRIS, and G. WEBER: Cancer Res. **23**, 1003 (1963).
WEBER, G., and H. P. MORRIS: Cancer Res. **23**, 987 (1963).

The Regulation of Enzyme Synthesis and it's Role in the Neoplastic Process

By H. C. Pitot

McArdle Laboratory, The Medical School, University of Wisconsin, Madison, Wisconsin, USA

With 8 figures

The environmental control of cellular metabolism is a topic which has fascinated biochemists throughout this century. However, it is only recently that our knowledge of the chemistry of living matter has grown to sufficient sophistication to allow us to study the mechanisms of cellular regulatory function. Perhaps the epitomy of the phenomena of cellular regulatory mechanisms is seen in the mammalian organism wherein a multicellular multiclonal organism must not only cope with the environment in which it lives but also must have delicately balanced mechanisms regulating an internal intracellular environment in which the cell colonies survive for their mutual benefit. Furthermore, unlike the unicellular organism whose only battle for survival is with the external elements, the mammalian organism consists predominantly of tissues and cells in the resting phase or interphase of their growth. In fact it is this maintenance of cells in specific positions in relation to one another and exhibiting homogeneous control of metabolism and growth which enables the multicellular organism to survive. It is certain aberrations in these homeostatic mechanisms which leads to aberrations in cellular function and growth which we have termed the disease, cancer. Thus, as has been stated before[1], neoplasia is a disease of evolution only existing as we know it in the multicellular organism characterized primarily, if not wholly, by aberrations in the control of cellular function and metabolism including cellular replication. This biological description of the disease, neoplasia, rightly leads us to the study of this process through a study of the abnormalities of mechanisms controlling genetic expression.

Studies on the control of genetic expression in mammalian liver

Studies on the regulation of enzyme synthesis and activity in mammalian tissues were initially stimulated by the work of KNOX[2] who described the dramatic increase in the rat liver enzyme, tryptophan pyrrolase, resulting from the administration of the amino acid, tryptophan, or the homone, cortisone. KNOX and his associates[3] further found that in the adrenalectomized animal the increase in enzyme levels resulting from these two stimuli were additive. Examination of intracellular amino acid pools indicated that only after the injection of tryptophan was the intracellular level of tryptophan dramatically increased. These studies were the first indication that the substrate of the enzyme and the hormone led to increases, in tryptophan pyrrolase activity by two different mechanisms. Later studies by GREENGARD, et al.[4], utilizing the antibiotic, actinomycin D, an inhibitor of DNA-directed RNA synthesis, showed that it was only the hormonal-stimulated increase in tryptophan pyrrolase that was inhibited by the antibiotic. The tryptophan-induced increase in the enzyme was not affected by the administration of this antibiotic to adrenalectomized animals. More recent work by SCHIMKE and his co-workers[5] showed that not only were the hormonal and substrate inductions in the adrenalectomized animal different in the ways mentioned above, but also the kinetics of induction was quite different. The hormone stimulated a complex third or fourth degree type of curve whereas tryptophan given to the adrenalectomized animal resulted in a virtual straight line increase in the level of the enzyme, the slope of the line being equal to the half-life of the enzyme. On this basis these workers suggested that the hormonal stimulation of enzyme level was a true enzyme induction resulting from an increased rate of synthesis of the enzyme while the tryptophan effect was due to the stabilization and prevention of breakdown of the enzyme by the usual process of intracellular protein turnover. Immunochemical studies confirmed these suggestions and thus demonstrated very nicely that the effects seen by GREENGARD with actinomycin D were definitely the effects on enzyme synthesis. These data, coupled with those of GREENGARD, also showed that in the adrenalectomized animal the messenger RNA template for this enzyme must be extremely stable since it apparently was able to function without any continued renewal.

As to the regulation of tryptophan pyrrolase at the enzyme level as differentiated from the above regulations which occur at either the transcriptional or translational level, WAGNER[6] has reported

Table 1. % *Inhibition of Tryptophan Pyrrolase Activity by Derivates of Nicotinic Acid*

Compound	Concentration of Inhibitor	
	5×10^{-4} M	2×10^{-4} M
Nicotinamide Mononucleotide	6%	—★
Nicotinic Acid	—	—
Deamino DPN	19	—
DPN	15	—
3-Acetylpyridine DPN	5	—
TPN	20	—
DPNH	55	30
TPNH	70	64

★ -denotes no measurable effect.

that the distal products of the pathway of the conversion of tryptophan to nicotinamide, i.e., DPN, TPN, nicotinamide mononucleotide and other related compounds, are capable of inhibiting

Table 2. *Regulation of Tryptophan Pyrrolase (TP) Synthesis by Nicotinamide in vivo in Adrenalectomized Rats* ★

	TP units g. Liver	DPM/TP/ Liver	DPM/mg Liver Prt.
Control	2	624	2595
+ Tryptophan	13	1140	2285
+ Tryptophan + Act D	15	1090	2310
+ Tryptophan + Nicotinamide	3	440	1980
+ Cortisone	17	17200	2400
+ Cortisone + Act D	4	940	2230
+ Cortisone + Nicotinamide	13	13100	2370

★ Rats given C^{14} leucine (10μC) 1 P 30′ prior to sacrifice. See ref. 5 for further details of method.

Nicotinamide (1mM) given with inducer. Actinomycin D (2 mg/kg) given 30′ priot to inducer.

the enzyme tryptophan pyrrolase. Studies in our laboratory by Dr. YOON SANG CHO have confirmed and extended these results of WAGNER as seen in Tab. 1. It may be noted that the most effective inhibitor of tryptophan pyrrolase is reduced triphosphopyridine nucleotide which even at $10^{-4}M$ gives a more than 50% inhibition of activity. Furthermore the activity curves seen in the presence of reduced TPN are sigmoidal in shape and resemble those characteric of so-called allosteric inhibition. As a further indication of the actual function of nicotinic acid derivatives in the regulation of tryptophan pyrrolase activity, the data of Tab. 2 should be considered. In this table it may be seen that utilizing the immunochemical methods described by SCHIMKE and his co-workers, it is possible to demonstrate in adrenalectomized animals a complete inhibition of the increase in tryptophan pyrrolase synthesis stimulated by tryptophan through the administration of nicotinamide. It is of interest that nicotinamide has little effect on the hormonal induction of the enzyme. The two-fold increase in the rate of tryptophan pyrrolase synthesis in the adrenalectomized animal may be demonstrated under the conditions of protein depletion followed by the administration of tryptophan.

Thus the enzyme tryptophan pyrrolase may serve as model for studies on the regulation of the control of genetic expression in mammalian systems. This enzyme exhibits both inductive and repressive control as well as feedback inhibition at the enzyme level. It now becomes of extreme interest to look at this system as well as other systems during the process of carcinogenesis as well as the ultimate product — the malignant cell itself.

Studies on the regulation of enzyme synthesis in cancer and carcinogenesis

Early studies carried out in the laboratory of Dr. EMANUEL FARBER by the author as a graduate student utilizing primary hepatocellular carcinomas produced by the feeding of ethionine or 3'-methyl-DAB indicated that in these primary lesions no significant induction of tryptophan pyrrolase could be demonstrated whereas the remaining liver could be significantly stimulated by the administration of tryptophan[7]. More critical studies were carried out later by the author as a postdoctoral fellow in the laboratory of Dr. VAN POTTER utilizing the highly differentiated hepatocellular carcinoma, the MORRIS 5123. Again this data showed rather conclusively that the induction of tryptophan pyrrolase was virtually nonexistent in this tumor[8]. A survey of a number of other highly differentiated hepatocellular carcinomas indicated this was also true. However, several highly differentiated hepatocellular carcinomas have now been demonstrated to possess some degree of inducibility of tryptophan pyrrolase [9,10]. The rather dramatic phenomenon that is seen in these lesions where it has been studied is that in the adrenalectomized animal absolutely no increase in tryptophan pyrrolase activity is noted in the tumor although the liver still responds as normal. Studies by CHO, et al.[9], as well as earlier studies by PITOT and MORRIS[8], showed that the failure of the induction of tryptophan pyrrolase was not due to abnormalities in the blood supply of the tumor compared to liver. As determined by radioactive tryptophan, the availability of the inducing agent to both tumor and liver was not significantly different. In line with these findings in hepatocellular carcinomas recent studies by KROGER and GREUER[11] on the hepatocarcinogen, N-nitroso-morpholine, indicate that administration of this material for 1 to 3 months results in complete loss of the substrate induction of tryptophan pyrrolase but no effect on the steroidal induction of this enzyme.

On the basis of the above experiments, coupled with those earlier reported by GREENGARD, it became of interest to see whether or not the induction of the enzyme in those tumors where this could be demonstrated was sensitive or resistant to the antibiotic actinomycin D. Studies reported by PITOT and CHO[12] with the Rueber hepatoma H-35 indicate that at a time at which the induction of this enzyme is resistant to actinomycin D in liver, it is completely sensitive in the tumor. These studies might then suggest that in those tumors where induction is still present, the neoplasm has lost the mechanism for the extended stabilization of the messenger RNA template for this enzyme. Furthermore it might be conjectured that in those tumors where no induction may be demonstrated the neoplasm has completely lost the ability to stabilize the messenger RNA template even for a period long enough for a single complete translation to occur.

It should be noted that studies by CHO in this laboratory on the purified enzyme from both the Rueber hepatoma H-35 and the Morris hepatoma 7793 have been unable to find any differences in kinetic properties of the tumor enzyme as related to substrate or substrate analogues of this enzyme and to its behavior on gel electrophoresis or in immunodiffusion. The tumor enzymes react to derivatives of nicotinic acid in a manner identical to that seen with the enzymes from normal liver. Thus, there does not appear to be any alteration in the structure of the enzyme itself indicating that the defects in control may not be secondary to structural mutations.

The synthesis of other enzymes in liver and their derangement in hepatocellular carcinomas has also been the subject of extensive recent investigation in this laboratory as well as in others. The enzyme, serine-threonine dehydrase, may be induced in normal liver to extremely high levels but in a series of hepatocellular carcinomas, both primary and secondary studied in this laboratory, the level of this enzyme was in most cases extremely low and could not be altered either by the administration of dietary protein or amino acids[12]. However in a few cases the enzyme was at extremely high levels in the tumors and could not be appreciably altered by various dietary manipulations. Adrenalectomy of the tumor-bearing host resulted in a lowering of the serine dehydrase level of these tumors, the Morris hepatoma 5123 and 7793, to control or below control levels. In two hepatomas, the Morris 7800 und Rueber H-35, the serine dehydrase was at relatively low levels in most instances and could be induced by the administration of either amino acids or dietary protein, but only in the intact host. Thus we see an extreme parallelism to that found in the case of tryptophan pyrrolase. When studies utilizing actinomycin D, such as that reported for tryptophan pyrrolase, were undertaken with this enzyme in two different types of hepatocellular carcinomas, the results obtained in Fig. 1 were seen. From this it is evident that in normal liver the induction of this enzyme is initially sensitive to the antibiotic and becomes resistant for about 6 to 8 hours, then reverting again to sensitivity. In contrast, in the Rueber hepatoma H-35, the induction of this enzyme is always sensitive to the antibiotic whereas in the hepatoma 5123 it appears to be always resistant. This is unlike the condition in the case of tryptophan pyrrolase where in all cases studied the induction of the enzyme appears to be sensitive to this antibiotic.

Other enzymes have also been studied in a similar manner. For example, the enzyme tyrosine-α-ketoglutarate transaminase may be induced by cortisone and this induction is sensitive to the antibiotic actinomycin. In contrast, the induction of this enzyme in tumors of adrenalectomized hosts is relatively resistant to the antibiotic. The mitochondrial enzyme, ornithine transaminase, exhibits a similar finding in being considerably more resistant to the antibiotic in the tumors than it is in liver. Recent studies in this laboratory have indicated that the enzyme glucokinase as well as the citrate cleavage enzyme, both of which are induced by

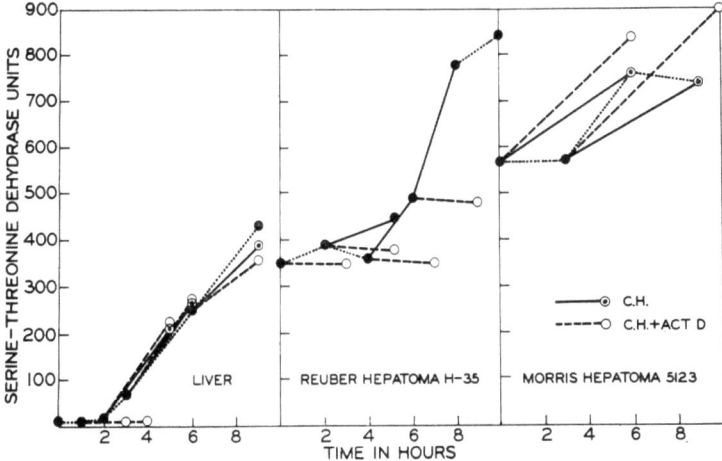

Fig. 1. The effect of Actinomycin D administration on the induction of serine-threonine dehydrase in liver, Rueber hepatoma H 35 and Morris hepatoma 5123. In all cases at 0 time a single dose of casein hydrolysate is administered by intubation to a group of rats. At intervals thereafter designated by the completely black circles a dose of Actinomycin (1 mg/kg) is given to a portion of this group. In most cases a second dose of casein hydrolysate is also given at this time. Four hours later the level of the enzyme is seen in the animals receiving Actinomycin as the open circles (and dashed lines) and those receiving a second dose of casein hydrolysate but no Actinomycin are designated as a circle with a dot in the center (solids lines). It is to be noticed that in the case of liver the induction is stopped at 0 and 1 hours by the administration of Actinomycin but not affected thereafter until between the 6 and 9 hour point. This is not actually shown on the graph but by 9 hours the system is again completely sensitive to the antibiotic. In contrast, at all times studied the induction of this enzyme in the H 35 is completely sensitive to the antibiotic whereas induction in the 5123 is completely resistant

glucose or glucose plus protein in normal liver are completely non-inducible in the tumors thus far studied. Again the question arises as to what is the effect of hepatocarcinogenesis on the control of enzyme synthesis, and does it bear any relation to the ultimate product — the neoplasm? Studies in this laboratory carried out largely by Dr. LIONEL POIRIER, and in collaboration with Drs. JAMES and ELIZABETH MILLER, have shown that the enzymes ornithine transaminase, histidase, glucokinase and the citrate cleavage enzyme lose their inducibility upon feeding of the carcinogen, 3'methyl-DAB. This loss in induction occurs very early in the case of some enzymes, especially ornithine transaminase. The results of this particular experiment are seen in Fig. 2. The loss of induction of this enzyme is very similar to that seen in the case of many of the final products. A similar finding is also seen with the enzyme, histidase, which has yet to be found inducible in any of

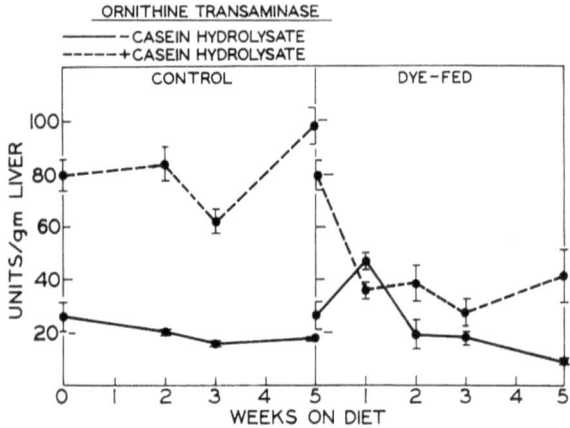

Fig. 2. The effect of feeding 3'-methyldimethylaminoazobenzene on the induction of hepatic ornithine transaminase. The dash line denotes the highest level of induction seen at the times indicated whereas the solid line is the control. The dye was fed at a level of 0.06% in a semi-synthetic diet

the hepatomas studied. Furthermore, in the case of the enzyme, serine dehydrase, although induction appears to be perfectly normal at the 3-week point of dye-feeding, this induction becomes resistant to actinomycin at a time at which it is normally sensitive in control liver. Thus, on the basis of the experiments carried out thus far, one might conjecture that very early during the process of carcinogenesis the entire liver cell population, or at least a substantial part thereof, appears to assume many of the characteristics of the ultimate product — the hepatocellular carcinoma, when the mechanisms controlling the synthesis of its enzymes are studied. Herein we may be dealing with some of the earliest alterations seen in the process of carcinogenesis. Preliminary studies have indicated that at the 3-week point placing the animal back onto a normal diet results in the reversion of most of these altered control mechanisms back to normal. However, if the animal is allowed to continue the diet until the 5-week point, a number of alterations in the control of enzyme synthesis do not appear to revert to normal at least after 2 weeks feeding of a normal diet, although malignant neoplasms rarely develop after this short period of dye feeding.

Studies on the mechanisms of the regulation of enzyme synthesis in liver and its alterations in liver cancer

The data presented thus far strongly suggest that one of the basic phenotypic alterations seen in the cancer cell is a dramatic alteration in the control of synthesis of many enzymes. It does not

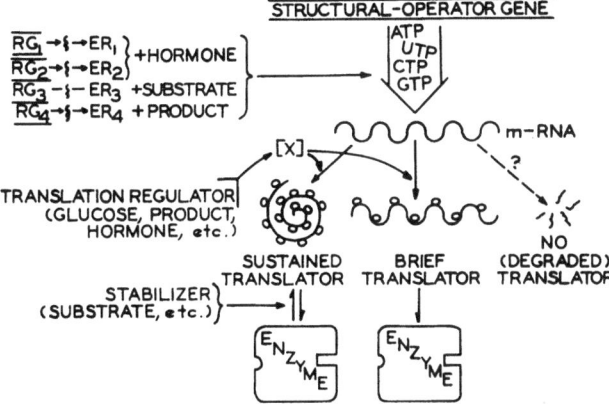

Fig. 3. A scheme for the regulation of genetic expression in multicellular organisms

appear that this alteration in control is the result of anatomical alterations in blood supply or any related conditions. Furthermore, studies such as those carried out with the enzyme, tryptophan pyrrolase, have not indicated any alteration in the structural features of the enzyme protein involved. Rather, it would appear that the mechanisms of these alterations lie in the actual area of the mechanism of the control of genetic expression at the translational and/or transcriptional steps.

Although the theoretical models suggested by microbial studies may have application in systems of multicellular organisms, these applications must be dramatically modified. Some of the modifications necessary are seen in Fig. 3. It should be noted that besides the usual operator and regulatory genes present in bacteria in the mammalian system the actual message itself may undergo one of three possible alterations. It may be degraded and never translated at all; it may be translated only briefly, such as is seen in most bacterial systems; or it may be stabilized and translated for an extended period of time, perhaps even for the lifetime of the individual. Based on the data of SCHIMKE and his co-workers, one may also argue that if the translator is sustained, at least one type of regulation may occur here and this appears to be a prevention of degradation of the product of the messenger RNA. Contral at the level of the enzyme by feedback appears to be somewhat similar in mammalian systems to that seen in microbial systems. The most distinctive feature of this scheme is, in fact, the sustained translator or, as we shall term it, the stable template which appears to be so characteristically found in mammalian tissues[13]. The data suggested above would argue for the fact that in neoplasms the stability of a number of messenger RNA templates is altered either in the direction of greater stability or in the direction of lesser stability. The question thus becomes, what is the mechanism of template stabilization and can this mechanism enter into a hereditary type of alteration such as we see in neoplasia?

In microbial systems several studies have been carried out on what appear to be stable messenger RNA templates. In particular, the study of AARONSON[14] in sporulating Bacillus cereus has suggested that the stable messenger RNA template here is in close association with the bacterial membrane. Since RNA is in many instances closely associated with membranes in mammalian systems, it would seem logical to argue that such a mechanism for stabilization may also occur in rat liver. Studies undertaken in this laboratory to see whether or not a messenger-like fraction in rat liver may be associated with the membrane of the cell have been carried out by Dr. LAMAR[15]. The results of these are seen in Fig. 4 and 5. The former figure shows an optical density and radioactive pattern of RNA isolated from the cytoplasm of rat liver after the administration of a low dose of actinomycin D. It may be noted that the optical density pattern is that usually seen minus the soluble RNA fraction which is removed by prior treatment with Sephadex-G 200. The radioactive pattern, however, is almost exclusively in the area of the 18 S peak. Furthermore, base composition of this fraction indicates that it is very much like that of DNA. Figure 5 shows the same fraction after the lysis of the ribosomes on the membrane and here again a peak of RNA having approximately the same sedimentation value as that in Fig. 4 may be noted. This RNA also has a base composition identical to that seen in DNA and it is relatively resistant to the action of exogenous

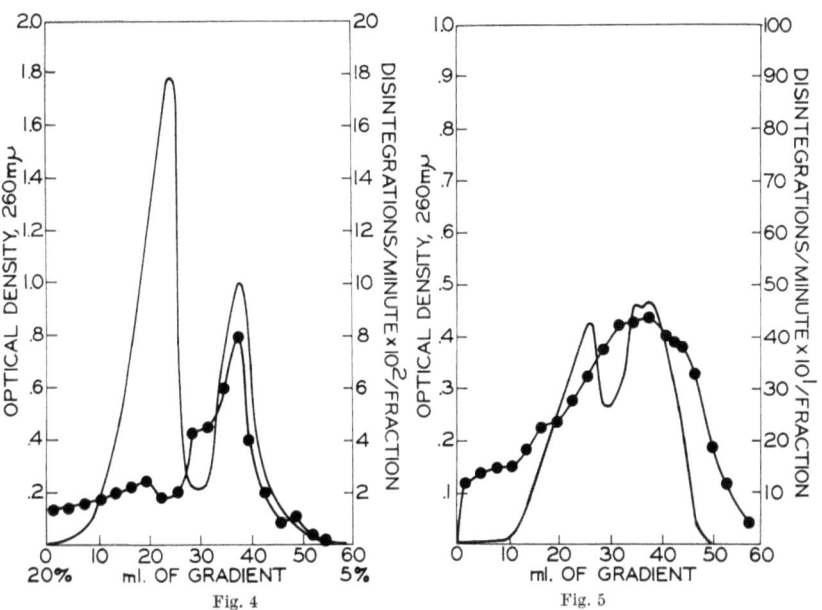

Fig. 4. Sucrose gradient pattern of cytoplasmic RNA from liver of animals treated with Actinomycin D. The solid line shows the optical density pattern and the line connected by solid dots shows the radioactivity of the fractions

Fig. 5. Sucrose gradient of RNA extracted from microsomal membranes after treatment with pyrophosphate. (See legend of Fig. 4)

nucleases. It would thus appear that this material in close association with the membrane of the cell exhibits a number of the characteristics usually reserved for so-called messenger RNA.

Studies in this laboratory by Dr. R. SUESS and G. BLOBEL[16] have indicated confirming the data of WEBB et al.[17] obtained *in vivo*, that polysomes will bind *in vitro* to the membranes of the endoplasmic reticulum of liver (see Fig. 6). This binding is of sufficient strength such that it is not altered by 24 hours of centrifugal force at 39000 (rpm) revolutions per minute. In Figure 6 is shown the amount of radioactive polysomes bound to membranes from both liver and highly differentiated hepatocellular carcinomas. It will be noted that the binding to liver membranes is considerably greater than that seen with the hepatoma membranes. The source of the polysomes themselves appears to not matter since either liver or highly or poorly differentiated carcinomas all give the same results with the same membrane preparation. Thus the alteration and binding characteristics appear to be a result of membrane alteration rather than polysome alteration. On the basis

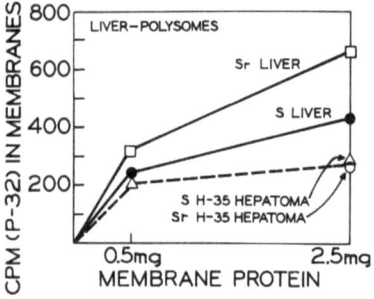

Fig. 6. Polysome binding to liver and hepatoma membranes of the endoplasmic reticulum. SR denotes membranes of the rough endoplasmic reticulum treated with chelating agents and S denotes membranes of the smooth endoplasmic reticulum. The measurement is of the radioactive P 32 labeled polysomes still in association with membranes layered over 2.0 molar sucrose after centrifugation at 150000 × g for 24 hours. Under these condotions all polysomes which are not bound to membranes sediment through the sucrose

of these findings the models seen in Fig. 7 and 8 are proposed as an explanation for template stability and its alteration in the neoplasms discussed[18]. The polysome unit is shown as a spiral resting on a membrane made up of a mosaic of chemical structures denoted by symbols having the signification O, D, W, etc. On the messenger RNA are regularly spaced boxes or attachment points presumably of certain base sequences which are in some way ,,complimentary" to certain structures on the membrane. Only 1 ribosome is represented in the polysome for the sake of clarity. In Fig. 8 the same scheme is depicted but now it will be noticed that the mosaic has been changed. It is suggested that this is what occurs in the neoplastic cell. The alteration in the mosaic of the membrane results in the situation that many of the attachment points on the RNA do not coincide with similar structures on the membrane, thus making for a looser association and a decreased template stability, e.g., serine dehydrase in the H-35. One may also imagine an increased template stability by having coincidence of the membrane structures with more attachment points than are normally seen in liver. Although in this model the association of attachment points with messenger RNA is depicted, an alternate model would suggest an association of the growing polypeptide chain with certain membrane structures. If this were true then one

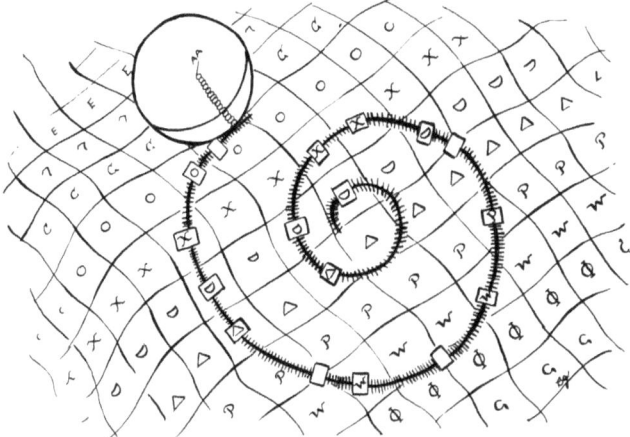

Fig. 7. Diagramatic representation of association between a polysome unit and the endoplasmic reticulum in liver. The messenger RNA template is represented as a coiled line with cross ties and boxed symbols representing nucleotide sequences having a structure capable of associating with a similarly designated symbol on the membrane surface which is represented by the background squares and symbols. Only one ribosome is shown associated with the template RNA for the purpose of clarity

might argue that enzyme induction and repression are involved with template stabilization. In fact enzyme induction itself may result from the induced stabilization of a template by the inducing molecule.

The question of the apparent hereditary transmission of the phenotypic characteristics noted in the cancer cell may also be considered in light of Fig. 7 and 8. In essence, the mechanism is that as suggested by the work of SONNEBORN in paramecia[19]. In this protozoan, environmental stimuli may alter the molecular arrangement of the mosaic pattern of the cell surface membrane of this organism. This altered mosaic may then be maintained for hundreds of cell generations without any concomitant genetic alterations. Such a condition may also occur in the mosaic of the membrane of endoplasmic reticulum, thus resulting in an altered template stability — such alteration being entirely extragenetic. As suggested earlier by PITOT and HEIDELBERGER[10] on the basis of purely theoretical grounds with regulatory circuits, the hypothesis here suggests

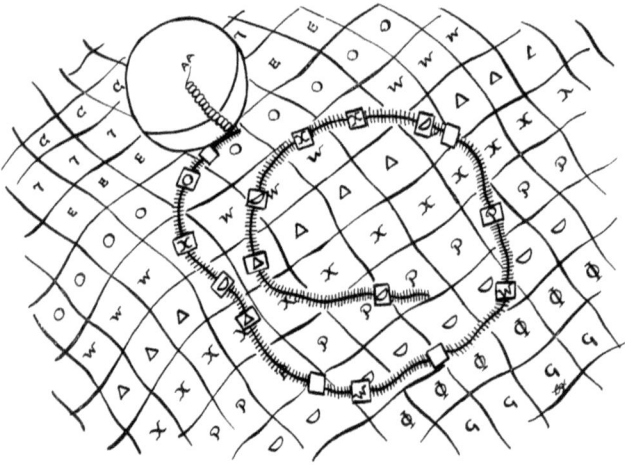

Fig. 8. Diagramatic representation of the association between a polysome unit and the endoplasmic reticulum in a hepatoma. Notice that the messenger RNA template is not altered in its sequence of boxed symbols. However, the mosaic pattern of the membrane has been altered in such a way that only a few of the like symbols of membrane and RNA are in a juxtaposition. Such an alteration could lead to a decreased stability of the association of RNA and membrane

that in the instances studied reversion of the neoplastic change is possible. Although reversion of neoplasms has been found to occur in nature in several instances[21], the controlled reversion of the neoplastic process has not yet been achieved. Until such a controlled reversion is achieved, the mechanisms mentioned here remain in the relatively theoretical realm.

References

[1] PITOT, H. C.: Cancer Res. **23**, 1474 (1963).
[2] KNOX, W. E.: Brit. J. exp. Path. **32**, 462 (1951).
[3] — Adaptive Enzymes in the Regulation of Animal Metabolism in Physalogical Adaptions, p. 107. Amer. Physiol. Soc. Washington, D. C., 1958.
[4] GREENGARD, O., M. A. SMITH, and G. ACO: J. biol. Chem. **238**, 1548 (1963).
[5] SCHIMKE, R. T., E. W. SWEENEY, and C. M. BERLIN. J. biol. Chem. **240**, 322 (1965).
[6] WAGNER, C.: Biochem. biophys. Res. Commun. **17**, 668 (1964).

[7] PITOT, H. C.: Bull. Tulane med. Fac. **19**, 17 (1959).
[8] —, and H. P. MORRIS: Cancer Res. **21**, 1009 (1961).
[9] CHO, Y. S., H. C. PITOT, and H. P. MORRIS: Cancer Res. **24**, 52 (1964).
[10] DYER, H. M., P. GUILLINO, and H. P. MORRIS: Cancer Res. **24**, 97 (1964).
[11] KRÖGER, H., u. B. GRENER: Hoppe-Seylers Z. physiol. Chem. **342**, 148 (1965).
[12] PITOT, H. C., and Y. S. CHO: Prog. exp. Tumor Res. (Basel) **7**, 158 (1965).
[13] — Metabolic Regulation in Metazoa. In TAYLOR, J. H. (ed.): Molecular Genetics, vol. **2**, New York: Academic Press, Inc. (In press).
[14] ARONSON, A. J.: molec. Biol. **13**, 92 (1965).
[15] LAMAR, C., M. PRIVAL, and H. PITOT: Cancer Res. (In press).
[16] SÜSS, R., G. BLOBEL, and H. C. PITOT: Biochem. biophys. Res. Commun. (In press).
[17] WEBB, T. E., G. BLOBEL, V. R. POTTER, and H. P. MORRIS: Cancer Res. **25**, 1219 (1965).
[18] PITOT, H. C., C. PERAINO, and C. LAMAR: Altered Template Stability in Rat Hepatomas in Developmental and Metabolic Control Mechanisms and Neoplasia, p. 413. Baltimore: Williams and Wilkins 1965.
[19] SONNEBORN, T. M.: Proc. nat. Acad. Sci. (Wash.) **51**, 915 (1964).
[20] PITOT, H. C., and C. HEIDELBERGER: Cancer Res. **23**, 1694 (1963).
[21] — Ann. Rev. Biochem., vol. 35 (In press).

Zur Frage tumorspezifischer Histone

Von H. LETTRÉ

Institut für experimentelle Krebsforschung der Universität Heidelberg

In Untersuchungen, welche wir 1960 durchgeführt haben, konnten wir die Ergebnisse von H. BUSCH bestätigen, daß in den Histonen von Tumorzellen eine Fraktion (RP2-L) auftritt, die in den Histonen normaler Zellen nicht zu finden ist. Wir verwendeten die Methode von DAVIS und BUSCH, um die Veränderungen des Histonspektrums von Tumorzellen unter der Einwirkung cytotoxischer Substanzen zu untersuchen, was für die Aufklärung ihres Wirkungsmechanismus von Bedeutung sein kann. HIDVÉGI u. Mitarb. [Brit. J. Cancer **17**, 377 (1963)] konnten in den Histonen aus Zellen des normalen Knochenmarks des Kaninchens die RP2-L-Fraktion nachweisen. Wir modifizierten die Methode von DAVIS und BUSCH so, daß wir das Zeitintervall zwischen der Injektion des als Indicator verwendeten radioaktiv markierten Lysins und der Tötung der Versuchstiere von 2 auf 6 Std ausdehnten. Hierbei konnten wir auch in den Histonen der normalen Leber die RP2-L-Fraktion nachweisen [Naturwissenschaften **53**, 134 (1966)]. Demnach scheint nur die Bildungsgeschwindigkeit der RP2-L-Fraktion bei verschiedenen Zellarten verschieden zu sein. Die Frage, ob die RP2-L-Fraktion ein spezifisches Kennzeichen maligner Zellen ist, bedarf einer Überprüfung unter Berücksichtigung des „turnovers" der Histone verschiedener Zellarten.

Allosterie-Effekte an Enzymen aus normalen und leukämischen Leukocyten

Von A. W. HOLLDORF

Biochemisches Institut der Universität Freiburg

Durch Induktion und Repression kommen Änderungen von Enzymaktivitäten in Zeiträumen von Minuten bis Stunden zustande. Wesentlich schneller können Enzymaktivitäten durch allosterische Hemmung oder Aktivierung verändert werden. Über Bedeutung und Verbreitung dieser bei Mikroorganismen recht gut untersuchten Regelvorgänge in normalen und in pathologischen Zellen von Säugern ist bisher viel weniger bekannt als über Induktion und Repression in diesen Zellen. Anknüpfend an Beobachtungen von PRAGER und BRYAN [Amer. Chem. Soc.; Abstr. of Papers, 145. Meeting (1963), S. 45] untersuchten wir Allosterieeffekte an der Aspartattranscarbamylase und an der Desoxycytidin-5'-monophosphat-Desaminase aus normalen und aus leukämischen Leukocyten. Es ergab sich dabei folgendes:

1. Die Aspartattranscarbamylase aus normalen wie aus leukämischen Leukocyten wird durch Cytidin-5'-monophosphat (CMP) allosterisch gehemmt. Durch 10^{-3} bzw. 10^{-2} M CMP wird das Enzym aus allen Gruppen von leukämischen Zellen (akute und chron. Myelosen, chron. Lymphadenosen) wesentlich weniger gehemmt als das Enzym aus normalen Zellen.

2. Die Desoxycytidin-5'-monophosphat (dCMP)-Desaminase aus allen Zelltypen wird durch Desoxyxytidin-5'-triphosphat (10^{-5} M) aktiviert. Diese Aktivierbarkeit ist beim Enzym aus Zellen von chronischen Leukosen wesentlich höher als bei dem Enzym aus normalen Leukocyten oder aus Zellen von akuten Myelosen.

3. Das gleiche Enzym aus allen Zelltypen wird durch Thymidin-5'-triphosphat (10^{-4} M) gehemmt, wobei sich keine Unterschiede zwischen den Enzymen aus den verschiedenen Zelltypen ergeben.

Zu den Ergebnissen ist folgendes zu bemerken: Es ergeben sich eindeutige Unterschiede im Verhalten der beiden Enzyme aus normalen und aus pathologischen Zellen. Ob diesen Erscheinungen jedoch eine biologische Bedeutung zukommt ist fraglich, da die für die Erzielung der Effekte erforderlichen Nucleotidkonzentrationen wesentlich über den stationären Konzentrationen dieser Metabolite in den Zellen liegen. Dies gilt auch für viele andere Allosterieeffekte. Andererseits sind an allosterisch beeinflußbaren Enzymen die allosterischen Bindungsorte meist wesentlich empfindlicher gegenüber exogenen Noxen (Schwermetalle, Hitze, Röntgenstrahlen usw.) als die katalytisch aktiven Zentren. Deshalb erscheint es gerechtfertigt, derartige Veränderungen oder gar den Verlust allosterischer Eigenschaften von Enzymen bei Entdifferenzierungsvorgängen zu untersuchen. Weiterhin ergibt sich die Frage, ob bei therapeutischen Eingriffen Veränderungen an allosterisch beeinflußbaren Enzymen erfolgen. Es ist denkbar, daß hier durch Chemotherapeutica noch bestehende celluläre Regelmechanismen zerstört werden, bevor der eigentliche therapeutische Effekt, z. B. die Hemmung der katalytischen Wirksamkeit eines Enzyms, erzielt wird.

Minderung der Euchromatinanteile in den Zellkernen von Tumorzellen

Von E. HARBERS

Institut für Medizinische Physik und Biophysik der Universität Göttingen

In somatischen Zellen ist von der Gesamtinformation der Desoxyribonucleinsäure (DNS) immer nur ein gewisser Anteil verfügbar entsprechend den spezifischen Leistungen der betreffenden Zellart. Zahlreiche Beobachtungen sprechen dafür, daß nur — oder weit überwiegend — die im Euchromatin enthaltenen DNS-Anteile für den Zellstoffwechsel zur Verfügung stehen, während die im dichtgepackten Heterochromatin enthaltene Information gar nicht — oder nur geringfügig — verwendet werden kann. Von FRENSTER et al.[1] wurde eine Methode beschrieben, die eine Fraktionierung des Zellkerninhaltes in eu- und heterochromatinreiche Fraktionen gestattet. Die Anwendung dieses Verfahrens ergab u. a., daß Actinomycin, dessen biologische Wirkung auf eine Komplexbildung mit der DNS zurückgeht, die DNS des Euchromatin sehr viel dichter besetzt als die des Heterochromatin; ferner zeigte das Euchromatin im Vergleich zum Heterochromatin eine wesentlich bessere Matrizenwirksamkeit für die RNS-Polymerase aus *Escherichia coli*[2]. In der Folgezeit wurde das Verhalten von Actinomycin ausgenutzt als Kriterium, um die optimalen Fraktionierungsbedingungen (vor allem die Dauer der erforderlichen Ultraschalleinwirkung) für Zellkerne aus verschiedenen Geweben der Ratte zu ermitteln und so auf das ungefähre Mengenverhältnis von Eu- und Heterochromatin rückschließen zu können. Aus den bisher vorliegenden Ergebnissen geht hervor, daß in drei untersuchten Transplantationstumoren (Walker-Carcinom und Yoshida-Sarkom der Ratte, Ehrlich-Ascitestumor der Maus) der Anteil der DNS im Euchromatin herabgesetzt ist gegenüber vier normalen Geweben der Ratte (Gehirn, Leber, Lunge und Nieren). Aus einer weiteren ersten abgeschlossenen Versuchsreihe geht hervor, daß es während der durch kontinuierliche Verabfolgung von Diäthylnitrosamin ausgelösten Cancerogenese in der Leber ebenfalls zu einer Minderung der Euchromatisierung kommt, die eine Einschränkung in der Verfügbarkeit der Information zur Folge haben sollte; in die gleiche Richtung weisen Ergebnisse histochemischer Untersuchungen von SANDRITTER[3]. Die geschilderten Befunde harmonieren mit den zahlreichen Beobachtungen, daß es während der Cancerogenese zu einem Verlust von Enzymen (Deletionen) kommen kann, die sich nunmehr vielleicht als Verschiebungen in der Verfügbarkeit der Information deuten lassen. Allerdings ist über den normalerweise wirksamen Kontrollmechanismus, der jeweils nach abgeschlossener Zellteilung Art und Ausmaß der Euchromatisierung regelt, bisher noch nichts bekannt. Die in den untersuchten Tumorzellen beobachtete herabgesetzte Euchromatisierung ist weiterhin noch von Interesse im Hinblick auf den Effekt von chemotherapeutisch wirksamen Stoffen, deren biologische Wirkung — wie z. B. bei den Actinomycinen — auf eine Wechselwirkung mit der DNS zurückgeht.

Literatur

[1] FRENSTER, J. H., V. G. ALLFREY, and A. E. MIRSKY: Proc. nat. Acad. Sci. (Wash.) 50, 1026 (1963).
[2] VOGT, M., u. E. HARBERS: (In Vorbereitung).
[3] SANDRITTER, W.: (Pers. Mitteilung).

Regulation der Thymidinkinase im Zellteilungscyclus

Von W. SACHSENMAIER

Institut für experimentelle Krebsforschung der Universität Heidelberg

Mit 1 Abbildung

Herr Prof. PITOT hat ein Modell diskutiert, das den malignen Charakter der Tumorzellen ursächlich auf eine gestörte Enzymregulation zurückführt. Dieses Modell basiert auf Untersuchungen über die Induzierbarkeit einiger katabolischer Enzyme in hochdifferenzierten „minimum deviation" Tumoren. Hieran knüpft sich zwangsläufig die Frage, welche Enzyme und welche Kontrollmechanismen spezifisch für die Steuerung des Zellwachstums verantwortlich sind; denn eine Störung gerade dieser Mechanismen muß offenbar für das unkontrollierte Wachstum der Tumoren verantwortlich sein.

In den letzten drei Jahren haben wir in Heidelberg als Fortsetzung einer am McArdle Memorial Institute for Cancer Research, Madison, begonnenen Arbeit versucht, einige Informationen über jene biochemischen Prozesse zu gewinnen, die spezifisch an der Mitoseregulation beteiligt sind. Als Studienobjekt verwenden wir ein für die Untersuchung biochemischer Vorgänge im Zellteilungscyclus besonders geeignet erscheinendes System, einen Schleimpilz, *Physarum polycephalum* (Klasse: Myxomycetae). Dieser Organismus bildet in seiner vegetativen Phase vielkernige Plasmodien, die etwa alle 9 Std natürlich synchrone Kernmitosen durchführen. Die Teilungssynchronie des Systems erlaubt es, Eingriffe mit Hemmstoffen bzw. mit Röntgen- und UV-Strahlung in definierten Stadien des Zellcyclus vorzunehmen. Aus den beobachteten Effekten können Rückschlüsse auf die mit der Mitose korrelierten Stoffwechselprozesse gezogen werden. Versuche mit Hemmstoffen der Ribonucleinsäure (RNS)- und Proteinsynthese haben gezeigt, daß als Vorbedingung zum Eintritt in die Mitose RNS bis etwa 2 Std vor Prophase und Protein bis etwa 20 min vor Prophase neu synthetisiert werden müssen[1]. Proteinsynthese ist offenbar auch noch während der frühen Mitosestadien erforderlich, um den normalen

Ablauf der weiteren Stadien bis zur Telophase zu gewährleisten. Diese Befunde stehen im Einklang mit der auch von anderen Autoren vertretenen Auffassung, daß der Ablauf des Zellcyclus weit-

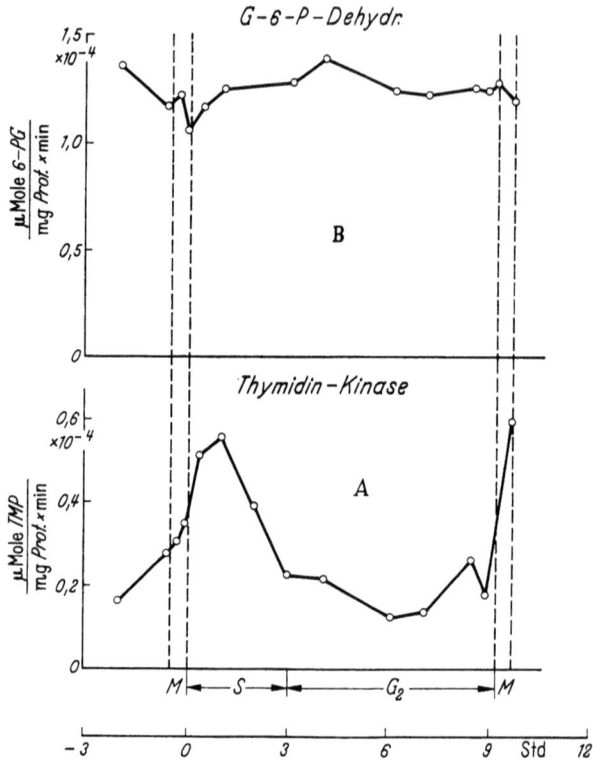

Abb. 1 a und b. Enzymaktivitäten in verschiedenen Stadien des Teilungscyclus synchroner Makroplasmodien von Physarum polycephalum. a) Thymidinkinase; b) Glucose-6-phosphatdehydrogenase. M Mitose (Prophase bis Telophase); S DNS-Syntheseperiode; G_2 prämitotische Ruhepause (aus [3])

gehend durch Genregulation gesteuert wird[2]. Hiernach wäre zu erwarten, daß bestimmte Enzyme in definierten Stadien des Zellcyclus periodisch induziert werden. Tatsächlich beobachten wir im Falle des am Desoxyribonucleinsäure (DNS)-Stoffwechsel beteiligten Enzyms Thymidinkinase einen plötzlichen Anstieg und Wiederabfall der Aktivität im Verlauf des Zellcyclus[3]. Das Maximum liegt kurz nach einer Mitose und fällt zeitlich mit dem Beginn der DNS-

Synthese bei diesem Organismus zusammen. Das zum Vergleich untersuchte Zwischenferment sowie einige andere orientierend von Herrn Dr. PETTE untersuchte Enzyme des Kohlenhydratstoffwechsels zeigen keine oder keine ausgeprägte Periodizität. Dies könnte bedeuten, daß nur die Aktivität solcher Enzyme periodisch verändert wird, die an periodischen Prozessen im Zellcyclus, wie DNS-Synthese und möglicherweise auch Mitose, beteiligt sind.

Herr v. FOURNIER konnte kürzlich im Rahmen seiner Doktorarbeit zeigen, daß der Aktivitätsanstieg der Thymidinkinase durch den Proteinsynthesehemmstoff Actidion unterdrückt wird, sofern dieser unmittelbar vor Beginn des erwarteten Anstiegs zugesetzt wird. Vorläufige Versuche mit Actinomycin lassen erkennen, daß auch durch Hemmung der RNS-Synthese der Aktivitätsanstieg verhindert werden kann, wobei der Hemmstoff allerdings schon zu einem etwas früheren Zeitpunkt, etwa 60 min vor dem erwarteten Aktivitätsanstieg, zugesetzt werden muß. Dies spricht dafür, daß RNS- und Proteinsynthese notwendige Voraussetzungen für den Enzymanstieg sind und die Thymidinkinase demnach durch periodische Induktion reguliert wird. Die unterschiedlichen Zeitgrenzen bis zu denen der Enzymanstieg gegen Actinomycin und Actidion empfindlich ist, könnten vielleicht für eine zeitlich getrennte Regulation auf der Ebene der „Transscription" und der „Translation" sprechen.

Unter Bedingungen, unter denen Actidion und Actinomycin den Enzymanstieg hemmen, wird auch der Eintritt in die Mitose verhindert. Dies läßt auf eine enge Koppelung der Regulation der Thymidinkinase mit der Regulation anderer, bislang noch weitgehend unbekannter Proteine schließen, die in spezifischer Weise mit dem Mitosegeschehen korreliert sind. Es wäre denkbar, daß die Thymidinkinase zusammen mit jenen noch unbekannten Enzymen über ein „Mitoseoperon" reguliert wird, das den Ablauf periodischer Vorgänge im Zellcyclus, wie DNS-Synthese und Mitose, kontrolliert. Der entscheidende Unterschied zwischen einer Normal- und einer Tumorzelle könnte nach dieser Vorstellung darin liegen, daß in der Normalzelle die periodische Aktivierung dieses „Mitoseoperons" durch extracelluläre Faktoren direkt oder indirekt, eventuell über die Zellmembran, blockiert werden kann. Die Tumorzelle hingegen wäre infolge einer Störung der für das Wachstum spezifischen Enzymregulation, etwa durch Veränderungen von Membraneigenschaften im Sinne des von Dr. PITOT diskutierten Modells, gegen diese extracellulären Regulatoren mehr oder weniger resistent.

Literatur

[1] SACHSENMAIER, W.: In 3. Wiss. Konf. d. Ges. Deutscher Naturforscher u. Ärzte, Semmering b. Wien, 1965 (Hrsg.: P. SITTE), S. 139 Berlin-Heidelberg-New York: Springer 1966.
[2] HOTTA, Y., and H. STERN: Proc. nat. Acad. Sci. (Wash.) **49**, 648 (1963).
[3] SACHSENMAIER, W., u. D. IVES: Biochem. Z. **343**, 399 (1965).

Stoffwechsel der Ribonucleinsäure in der Leber während der Applikation von N-Nitroso-morpholin

Von H. KRÖGER

Biochemisches Institut der Universität Freiburg

Wir befassen uns seit einiger Zeit mit den Stoffwechselveränderungen, die in der Leber auftreten, wenn Normalzellen sich in Tumorzellen verwandeln. Wir erreichen die experimentelle Carcinogenese mit N-Nitroso-morpholin, dessen stark carcinogene Wirkung von DRUCKREY et al.[1] festgestellt wurde. In früheren Untersuchungen konnten wir zeigen, daß die Substratinduktion der Tyrosin-α-Ketoglutarat-Transaminase nach 20 Tagen, die der Tryptophan-Pyrrolase nach 70 Tagen Fütterung mit N-Nitrosomorpholin herabgesetzt ist[2]. Die Cortisoninduktion der beiden Enzyme wurde nicht beeinflußt.

Wir haben nun nach dem Grund der verminderten Substratinduktion gesucht. Vor allem interessierte uns die Stabilität der Messenger-RNS (M-RNS); denn von PITOT et al.[3] wurde in Hepatomen eine herabgesetzte Stabilität dieser RNS-Art gefunden. Wir haben für die Tryptophan-Pyrrolase ein Verfahren eingesetzt, mit dem man ohne Actinomycin Aussagen über die Stabilität der M-RNS machen kann[4]. Mit dieser Methode konnten wir zeigen, daß die Halblebenszeit der M-RNS für die Tryptophan-Pyrrolase 4 bis 5 Std beträgt. In der präcancerösen Leber von Tieren, die 170 Tage N-Nitroso-morpholin erhalten hatten, hat die durch Cortison aufgebaute M-RNS die gleiche Stabilität wie in der normalen Leber.

Weiterhin haben wir die RNS-Synthese in der präcancerösen Leber studiert. Dazu wurde den Tieren Orotsäure-6-C^{14} intraperitoneal injiziert und die RNS aus der Leber nach dem Verfahren von GEORGIEV et al.[5] isoliert. Es zeigte sich, daß die RNS-Synthese schon bald nach Fütterungsbeginn gesteigert ist. Diese Veränderungen betreffen im wesentlichen die Kernfraktion.

In weiteren Untersuchungen fanden wir, daß die Synthese von spezifischer M-RNS während der Substratinduktion von Tyrosin-

α-Ketoglutarat-Transaminase und Tryptophan-Pyrrolase herabgesetzt ist. Nach den bisher vorliegenden Ergebnissen greift die carcinogene Substanz N-Nitroso-morpholin schon sehr bald nach Fütterungsbeginn in die Stoffwechselregulation der Leber ein.

Literatur

[1] DRUCKREY, H., R. PREUSSMANN, D. SCHMÄHL und M. MÜLLER: Naturwissenschaften 48, 134 (1961).
[2] KRÖGER, H. ,u. B. GREUER: Hoppe-Seylers Z. physiol. Chem. 342, 148 (1965).
[3] GEORGIEV, G. P., and V. L. MANTIEVA: Biokhimiya 25, 103 (1960).
[4] KRÖGER, H., J. PHILIPP und A. WICKE: Biochem. Z. 344, 227 (1966).
[5] PITOT, H. C., C. PERAINO, A. N. PRIES, and L. KENNAN: Advances Enzyme Regulation 3, 359 (1965).

Untersuchungen über den Nucleinsäurestoffwechsel in Zellkernen und Mitochondrien von Morris-Hepatomen

Von D. NEUBERT

Pharmakologisches Institut der Freien Universität Berlin

Mit 3 Abbildungen

Wir haben in Zusammenarbeit mit Dr. H. P. MORRIS vom National Cancer Institute, Bethesda, Untersuchungen über den Nucleinsäurestoffwechsel isolierter Zellkerne und Mitochondrien von Hepa-

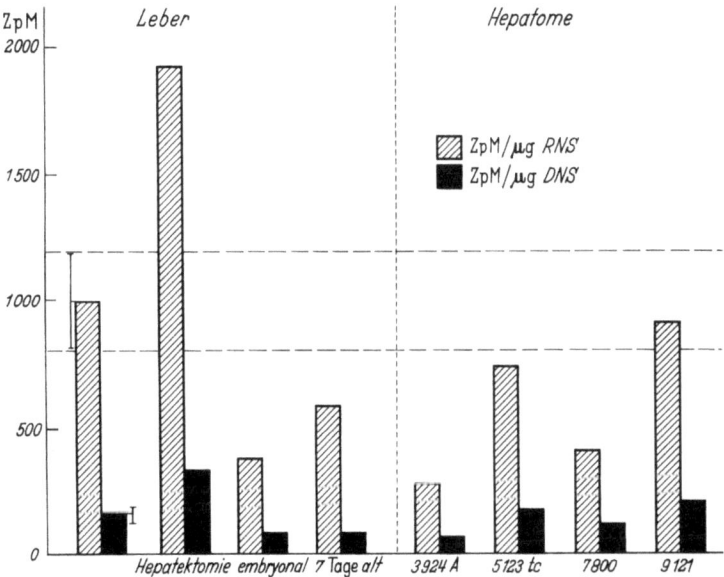

Abb. 1. Einbau von UTP-H^3 in die RNS isolierter Zellkerne. 5 min Inkubation, pH 8,0, 124 μM UTP, 1000 μM ATP, GTP, CTP. Die Zellkerne wurden durch Abtrennung in 2,2 M Saccharose rein dargestellt Die Wachstumsrate der aufgeführten Hepatome nimmt von li. nach re. ab (3924A etwa 12 Tage, 9121 etwa 120 Tage)

tomen durchgeführt und die enzymatischen Aktivitäten mit denen von Leberzellorganellen verglichen.

1. Isolierte Zellkerne: Es ist seit längerer Zeit bekannt, daß in den Zellkernen einiger schnellwachsender Gewebe (z. B. Lebergewebe nach partieller Hepatektomie) eine erhöhte *RNS-Polymerase*-Aktivität nachgewiesen werden kann. Die isolierten Zellkerne aus verschiedenen Morris-Hepatomen zeigen dagegen keinen gegenüber Lebergewebe erhöhten Einbau von UTP in die RNS (Abb. 1). Bemerkenswerterweise ist auch das Ausmaß der RNS-Synthese in den Leberzellkernen von wachsenden Ratten verglichen mit dem von ausgewachsenen Tieren nicht erhöht, sondern eindeutig erniedrigt. Da die *Proteinsynthese*-Aktivität sowohl bei Hepatomen[1] als auch bei wachsendem Lebergewebe deutlich erhöht ist, scheint eine gewisse Diskrepanz zu bestehen. Dies mag einmal damit erklärt werden, daß bei unseren Versuchen möglicherweise nicht nur m-RNS markiert wurde, sondern zum Teil auch r-RNS. Außerdem könnte aber auch die Stabilität von m-RNS bei den Hepatomen erhöht sein — wie PITOT et al.[2] es für einige Fermente gezeigt haben —, so daß für eine erhöhte Proteinsynthese keine gesteigerte RNS-Synthese notwendig wäre. Die Befunde ergeben, daß eine erhöhte Wachstumsrate durchaus nicht immer mit einer gesteigerten RNS-Polymerase-Aktivität im Zellkern einhergehen muß, sondern daß diese Aktivität sogar vermindert sein kann.

2. Isolierte Mitochondrien: Besonders eingehend haben wir uns mit dem Mitochondrienstoffwechsel beschäftigt, nachdem wir vor einigen Jahren eine RNS-Polymerase-Aktivität auch in Warmblüter-Mitochondrien nachweisen konnten[3,4]. Bei isolierten Mitochondrien von Morris-Hepatomen war die *RNS-Polymerase*-Aktivität gegenüber Mitochondrien aus normalem Lebergewebe regelmäßig erhöht. Die *Proteinsynthese*-Aktivität (Tabelle) der Mitochondrien aus den Tumoren entsprach jedoch vollkommen der von Lebergewebe. Im Prinzip ähnliche — sonst allerdings von unseren abweichende — Befunde konnten GRAFFI et al.[5] an anderen Impftumoren erheben. Es scheint also eine Veränderung im Nucleinsäurestoffwechsel von Tumormitochondrien vor-

Tabelle. *Einbau von C^{14}-Leucin in Proteine isolierter Mitochondrien*

	ZpM/mg Protein (10 min Inkubation)
Leber (erwachsene Tiere) ..	168 ∓ 30
Leber (embryonal)	250
Hepatom 3924 A........	180
5123 tc	140 ∓ 30
7800..........	130 ∓ 30
9121..........	196
7787..........	278
Yoshida-Sarkom	170

Der gemessene C^{14}-Leucineinbau ist vollkommen RNase-resistent.
Chloramphenicol hemmt die Reaktion zu 60 bis 80%.

zuliegen, die auch bei anderen Tumoren wie Jensen-, Walker-, Yoshida-, Shay-Chloro- oder DS-Tumoren der Ratte nachzuweisen ist[6]. Mitochondrien enthalten auch eine kleine Menge DNS, die sich mit radioaktivem Thymidin markieren läßt[7]. Durch ihre Markierungsrate nach Gabe von H^3- oder C^{14}-Thymidin läßt sich die mitochondriale DNS klar von der Zellkern-DNS abgrenzen. Kürzlich konnten wir auch in vitro eine *DNS-Polymerase*-Aktivität in

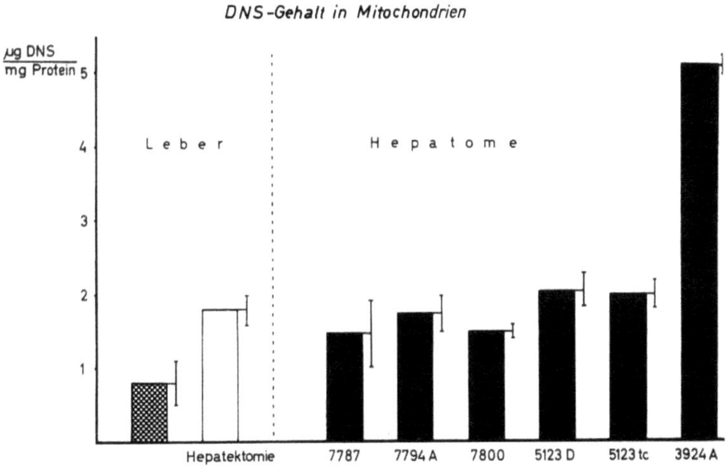

DNS-Gehalt in Mitochondrien

Abb. 2. DNS-Gehalt in Mitochondrien. Die durch Flotieren im Saccharose-Gradienten gereinigten Mitochondrien wurden fünfmal mit kalter, 0,2 n Perchlorsäure und anschließend mit Alkohol/Äther, Benzol und Äther gewaschen. Die DNS wurde mit 0,5 n Perchlorsäure (15 min bei 75°) zweimal extrahiert und nach der Burton-Methode bestimmt. Proteinbestimmung mit der Biuret-Methode

Mitochondrien nachweisen[8]. Der *DNS-Gehalt* von Tumormitochondrien liegt ausnahmslos höher als der von normalen Mitochondrien. Allerdings haben wir auch nach partieller Hepatektomie einen Anstieg des DNS-Gehaltes gefunden (Abb. 2). Wir sind dabei, zu prüfen, ob sich die DNS in Tumormitochondrien auch in ihren Eigenschaften von der normaler Zellen unterscheidet.

Nach einmaliger Markierung mitochondrialer DNS mit H^3-Thymidin in vivo haben wir die *Geschwindigkeit des Absinkens der spezifischen Aktivität* über einen Zeitraum von 5 bis 30 Tagen untersucht. Die ,,Halbwertzeit" — wenn man den Ausdruck einmal benutzen darf — der mitochondrialen DNS beträgt bei Lebergewebe erwachsener Ratten 8 bis 9 Tage. Dieser Umsatz entspricht möglicherweise einem Umsatz der Hauptkomponenten der Mitochondrien. Die „Halbwertzeit" der spezifischen Aktivität mitochondrialer DNS entspricht bei den untersuchten Hepatomen etwa den Werten, die für nichtwachsendes Lebergewebe gefunden wurden

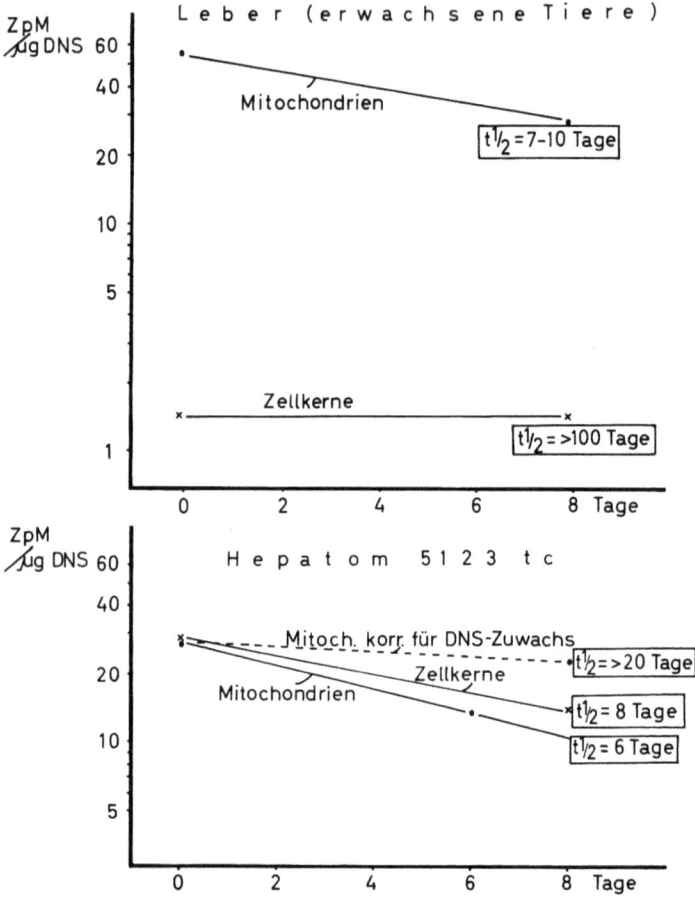

Abb. 3. Spezifische Aktivität der DNS in Zellkernen und Mitochondrien nach einmaliger Gabe von H^3-Thymidin. Die weiblichen Buffalo-Ratten erhielten zum Zeitpunkt Null 1 mC/kg (6C/mmol) H^3-Thymidin intravenös injiziert. 90 min später wurde den Tieren eine Dosis von 400 mg/kg inaktives Thymidin intraperitoneal verabreicht. Verschiedene Zeiten nach der Thymidingabe wurden die Zellkerne und Mitochondrien aus der Leber und dem Hepatom durch Gradientenzentrifugation rein dargestellt und die spezifische Aktivität der DNS gemessen. Die wiedergegebenen Kurven geben Mittelwerte aus zehn Versuchen mit Lebergewebe und vier Versuchsserien mit Hepatomen wieder. Ähnliche Kurvenverläufe wurden mit vier weiteren Morris-Hepatomen erhalten.

(Abb. 3). Im Gegensatz zur nichtwachsenden Leber hat jedoch die Größe des Tumors und damit aller Wahrscheinlichkeit nach auch die Gesamtzahl der Mitochondrien pro Tumor während der Versuchszeit zugenommen. Als Maß mag die Verdünnung der radioaktiv markierten Zellkern-DNS durch neugebildete inaktive DNS dienen. Ein Teil der Abfallrate der spezifischen Aktivität ist daher auch bei der Mitochondrienfraktion der Hepatome auf eine Verdünnung der Radioaktivität durch den DNS-Zuwachs während der Versuchsperiode zu erklären. Daraus ergibt sich, daß die mittlere „Lebenszeit" der Mitochondrien-DNS in den Hepatomen ganz wesentlich länger ist als bei der Leber, ja daß nach 8 Tagen noch der größte Teil aller markierten Mitochondrien vorhanden ist. Das schnellwachsende Gewebe deckt den erhöhten Bedarf an Mitochondrien anscheinend nicht durch eine gesteigerte Neubildungsrate, sondern vor allem durch eine Verlängerung der „Lebenszeit" der vorhandenen Mitochondrien. Hierzu paßt auch der erwähnte Befund, daß die Proteinsynthese in den Hepatommitochondrien nicht gesteigert ist.

Selbstverständlich kann dieser Anpassungsmechanismus nur begrenzt funktionieren, nämlich so lange wie die Wachstumsrate die Halbwertzeit der Mitochondrien nicht überschreitet. Bei einem schnelleren Wachstum muß dann auch die Neubildungsrate von Mitochondrien erhöht sein (was offenbar unter bestimmten Bedingungen der Fall ist), oder es muß — wenigstens während der schnellen Wachstumsphasen des Tumors — ein Mitochondriendefizit resultieren.

Literatur

[1] WAGLE, S. R., H. P. MORRIS, and G. WEBER: Cancer Res. **23**, 1003 (1963).
[2] PITOT, H. C., C. PERAINO, C. LAMAR, and A. L. KENNAN: Proc. nat. Acad. Sci. (Wash.) **54**, 845 (1965).
[3] NEUBERT, D., R. TIMMLER und H. HELGE: 28. Tag. dtsch. Pharmakol. Ges. 1964.
[4] —, H. HELGE und H.-J. MERKER: Biochem. Z. **343**, 44 (1965).
[5] GRAFFI, A., G. BUTSCHAK, and E. J. SCHNEIDER: Biochem. biophys. Res. Commun. **21**, 418 (1965).
[6] Gemeinsam mit Prof. Dr. BROCK, Brackwede, durchgeführte Untersuchungen.
[7] NEUBERT, D., H. HELGE und R. BASS: Naunyn-Schmiedebergs Arch. exp. Path. Pharmak. **252**, 258 (1965).
[8] SCHMIEDER, M., u. D. NEUBERT: 7. Frühjahrstag. dtsch. Pharmakol. Ges. 1966.

Diskussion

HEIDELBERGER (Madison/Wisconsin): I would like to make one or two remarks about Professor WARBURG's very interesting talk. I should like to say that I am a biochemist from the United States, who is in your country not at my own suggestion but at the invitation of your society. Now I just would like to make it very clear that all of the biochemists in the United States, particularly those working in cancer research have a tremendous admiration for the really enormous contribution that Professor WARBURG has made not only to biochemistry as a whole but to cancer research in general. This contribution has really been gigantic. However while we have recognized the importance of these contributions and the magnificence of his experimental techniques, nevertheless there is some difference in interpretation of his results. And these differences in interpretation are not limited just to the United States but also occur to some extent in this country. I would like to suggest and emphasize that honest differences in scientific opinion do not represent in any way enmity. I think in this era of history enmity in science as well as enmity between our two countries is a thing of the past.

I would now just like to make one remark about the work of Dr. DEAN BURK which won for him the DOMAGK Prize and which Professor WARBURG indicated on one of his slides in which glycolysis was plotted against the growth rate of hepatomas and a more or less straight line was presented. Now, unfortunately the individual points were not included in his graph. However, I had the opportunity to hear Dr. BURK present his results last year in the United States. If one takes the value for normal liver it was given at about 0.6, and the values for several of the very slow growing MORRIS-hepatomas were given at about 1.2 and then other tumors increased proportionally. Now the value of 0.6 for liver is extremely low. Most people, I understand, find values of about 1.0. So in this region we have several very slow growing tumors, which nevertheless are tumors and eventually will kill the animals, where the anaerobic fermentation rate appears to be not significantly different from liver. I would certainly hesitate to attribute a qualitative difference between 0.6, a very low value for liver, and about 1.0 which the slow growing tumors have.

BUSCH (Houston/Texas): The resolution of the data of DEAN BURK and of other data relating to glycolysis of course is not difficult. WEBER and MORRIS have published a long series of excellent experiments, perhaps 10 or more full publications on the question of glycolysis in "Cancer Research" and other American journals. They pointed out that although there was a relationship between growth rate and glycolysis that in fact there were some significant aberrations, i.e., some hepatomas had low glycolytic activity, i.e., lower than normal liver. Rapidly growing tumors had in general high glycolysis. Unfortunately, the work of Dr. BURK has not been published in an edited journal. With respect to Professor WARBURG's excellent studies, as noted in „Biochemistry of Cancer Cell", attention was called to the poor blood supply to neoplastic tissues as relating to glycolysis in vivo. As Prof.

Diskussion

WARBURG reported (1923 to 1928) the blood supply to tumors may account for the high glycolysis in vivo. In our own experiments years ago with radioactive pyruvate, radioactive glucose and radioactive acetate we did find some differences in tumors and other tissues. These differences, however, do not reflect enzymatic capacity since the tissues studied have very similar enzymatic capacities. These activities that differ in vivo may very well reflect the physiological state of the tissue and its environment. In short, glycolysis is an energy source and unless it is specifically related to nuclear function, it is probably unimportant to the cancer problem.

WARBURG (Berlin): Herr HEIDELBERGER hat in der Diskussion zu meinem Vortrag hervorgehoben, daß die Kurve, die nach DEAN BURK den Zusammenhang zwischen Gärung und Wachstumsgeschwindigkeit der Morris-Tumoren darstellt, *nicht* durch den Nullpunkt geht, woraus folgen würde, daß die langsamen Morris-Tumoren wachsen, ohne zu gären; daß es also Tumoren gibt, die nicht gären; daß also die letzte Ursache des Krebses nicht die Anaerobiose der Krebszellen sein kann.

Ich verstehe nicht, warum Herr HEIDELBERGER heute diese alten Einwände wiederholt, die von VAN POTTER ausgingen und längst widerlegt sind, sowohl durch die *in-vitro*-Experimente von DEAN BURK, als auch durch die *in-vivo*-Experimente von PIETRO GULLINO. (Vergl. auch SILVIO FIALA, [Naturwissenschaften **53**, 228 (1966)], wo für die langsamsten Morris-Tumoren neben Atmungsabfall und Gärungsanstieg auch die Verminderung der Mitochondrien nachgewiesen ist.)

HEIDELBERGER: I quoted the data of DEAN BURK because he is your disciple, and because these experiments won him the DOMAGK Prize.

WARBURG: Herr BUSCH hat gegen die Experimente von GULLINO eingewendet, daß die im Ovar wachsenden Morris-Tumoren langsamer mit Blut durchströmt werden als die Leber, und daß dieser Umstand die Milchsäurebildung erklärt. Die Anaerobiose der Tumoren — die Gegenstand dieser Diskussion ist — wird also hier durch den merkwürdigen Einwand widerlegt, daß die im Ovar wachsenden Morris-Tumoren als Anaerobier wachsen.

Mehr erfreut es mich zu hören, daß nunmehr von vielen Seiten Hemmungen von Atmungsfermenten und Zunahmen von Gärungsfermenten in Tumoren gefunden werden, eine willkommene Ausarbeitung der Tatsache, daß bei der Entstehung der Tumoren die Atmung sinkt und die Gärung steigt. Das Ferment Tryptophan-Pyrrolase wird in diesem Zusammenhang besonders häufig erwähnt, ohne daß gesagt wird, daß es ein sauerstoffübertragendes Ferment mit der Wirkungsgruppe Eisen ist. Dies gilt besonders für den Vortrag von Herrn PITOT. (Vergl. auch die Arbeit von GOODFRIEND, SOKOL und KAPLAN [J. molec. Biol. **15**, 18 (1966)], die für den von uns entdeckten Stoffwechselumschlag *beim Wachstum* der embryonalen Hühnerzellen zeigten, daß die Milchsäuredehydrogenase zunimmt.) Alle diese Arbeiten über die Fermente der Krebszellen werden aus unerfindlichen Gründen stets mit den Buchstaben ,,DNS" und ,,RNS" verknüpft, obwohl die Verbindung mit dem Wort ,,Anaerobiose" konstruktiver wäre.

76   Diskussion

DITTMANN (Homburg/Saar)*: Die Nettowachstumsgeschwindigkeit normaler Leber ist null, während selbst die Morris-Hepatome „minimaler Abweichung" eine endliche Wachstumsgeschwindigkeit haben. Auf absoluter Basis sind die Unterschiede der anaeroben Gärung tatsächlich nicht hoch; aber relativ zueinander sind die $Q_{CO_2}^{N_2}$-Werte selbst bei den sehr langsam wachsenden Hepatomen im Mittel zwei- bis fünfmal so hoch wie bei normaler Leber. Ich finde es an der Arbeit von BURK und WOODS interessant, wie sie gezeigt haben, daß schon eine kleine absolute Zunahme der anaeroben Milchsäurebildung ein essentieller Faktor bei der Entdifferenzierung einer Leberzelle zu einer Hepatomzelle ist.

HECKER (Heidelberg): I would like to ask a short question in connection with the talk of Dr. BUSCH referring to the cancer polyoperon V. As I understood your talk you are suggesting the presence of this genom in the normal cell, but in a repressed state. Is this correct?

BUSCH: With respect to Dr. MANDEL's comments on the polymerase activity, we have not studied the RNA-polymerase of the nucleus, but we have studied this activity in nucleoli. Dr. Ro and myself reported that the nearest neighbor frequency of the product is different in the nucleoli of tumor cells from that of liver cells. This study suggested different templates function differently in these nucleoli. We do not suspect, as Prof. MANDEL has pointed out, that there is any difference in the polymerases. We believe the templates differ.

With respect to Dr. LETTRE's very interesting contribution I would like to say that the RP₂L-story is an interesting old story. We were very interested in what are the components of RP₂L and about 5 years ago we began studies, which we reported partially in "Cancer Research" which suggested that RP₂L contains a number of components. Accordingly Dr. HNILICA and I published an extensive report on the amino acid composition, electrophoretic mobilities and other characteristics of the histones of neoplastic cells (BUSCH, Histones and Other Nuclear Proteins), and following up the work that was done by BUTLER's group, we did not find that the histones are significantly different in the tumor and other tissues. Histones are now defined as the proteins extractible with acid from a purified deoxyribonucleoprotein. RP₂-L is a mixture of histones and acidic proteins, and recently we have been concerned with the nucleolar proteins which may be important to the RP₂L problem (Cancer Research, May, 1966). One of the interesting facts about the nucleolar proteins is, that they are divisible into 3 classes; one of these are the nucleolar histones. There are definitely histones in the nucleoli, and in tumors the nucleoli are large by comparison with some other tissues, so the amount of histones is greater. What is immensely greater is the amount of an acidic protein. We know very little even about solubilization of such proteins. There are many of them and they include in part a number of nucleolar enzymes. The first extensive report on nucleolar

---

* Diese Bemerkung konnte aus Zeitmangel nicht gesprochen werden; sie wurde schriftlich vorgelegt.

enzymes is that of SIEBERT et al. from our laboratory (JBC, January, 1966). We know now that there are perhaps 17 or 20 nucleolar enzymes. Dr. CHAKRAVORTY has shown that the amount of ribonuclease-inhibitor is great in the tumors. Thus, there are many proteins some of which are important to the RP_2L question which remains to be studied.

About the cancer gene, one wonders why any gene is repressed. We are sure, many genes are repressed and most people agree with the old interpretations of ALLFREY and MIRSKY, who pointed out that in most cells perhaps 97% of the genome is suppressed. There is a tremendous difference between the cell of the retina, the cell of the myocardium, the cell of the kidney and so on. These differences may be accountable for by the very large amount of genes that must be carried by mammalian cells in order to function. A cancer cell may represent a last stage of survival of a living cell.

Dr. SCHERRER and Dr. DARNELL were pioneers in the field of rapidly synthesized RNA, and they were the ones who pointed out that in HeLa-cells there was a rapid synthesis of GC-rich RNA. Later OKAMURA in our laboratory pointed out that this rapidly synthesized RNA in a nontumor tissue could be very AU-rich. HAREL in France pointed out the same kind of thing.

HOLZER (Freiburg i.Br.): Dr. PITOT, you mentioned many cases in which you think that a change or a difference in regulation mechanisms is characteristic for a tumor as compared to a normal tissue. My question is: Could it not be that the fundamental metabolism of a tumor is changed and that the difference in regulation is nothing but a consequence of this change ? Glycolysis of tumors, for instance, differs from that of normal tissue. This could be the reason for differences in regulation. As you mentioned, glucose induction of glucokinase does not work in tumors. Could it not be that the rapid glucose metabolism in the tumor leads to smaller steady state concentrations of glucose and thus prevents the induction ? The induction mechanism itself might well be intact. In many other cases you mentioned, one could speculate too, that it is a change in metabolism rather than a damage of or a difference in a regulation mechanism which is characteristic for tumors.

PITOT (Madison/Wisconsin): In answer to your question, the fundamental metabolism of a neoplasm *is* changed, however, the data we presented suggest that this change is the result of altered control mechanisms, not their cause. The strongest piece of evidence for this is that mechanisms controlling enzyme synthesis in hepatomas are quite different from one hepatoma to another. Thus, the alteration in controls that we see are unique to each neoplasm studied. By each neoplasm I mean the transplanted tumor line. Any one line of tumor, such as the 5123, Rueber H35, etc., breed true and maintain their set of altered control mechanisms from generation to generation. If, as your question suggests, these alterations are the result of altered metabolism, and thus altered metabolic products, each tumor should have a virtually unique metabolism, different from any other tumor or the normal situation. If one considers glycolysis alone, this is obviously not true, since according to Dr. WARBURG's theory, all tumors are characterized by the glycolytic lesion. Thus, although one defect, that of glucokinase, in certain hepatomas might suggest that it is a result of altered glucose metabolism, this is not so in at least one other hepatoma, the MORRIS 7787, which has a high level of glucokinase that does show some slight response to environmental stimuli.

HILZ (Hamburg): One question to Dr. PITOT: You compared all your tumor events with normal liver as far as I saw. If you would compare your results with rapidly growing tissues you would probably have some changes in regulation phenomena. How far are your effects specific for tumor tissues ?

PITOT: The pattern of effects that we described here in a number of hepatocellular carcinomas are, as far as we can tell, unique to the neoplasms. Rapidly growing liver, such as regenerating liver, exhibits all of the normal control mechanisms found in resting liver. Fetal liver, which is only half liver because of the high content of bone marrow present in the tissue, exhibits very little response to environmental stimuli, but at the same time never shows the extremely high, nonchangeable levels of inducible enzymes such as tyrosine transaminase in many highly differentiated hepatomas or serine dehydrase in, for example, the Morris 5123 hepatoma. Thus, it appears that the effects that we have demonstrated may be entirely unrelated to growth rate per se.

MANDEL (Strasbourg): A very short comment: I think we have to be very careful in the interpretation of several data in cancer cells. We found in rat liver from animals on a poor protein diet tremendous changes in polysomes, in the attachment of m-RNA to the ribosomes and that there is also a derepression of RNA-polymerase activity. In cancer cells we have a kind of poor protein diet or a kind of malnutrition, because the synthesis of many amino acids decreases. Glycolysis is the main energy source, not the citric acid cycle. I think you should keep in mind that some of the effects may be the effects of troubles in nutrition of the cells, in which we found some troubles in the attachment of m-RNA to ribosomes and also the derepression of RNA-polymerase.

PITOT: I agree with Dr. MANDEL wholeheartedly on this point. In order to investigate such a phenomenon, Dr. RUDOLF SÜSS, working in our laboratory, has been able to show that polysomes isolated from tumors are bound *in vitro* to membrane components considerably less than those from liver. This defect appears to reside in the membrane components rather than in the polysomes themselves. Dr. WEBB and his associates, working in Dr. POTTER's laboratory, have also investigated polysome patterns *in vivo* in liver and several hepatomas. They find, like Dr. MANDEL, it is possible to alter the polysome pattern of liver by various dietary and other manipulations, but the polysome pattern of hepatomas cannot be altered by similar manipulation, again suggesting defects in the regulation of the protein synthesizing mechanism of the neoplasms.

Alkylation of Nucleic Acids and Carcinogenesis

By P. N. MAGEE

Toxicology Research Unit, Medical Research Council Laboratories, Carshalton, Surrey, Great Britain

With 5 Figures

The idea that the induction of cancer might follow mutation of somatic cells was first put forward by BOVERI in 1914[1]. He suggested that the great frequency of mitotic abnormalities in cancers indicates an abnormal chromosome constitution which might be the primary cause of malignant change in the cells. The presence of visible chromosomal abnormalities is not, however, an essential feature of malignancy and the ideas of BOVERI have been critically reviewed by KOLLER[2] in the light of more recent cytological analysis. KOLLER points out that primary tumours always contain some cells of normal karyotype and that these may be the only type, particularly during the early stages of tumour development. Furthermore, there is evidence that mitotic abnormalities and cell heterogeneity in tumours may be the result of defective nutritional and blood supply within the tumour mass and thus a secondary effect rather than the primary cause of the malignant change. It seems, therefore, that morphologically detectable chromosome abnormalities have not been established as the cause of the neoplastic transformation but this does not exclude chromosomal changes which cannot be detected microscopically but which may occur in the induction of point mutations.

Other theories of carcinogenesis have been advanced which may be included in the category of somatic mutation but which postulate altered activity of extrachromosomal genetic factors. DARLINGTON[3] postulated three classes of cytoplasmic particle, the hereditary plasmagenes, the naturally infectious viruses and an intermediate class, the pro-viruses. These particles were regarded as interchangeable and they all might be involved in the production of cancer. DARLINGTON concluded: "The origin of cancer can therefore be ascribed to mutations in cytoplasmic determinants, indifferently infectious or non-infectious, which make themselves visible by causing the resumption of growth". An attractive aspect of this hypothesis is its suggestion of a similar basic mechanism by which a somatic mutagen and a virus might induce cancer.

Whatever the nature of the initial event in carcinogenesis it seems that there must be some change in the heritable properties of the cells. This conclusion is difficult to escape because tumours can be transmitted by the innoculation of very few or even single cells[4,5,6] which shows that the cells are irreversibly neoplastic and that this property is transmitted to the descendants. The simplest explanation of a hereditary cellular change is that there is some change in the genetic material which is now generally accepted to be composed of nucleic acids. The change could involve a direct action on the chromosomal DNA by the carcinogen or an indirect action on nuclear histones or via some cytoplasmic interaction. The latter possibility is implicit in the extrachromosomal mutational hypotheses[3] discussed above and also in the more recent suggestions of JACOB and MONOD[7] which have been developed by PITOT and HEIDELBERGER[8]. These authors have proposed that carcinogenesis does not necessarily involve modification of DNA but might result from changes in repressor or other regulatory molecules outside the nucleus which could be either protein or RNA.

Aspects of the somatic mutation hypothesis have been widely discussed[9,10,11,12] with widely differing conclusions (PULLMAN[13]). As CLAYSON[14] has pointed out, the hypothesis is difficult to assess because of its vagueness and the different ways in which it has been interpreted. It is also difficult to design critical experiments to test the hypothesis. A major criticism has arisen from the lack of general correlation between the mutagenic and carcinogenic action of different groups of chemical agents[9]. A notable example of this is the failure to demonstrate unequivocal mutagenic activity in the usual test systems by the polycyclic hydrocarbons and by the aromatic amines and azo compounds[14], all of which are powerful carcinogens. It must be emphasised, however, that this failure to demonstrate correlation between the two properties may have been due to the inadequacy of the methods used[9].

On the other hand some active mutagens are carcinogenic, such as urethane, the biological alkylating agents[14] and many nitroso compounds[15]. Again, however, there may be lack of correlation of the two properties since other powerful mutagens such as formaldehyde, nitrous acid and hydroxylamine have not been reported as carcinogenic. Among these various agents the nitroso compounds stand out because they are among the most powerful mutagenic and carcinogenic agents known (reviews by MAGEE and SCHOENTAL[16], and MAGEE and BARNES[17]).

The first report of a chemical compound with mutagenic activity was by AUERBACH and ROBSON in 1946[18] who showed that mustard

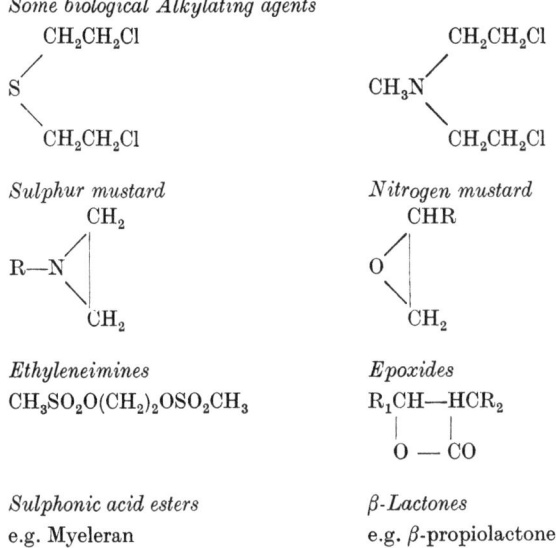

Fig. 1. Some typical biological alkylating agents

gas could induce mutations in *Drosophila melanogaster*. Since then several of the biological alkylating agents have been shown to be mutagenic and also carcinogenic. Some typical biological alkylating agents are shown in Fig. 1. Although many of these compounds are undoubtedly carcinogenic, their potency varies and, as a group, they cannot be regarded as powerful carcinogens[11,19]. β-propiolactone is an effective carcinogen by repeated subcutaneous[20] or cutaneous application[21].

The biological alkylating agents are reactive chemicals and are thought to owe their biological activity to alkylation of receptor sites in essential cellular components[22]. This conclusion is based on extensive studies of the reactions of the alkylating agents with proteins, nucleic acids, amino acids and other molecules *in vitro* (see Ross[23]). The evidence for alkylation *in vivo* is much less extensive and will be further discussed below.

The reaction of the alkylating agents with nucleic acids *in vitro* has been extensively studied by many workers, notably LAWLEY and BROOKES[19]. Using first sulphur mustard and then a range of other agents, it was shown that the most reactive sites of alkylation of both nucleic acids were the ring nitrogen atoms of the bases. Of these, considerably the most reactive was the N-7 of guanine, followed by the N-3 of the adenine moities and then other sites to a smaller extent. LAWLEY and BROOKES have suggested mechanisms by which alkylation on N-7 of the guanine bases of DNA could be responsible for the mutagenic action of the alkylating agents[19]. These possibilities will be discussed below.

Since it is now generally accepted that the genetic material of the cell is DNA it is reasonable to assume that mutagenesis involves a change in the structure of this nucleic acid. Various molecular mechanisms of mutations have been discussed by FREESE[24]. With carcinogenesis, however, no such key role for DNA can be assumed unless some form of mutational change is postulated. Clearly such a process might be involved in carcinogenesis by the alkylating agents and this suggestion was made explicitly by HADDOW in 1958[25] who stated: — "Whatever the precise chemical mechanism of the action of carcinogens, there can be little doubt of the importance of their combination with genetical material or its precursors, if (as seems likely) this is essential for those at least which function through biological alkylation."

The remainder of this paper will be concerned with a consideration of this hypothesis in the light of recent work with carcinogenic nitroso compounds and other agents which have been shown to alkylate nucleic acids in the organs of intact animals *in vivo*. Unpublished work done in collaboration with Dr. VALDA CRADDOCK, MR. P. SWANN, Dr. B. TERRACINI and Mr. D. LEAVER will be included.

The carcinogenic action of the simplest nitrosamine, dimethylnitrosamine (Fig. 2) was discovered by MAGEE and BARNES in 1956[26] and since than a very large number of other nitrosamines have been found to be carcinogenic by DRUCKREY, SCHMÄHL, PREUSSMANN and their colleagues who have carried out very extensive studies

Alkylation of Nucleic Acids and Carcinogenesis 83

on the relation of chemical structure to carcinogenic activity[27]. As well as the nitrosamines, certain nitrosamides such as N-nitrosomethylurea[23] (Fig. 2) and N-nitrosomethylurethane[29] are also very powerfully carcinogenic. Aspects of carcinogenesis by the nitroso

$$\text{Dimethylnitrosamine} \quad \begin{array}{c} CH_3 \\ \end{array}\!\!\!\!\diagdown\!\!\!\!\begin{array}{c} \\ NNO \\ \end{array}$$
$$CH_3\!\!\diagup$$

$$\text{Nitrosomethylurea} \quad \begin{array}{c} CH_3 \\ \end{array}\!\!\!\!\diagdown\!\!\!\!\begin{array}{c} \\ NNO \\ \end{array}$$
$$O\!=\!C\!\!\diagup$$
$$|$$
$$NH_2$$

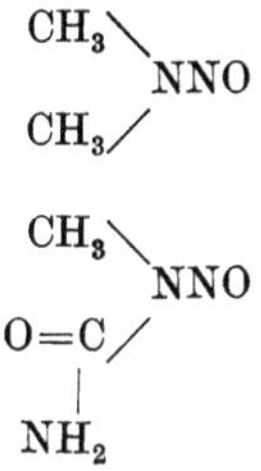

Fig. 2. Chemical structures of dimethylnitrosamine and N-nitrosourea

compounds have been discussed in several reviews[16,17,27,30,31,32]. As mentioned above, and in contrast with some other types of powerfully carcinogenic agents, the nitrosamines are equally powerful as mutagens and this strong correlation of mutagenic and carcinogenic properties makes them very useful agents for the study of possible relationships between the two processes.

Pathology of acute injury and carcinogenesis by nitroso compounds

In this section a brief account of the pathology of the lesions induced by the carcinogenic nitroso compounds will be given since this forms an essential background to the biochemical experiments to be described later. Dimethylnitrosamine is a quite highly toxic compound with a marked specificity for the liver. In doses of the order of 20 to 40 mg per kg body wt., which are about the median lethal level, it induces severe centrilobular haemorrhagic necrosis of the liver in the rat and in all other species of laboratory animal tested[33]. Lesions do occur elsewhere such as haemorrhages into the intestine, the lungs and lymph nodes and there may be extensive haemorrhagic ascites. In animals which survive the acute poisoning the liver regenerates and appears to recover almost completely. When dimethylnitrosamine was fed to rats at a level of about 50 parts per million in a normal diet malignant tumours of the liver were induced[23]. Feeding dimethylnitrosamine to rats at higher dietary levels for shorter periods e.g. 100 parts per million for one month or at 200 parts per million for one week did not induce liver cancer, but a high incidence of tumours of the kidney was obtained[31]. Tumours of the kidney also arose in some rats which had survived one dose only of dimethylnitrosamine[31], at about the median lethal level, which is of special interest since the compound is rapidly metabolised and does not persist in the body. The other group of nitroso compounds, the nitrosamides, as exemplified by N-nitrosomethylurethane and N-nitrosomethylurea are less stable than dimethylnitrosamine and induce lesions at the site of application and to a greater or less extent in distant organs. N-nitrosomethylurea is an extraordinarily effective carcinogen and must be among the most powerful known. Its carcinogenic action was discovered by DRUCKREY, PREUSSMANN, SCHMÄHL and MULLER[28] who induced cancer of the stomach in rats by continued administration of the compound in the drinking water. In later work, tumours in a variety of organs were induced in the rat by single intravenous injections of N-nitrosomethylurea[35] and, with repeated intravenous injections, even tumours of the brain[36]. Very recently, GRAFFI and HOFFMANN[37] have reported carcinogenesis after local application to the skin of mice and state that the potency of N-nitrosomethylurea is as great or even greater than that of strong polycyclic hydrocarbons. Recent work at CARSHALTON has shown that squamous carcinomas of the stomach, tumours of the large and small intestine and of the kidney and also of the skin can be induced by single oral doses of N-nitrosomethylurea in the rat[33]. DRUCKREY and his coworkers have tested a very large number of nitrosamines for carcinogenic activity and their results are summarized in the various review articles cited above. The induction of cancer of the rat bladder by dibutylnitrosamine[39] and of the oesophagus by unsymmetrical nitrosamines[40] are of particular interest.

Table 1. *Species in which the induction of liver tumours by dimethyl and diethylnitrosamine have been reported*

Dimethylnitrosamine	Diethylnitrosamine
Rat[26]	Rat[66]
Mouse[63]	Mouse[67]
Hamster[64]	Hamster[68]
Rainbow Trout[65]	Guinea Pig[69]
	Rabbit[70,71]
	Dog[72]
	Monkey[73]
	Aquarium Fish[74]

Diethylnitrosamine has been tested for carcinogenicity in more species than any other. Those in which it has induced liver tumours are listed in Tab. 1, together with those susceptible to hepatocarcinogenesis by dimethylnitrosamine.

In summary, it appears that tumours can be induced in any organ of the body by the appropriate nitroso carcinogen and that no species has so far been reported resistant to these carcinogens. With some nitrosamines repeated dosage is not necessary and tumours have been induced in a variety of organs by single doses only.

Decomposition of carcinogenic nitroso compounds in the body

The dialkylnitrosamines are stable at neutral pH but some have been shown to be rapidly decomposed after injection into animals[41,42,43]. Most of the experimental work reported has been done with dimethyl and diethylnitrosamine, both of which are rapidly metabolised. The organ mainly concerned with the metabolism of the nitrosamines is the liver but smaller degrees of decomposition may occur in the kidneys and perhaps other organs in the rat. The decomposition of these nitrosamines is enzymically catalysed[44,45] the enzyme systems involved having some similarities to the drug-metabolizing enzymes described by BRODIE and his co-workers[46]. It is emphasised that a dose of dimethylnitrosamine adequate to induce tumours of the kidney in some of the survivors appears to be completely metabolised within 24 hr. since none of the carcinogen can be detected in the bodies of rats at this time after treatment. The rapid metabolism of dimethylnitrosamine has been confirmed in studies using the compound labelled with radioactive carbon in the methyl groups[41,43].

The nitrosamides are less stable compounds and probably decompose rapidly in the tissues. The rate of decomposition of N-nitrosomethylurethane at neutral pH is greatly increased by the presence of sulphydryl compounds such as cysteine[17]. These compounds are, of course, widely used in organic chemistry for the preparation of diazomethane which they yield at alkaline pH.

This capacity to decompose with the formation of a powerful alkylating agent has led to a hypothesis to explain the mechanism of cellular injury and carcinogenesis by the nitroso compounds. Since diazomethane is a very toxic and irritant material, a powerful mutagen and a methylating agent, it has been suggested that this might be the active intermediate in the biological action of the nitroso compounds. With the more stable nitrosamines the alkylating agent would only be formed in organs containing the necessary enzyme systems for their decomposition but with the less stable nitrosamides the intermediate would be formed in the tissues without the necessity of enzyme action.

The alkylation hypothesis of nitrosamine carcinogenesis

Evidence that alkylation of cellular components does occur in animals treated with nitroso carcinogens has been obtained from experiments using nitrosamines labelled with radioactive tracers

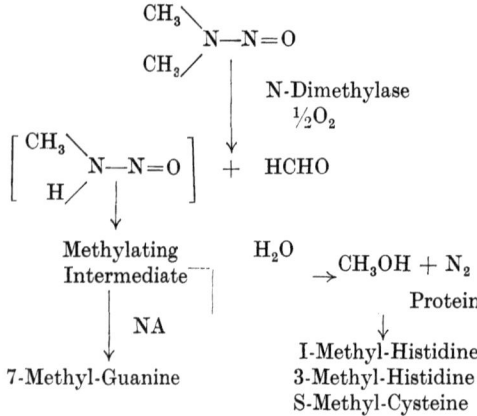

Fig. 3. A suggested metabolic scheme for dimethylnitrosamine

in their alkyl groups[48,49]. Much of this work has been done with [^{14}C] dimethylnitrosamine. Since the postulated methylating intermediate would be highly reactive it would be extremely difficult to detect. Its presence can be inferred, however, from the finding of excess amounts of methylated cellular components in which the radioactivity is derived from the injected dimethylnitrosamine. Following the injection of labelled dimethylnitrosamine into rats and other species the nucleic acids of the livers and some other organs[49,50] were found to be radioactive and much of this radioactivity was in 7-methylguanine. Since this methylated base is only present in very small amount, or absent, in the nucleic acids of untreated animals and since it is the major excepted product of methylation of nucleic acids[19], it is reasonable to conclude that an active methylating intermediate must have been formed metabolically in the liver from the injected dimethylnitrosamine. A suggetted metabolic scheme for dimethylnitrosamine is shown in Fig. 3. It can be seen that the greater part of the liberated methyl groups is oxidised to formaldehyde or methanol and further to CO_2 which is expired. Both DNA and RNA are methylated as are liver proteins and it must be emphasised that other cellular components must very probably be methylated as well.

Although there is good evidence that alkylation of cellular components does occur in some organs of animals treated with nitroso carcinogens it remains to be established whether such alkylations are related to the induction of cancer and, if so, which cellular receptors are critically involved. There is much indirect evidence that alkylation may be important from the very extensive studies on the relation of chemical structure to biological activity by DRUCKREY, PREUSSMANN, SCHMÄHL and their co-workers[27]. This group has tested a very large number of different nitroso compounds and found that there is a good general correlation between nitrosamines which could yield an alkylating intermediate and those which are carcinogenic although there are some discrepancies. The objection has been made that some heterocyclic nitrosamines are powerfully carcinogenic but would not be excepted to undergo metabolic opening of the heterocyclic ring to give an alkylating intermediate[30]. Only further experimental work can resolve this question.

Possible carcinogenic decomposition products of the nitroso compounds

Several other possibly carcinogenic products might be formed from the nitroso compounds[75] within the target cells. Some of these are listed in Tab. 2. The alkylating intermediate has been discussed above and will be considered further later. The aldehyde is unlikely to be an active intermediate, even if formed metabolically within the cells, since many drugs which are not hepatotoxic or carcinogenic are known to yield aldehydes on

Table 2. *Known and possible breakdown products of nitroso carcinogens*[75]. *All are mutagens*

Aldehyde
Nitrous Acid
Hydroxylamine derivative
Hydrazine derivative
Alkylating agent

oxidative dealkylation[46]. Nitrous acid might be formed from nitroso carcinogens in the body and it must be seriously considered as a possible carcinogenic intermediate. It is not carcinogenic on prolonged feeding to rats [51] but its behaviour might be different if released inside the cells. It is a very effective mutagen and is thought to act by deamination of nucleic acid bases[21]. Extensive deamination of nucleic acids in mammalian cells *in vivo* would not be expected because this reaction is highly dependent on pH and becomes virtually absent above pH 6,5 *in vitro* and in the yeast cell *in vivo*[52]. On the other hand, gene mutations by nitrous acid are thought to occur after deamination of only one nucleic acid base[53] and no experimental methods at present available would be sensitive enough to detect this. An attempt has recently been made, however, to detect any deamination that may have occurred in the DNA and RNA of livers of rats treated with dimethylnitrosamine. The nucleic acids were hydrolysed and analysed for xanthine and hypoxanthine, the expected deamination products of guanine and adenine respectively. The experiments were complicated by the occurrence of artefactual deamination of the bases, particularly adenine, during the process of extraction and hydrolysis of the nucleic acids. An attempt was made to circumvent this problem by using rats whose nucleic acids had been pre-labelled with [^{14}C] and tritium respectively. Animals pre-labelled with the one isotope were treated with dimethylnitrosamine and those pre-labelled with the other served as controls. Immediately after removal from the animal the livers were pooled and a single mixed preparation of the nucleic acids was made which was hydrolysed and examined by ion-exchange chromatography as usual, using differential scintillation radio assay. Using these methods it was not possible to detect any significant increase in the amount of xanthine or hypoxanthine in hydrolysates from the nitrosamine treated animals over that from the controls[54]. Although a very small degree of nucleic acid deamination in the carcinogen treated rats could not be excluded it must have been much less than the degree of metylation, which was confirmed. Similar attempts were made to find N^6-hydroxycytosine, the expected reaction product of hydroxylamine with nucleic acids, but this could not be detected[54]. Reduction of nitrosomorpholine to the corresponding hydrazine derivative has recently been reported to occur on anaerobic incubation of the nitrosamine with liver microsomal preparations *in vitro*[55]. Since hydrazine and some of its derivatives are known to be mutagenic and carcinogenic, the possible formation of such derivatives from nitroso carcinogens must be seriously considered.

Quantitative studies of alkylation of nucleic acids in vivo by nitroso carcinogens and other alkylating agents

As mentioned above, several of the biological alkylating agents are proven carcinogens but, as a class, their potency is considerably less than that of the nitroso compounds. If therefore, methylation of nucleic acids or other cellular components is related to the induction of cancer by dimethylnitrosamine and N-nitrosomethylurea, other methylating agents which are not nitroso compounds might be expected to be equally active. Dimethylsulphate, for example, is known to methylate nucleic acids effectively *in vitro*, predominantly on the N-7 position of guanine. If it could penetrate to the critical intracellular sites for carcinogenesis the alkylation hypothesis would predict that it should be an active carcinogen. DRUCKREY and his co-workers[56] have recently shown that dimethylsulphate is locally carcinogenic after repeated subcutaneous injections in the rat, inducing sarcomas with metastases. Recent work at CARSHALTON, however, has shown that this compound is only very weakly active, if at all, after repeated administration by other routes[57]. Rats and mice were given repeated oral, intraperitoneal and rectal injections of dimethylsulphate and some rats also received single intravenous doses. The only tumours observed were a squamous carcinoma of the stomach with widespread metastases in a mouse after repeated oral dosage and a brain tumour in a rat which had received one intravenous injection and several by the intraperitoneal route[57]. These results are clearly in sharp contrast with those obtained with N-nitrosomethylurea and an attempt to explain this difference was made. Rats were injected intravenously with [^{14}C] dimethylsulphate and [^{14}C] nitrosomethylurea and the extent of alkylation of the DNA and RNA in a variety of organs was determined. The dose of nitrosomethylurea was known to be carcinogenic and that of dimethylsulphate was close to the lethal level. It was found that the methylation of nucleic acids after injetion of dimethylsulphate was considerably lower than that after nitrosomethylurea in all the organs studied[58]. The results with rat kidney are shown in Fig.4 where they are compared with the methylation after treatment with dimethylnitrosamine and also with methyl methanesulphonate. Clearly methylation after injection of the two nitrosamines is considerably higher than after treatment with dimethylsulphate and this is reflected by the incidence of tumours. It is again emphasised that the dose of dimethylsulphate was almost lethal and appreciably higher levels of methylation could not therefore be obtained

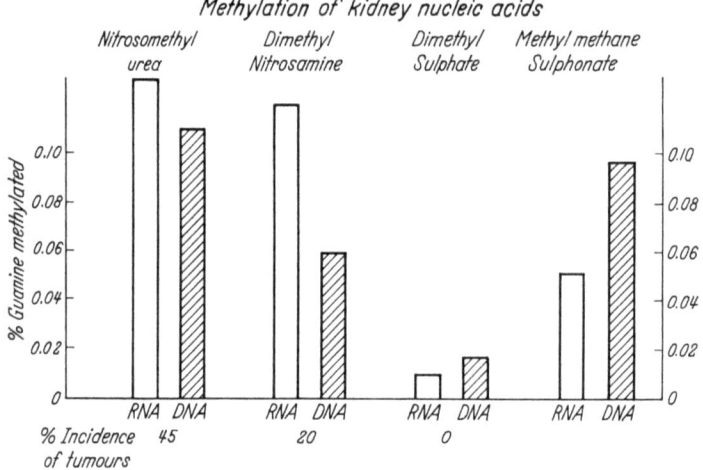

Fig. 4. Degree of methylation of kidney nucleic acids and incidence of renal tumours in rats given single doses of nitroso carcinogens and alkylating agents. Dimethylnitrosamine, 30 mg/kg orally; Nitrosomethylurea, 90 mg/kg intravenously; Dimethylsulphate, 80 mg/kg intravenously; Methyl methanesulphonate, 120 mg/kg intravenously. The incidence of tumours with nitrosomethylurea is according to DRUCKREY et al. (1963)[35]. Experiments are in progress to test the carcinogenic activity of methyl methanesulphonate under these conditions

without killing the animals. The results with methyl methanesulphonate are interesting because, with this compound, the levels of methylation were similar to those with the nitroso compounds. It is not known whether methyl methanesulphonate is carcinogenic to rat kidney under these conditions of single rather high intravenous dosage but experiments are in progress at CARSHALTON to test this. Since the alkylation hypothesis would certainly predict that some kidney tumours should be induced under these conditions the results of these experiments will be of some interest.

It is concluded that if alkylation of nucleic acids is significant for carcinogenesis its quantitative degree may be critical. By virtue of releasing an alkylating intermediate within the cells, the nitroso carcinogens may be able to attain this critical level of alkylation and therefore prove more effective than some other alkylating agents which cannot achieve this.

The fate of the nucleic acid methyl groups derived from dimethylnitrosamine in vivo

In rats treated with dimethylnitrosamine in doses (about 30 mg per kg body wt.) adequate to induce liver necrosis in all and kidney tumours in about 20% of the survivors, methylation of both nucleic acids in liver and kidney reaches a maximum level at about 6 hr after intraperitoneal injection. In the liver these levels correspond to methylation of about 1% of the guanines of the RNA and about 0,5% in the DNA: in the kidney the levels are about 0,1% and 0,05% respectively. The level in the liver RNA begins to fall sharply at about 12 hr after the injection and that in the DNA perhaps even earlier[59]. The liver, of course, undergoes a severe centrilobular necrosis and some of this rapid loss of the methyl groups must reflect early breakdown of cells. The level of methylation in the kidney nucleic acids also falls with time but much more slowly than in the liver. It cannot be said with certainty for how long the methyl groups derived from dimethylnitrosamine remain in the nucleic acids of these rat organs and this requires further study. Technical problems arise in the detection and estimation of minute quantities of 7-methylguanine in the nucleic acids even when very sensitive radiochemical methods are used.

The rapid loss of the methyl groups from the nucleic acids of the dimethylnitrosamine-treated rats raises the question whether this represents a simple demethylation, whether the 7-methylguanine is being removed intact or, of course, both. 7-methylguanine has been known to be a normal component of human urine since the end of the last century[60] and recently it has also been found in rat urine[61]. If the methylated base is removed from the nucleic acids of rats treated with dimethylnitrosamine it might be expected to appear in greater amounts after treatment of the animals with the carcinogen. Furthermore, if the nitrosamine is radioactively labelled, the urinary 7-methylguanine would also be expected to be tabelled. This has been shown to be the case[62]. A rat was treated with the usual dose of dimethylnitrosamine which, in this experment, had considerably higher specific radioactivity than usual. The urine was collected for the first five hours after the injection and the urinary purines were isolated and analysed chromatographically. The results are shown in Fig. 5. It can be seen that the urinary 7-methylguanine is heavily labelled, strongly suggesting that it has been derived from the methylated nucleic acids since labelled 7-methylguanine was not detected in the acid soluble fraction of rat liver at the same time after injection of dimethyl-

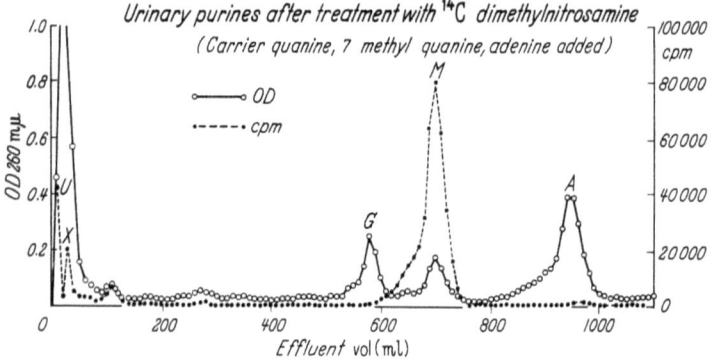

Fig. 5. Ion-exchange chromatography (Dowex-50 [H÷]) of urinary purines from a rat treated with [^{14}C]dimethylnitrosamine, 30 mg/kg, 180 μc. The urine was collected during five hours after injection, not quantitatively
U, uric acid; X, xanthine; H, hypoxanthine; G, guanine; M, 7-methylguanine

nitrosamine in earlier experiments[49]. At this stage there is little microscopical evidence of cellular injury in the liver but it cannot be said with certainty whether the labelled urinary 7-methylguanine is derived from the nucleic acids by some sort of excision mechanism or whether it simply represents general breakdown of the nucleic acid as part of the impending necrosis of the liver. An attempt was made to determine whether any of the urinary 7-methylguanine was derived from DNA. In these experiments rats were used which had their DNA pre-labelled in some organs by giving repeated injections of [^{14}C] formate in the neonatal period and then allowing them to attain an age of about 1 year when their livers were shown to contain radioactivity in the DNA, but virtually none in the RNA or protein. These animals were injected with [^{3}H] dimethylnitrosamine and their urine collected over a period of two days. The urinary 7-methylguanine was heavily labelled with tritium and also contained a small, but definite amount of ^{14}C label. This indicates that some of the urinary 7-methylguanine in the nitrosamine treated rat must have come from methylated DNA but again it may have been released by general cellular injury[62].

Conclusions

From the above it can be concluded that the alkylation hypothesis of nitrosamine carcinogenesis is an attractive one but that it is very far from being established. There is no doubt that alkylation of the genetic material does occur in the somatic cells of the nitrosamine-treated animals which is an experimental confirmation of the suggestion made by HADDOW in 1958[25] and, to this extent, gives support to mutational hypotheses of cancer. BROOKES and LAWLEY[19] have discussed mechanisms at the molecular level by which alkylating agents may induce genetic mutations. They suggest two possibilities. In the one, alkylation on the 7-position of guanine is suggested to cause anomalous base-pairing, thymine pairing with the ionized 7-methylguanine. In the other, depurination with splitting out of 7-methylguanine from the DNA chain is suggested to induce mutation by deletion. Grosser deletions of genetic material might also be produced by chemical degradation of the alkylated DNA or by the action of specific deoxyribonucleases. The results described in this paper are consistent with the occurrence of both these mutational mechanisms in the cells of somatic organs in which cancer has been induced by a single dose of a nitroso carcinogen.

References

[1] BOVERI, T. H.: Zur Frage der Entstehung maligner Tumoren. Jena: Fischer 1914.
[2] KOLLER, P. C.: In Cell Physiology of Neoplasis. Austin: University of Texas Press 1960.
[3] DARLINGTON, C. D.: Brit. J. Cancer 2, 118 (1948).
[4] FURTH, J., and M. C. KAHN: Amer. J. Cancer 31, 276 (1937).
[5] KLEIN, E.: Exp. Cell Res. 8, 213 (1955).
[6] FISHER, B., and E. R. FISHER: Science 130, 918 (1959).
[7] JACOB, F., and J. MONOD: Cold Spr. Harb. Symp. Quart. Biol. 26, 193 (1961).
[8] PITOT, H. C., and C. HEIDELBERGER: Cancer Res. 23, 1694 (1963).
[9] BURDETTE, W. J.: Cancer Res. 15, 201 (1955).
[10] ROUS, P.: Nature (Lond.) 183, 1357 (1959).
[11] BURNET, F. M.: Brit. med. J. 1, 779 (1957).
[12] KAPLAN, H. S.: Cancer Res. 19, 791 (1959).

[13] PULLMAN, B.: J. cell comp. Physiol. **64**, 91 (1964).
[14] CLAYSON, D. B.: Chemical Carcinogenesis. London: J. & A. Churchill 1962.
[15] PASTERNAK, L.: Arzneimittel-Forsch. **14**, 802 (1964).
[16] MAGEE, P. N., and R. SCHOENTAL: Brit. med. Bull. **20**, 102 (1964).
[17] —, and J. M. BARNES: Advanc. Cancer Res. (In press).
[18] AUERBACH, C., and J. M. ROBSON: Nature (Lond.) **157**, 302 (1946).
[19] BROOKES, P., and P. D. LAWLEY: J. cell. comp. Physiol. **64**, 111 (1964).
[20] WALPOLE, A. L., D. C. ROBERTS, F. L. ROSE, J. A. HENDRY, and R. F. HOMER: Brit. J. Pharmacol. **9**, 306 (1954).
[21] ROE, F. J. C., and O. M. GLENDENNING: Brit. J. Cancer **10**, 357 (1956).
[22] PETERS, R. A.: Nature (Lond.) **159**, 149 (1947).
[23] ROSS, W. C.: Biological Alkylating Agents. London: Butterworth 1962.
[24] FREESE, E.: In Molecular Genetics. Part I., ed. by Taylor, J. H. New York: Academic Press 1963.
[25] HADDOW, A.: Brit. med. Bull. **14**, 79 (1958).
[26] MAGEE, P. N., and J. M. BARNES: Brit. J. Cancer **10**, 114 (1956).
[27] DRUCKREY, H., R. PREUSSMANN, and D. SCHMÄHL: Acta Un. int. Cancr. **19**, 510 (1963).
[28] DRUCKREY, H., R. PREUSSMANN, D. SCHMÄHL und M. MÜLLER: Naturwissenschaften **48**, 165 (1961).
[29] SCHOENTAL, R.: Nature (Lond.) **188**, 420 (1960).
[30] ARCOS, J. C., and M. ARCOS: Fortschr. Arzneimittel-Forsch. **4**, 407 (1962).
[31] WEISBURGER, J. H., and E. K. WEISBURGER: Clin. Pharmacol. Ther. **4**, 110 (1963).
[32] MILLER, E. C., and J. A. MILLER: Pharmacol. Rev. **18**, 805 (1966).
[33] BARNES, J. M., and P. N. MAGEE: Brit. J. industr. Med. **11**, 167 (1954).
[34] MAGEE, P. N., and J. M. BARNES: J. Path. Bact. **84**, 19 (1962).
[35] DRUCKREY, H., D. STEINHOFF, R. PREUSSMANN und S. IVANKOVIC: Naturwissenschaften **50**, 735 (1963).
[36] —, S. IVANKOVIC und R. PREUSSMANN: Naturwissenschaften **51**, 144 (1964).
[37] GRAFFI, A., and F. HOFFMANN: Acta biol. med. germ. **16**, K 1 (1966).
[38] LEAVER, D. D., P. F. SWANN, and P. N. MAGEE: Unpublished work.
[39] DRUCKREY, H., R. PREUSSMANN, S. IVANKOVIC, C. H. SCHMIDT, H. D. MENNEL und K. W. STAHL: Z. Krebsforsch. **66**, 280 (1964).
[40] — —, G. BLUM, S. IVANKOVIC und J. AFKHAM: Naturwissenschaften **50**, 100 (1963).
[41] DUTTON, A. H., and D. F. HEATH: Nature (Lond.) **178**, 644 (1956).
[42] MAGEE, P. N.: Biochem. J. **64**, 676 (1956).
[43] HEATH, D. F.: Biochem. J. **85**, 72 (1962).
[44] MAGEE, P. N., and M. VANDEKAR: Biochem. J. **70**, 600 (1958).
[45] BROUWERS, J. A. J., and P. EMMELOT: Exp. Cell Res. **19**, 467 (1960).
[46] BRODIE, B. B., J. R. GILLETTE, and B. N. LA DU: Ann. Rev. Biochem. **27**, 427 (1958).
[47] SCHOENTAL, R.: Nature (Lond.) **192**, 670 (1961).
[48] MAGEE, P. N., and T. HULTIN: Biochem. J. **83**, 106 (1962).
[49] —, and E. FARBER: Biochem. J. **83**, 114 (1962).
[50] LEE, K. Y., W. LIJINSKY, and P. N. MAGEE: J. nat. Cancer Inst. **32**, 65 (1964).

[51] DRUCKREY, H., D. STEINHOFF, H. BEUTHNER, H. SCHNEIDER und P. KLÄRNER: Arzneimittel-Forsch. **13**, 320 (1963).
[52] LOCHMANN, E. R., u. W. STEIN: Z. Naturforsch. **18 b**, 809 (1963).
[53] GIERER, A., and K. W. MUNDRY: Nature (Lond.) **182**, 1457 (1958).
[54] CRADDOCK, V. M., and P. N. MAGEE: Biochem. J. (In press).
[55] SÜSS, R.: Z. Naturforsch. **20 b**, 714 (1965).
[56] DRUCKREY, H., R. PREUSSMANN, N. NASHED und S. IVANKOVIC: Z. Krebsforsch. **68**, 103 (1966).
[57] TERRACINI, B., and P. N. MAGEE: Unpublished work.
[58] SWANN, P., V. M. CRADDOCK, and P. N. MAGEE: Second Meeting Federat. Europ. Biochem. Soc. p. 322 (1965).
[59] CRADDOCK, V. M., and P. N. MAGEE: Biochem. J. **89**, 32 (1963).
[60] KRÜGER, M., u. G. SALOMON: Z. physiol. Chem. **24**, 364 (1898).
[61] MANDEL, L. R., P. R. SRINIVASAN, and E. BOREK: Nature (Lond.) **209**, 586 (1966).
[62] CRADDOCK, V. M., and P. N. MAGEE: Unpublished work.
[63] TAKAYAMA, S., and K. OOTA: Gann **54**, 465 (1963).
[64] TOMATIS, L., P. N. MAGEE, and P. SHUBIK: J. nat. Cancer Inst. **33**, 341 (1964).
[65] HALVER, J. E., C. L. JOHNSON, and L. M. ASHLEY: Fed. Proc. **21**, 390 (1962).
[66] SCHMÄHL, D., R. PREUSSMANN und H. HAMPERL: Naturwissenschaften **47**, 89 (1960).
[67] —, C. THOMAS und K. KÖNIG: Naturwissenschaften **50**, 407 (1963).
[68] HERROLD, K. M., and L. J. DUNHAM: Cancer Res. **23**, 773 (1963).
[69] DRUCKREY, H., u. D. STEINHOFF: Naturwissenschaften **49**, 497 (1962).
[70] SCHMÄHL, D., u. C. THOMAS: Naturwissenschaften **52**, 165 (1965).
[71] RAPP, H. J., J. H. CARLETON, C. CRISLER, and E. M. NADEL: J. nat. Cancer Inst. **34**, 453 (1965).
[72] SCHMÄHL, D., C. THOMAS und G. F. SCHELD: Naturwissenschaften **51**, 466 (1964).
[73] O'GARA, R. W., and M. G. KELLY: Proc. Amer. Ass. Cancer Res. **6**, 50 (1965).
[74] STANTON, M. F.: J. nat. Cancer Inst. **34**, 117 (1965).
[75] HEATH, D. F., and A. DUTTON: Biochem. J. **70**, 619 (1958).

Zur Frage der Wechselwirkung zwischen Nucleinsäuren und aromatischen Kohlenwasserstoffen und Aminen

Von H. DANNENBERG

Max-Planck-Institut für Biochemie, München

Mit 5 Abbildungen

Im Sinne der Mutationstheorien der Krebsentstehung hat man in den letzten Jahren, wie bei anderen carcinogenen Faktoren,

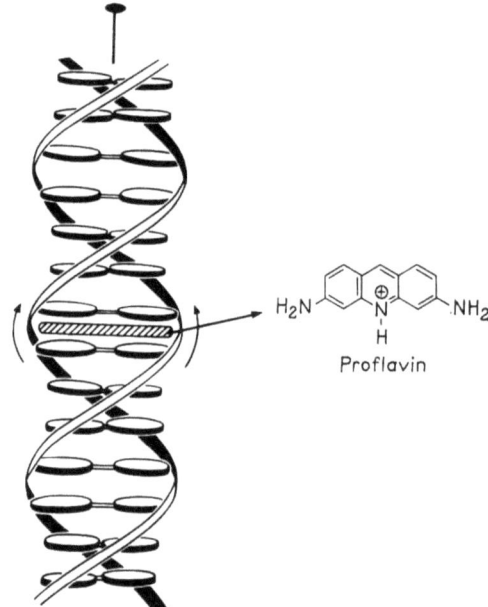

Abb. 1. Intercalarer Einbau von Proflavin in doppelsträngige DNA

auch bei den krebserzeugenden aromatischen Kohlenwasserstoffen[1] und Aminen nach Wechselwirkungen mit dem genetischen Material, den Nucleinsäuren und vor allem den Desoxyribonucleinsäuren (DNA) gesucht. Ausdruck einer derartigen Wechselwirkung können auch der hemmende Einfluß derartiger Verbindungen auf die Replikation gewisser Viren[2] und Bakteriophagen[3] und der verminderte Einbau von Thymidin in die DNA verschiedener Gewebe der Ratte[4] sein. Der molekulare Mechanismus dieser Wirkungen ist noch unbekannt.

Für mutagene Acridine[5] wie z. B. Proflavin sind von L. S. LERMAN[6] experimentelle Befunde beigebracht worden, nach welchen diese flachen, eben gebauten aromatischen Basen zwischen Basenpaare der doppelsträngigen DNA eingebaut werden (vgl. dazu aber [7-9]). Ein derartiger Einbau (s. Abb. 1) soll, möglicherweise bei der Replikation, zu einer bleibenden Veränderung einer DNA führen.

Dieser Mechanismus ist von E. BOYLAND und B. GREEN[10] (vgl. auch [11,12]) auch für die krebserzeugenden aromatischen Kohlenwasserstoffe, deren wirksamste Vertreter eine sehr ähnliche Molekularform wie die durch Wasserstoffbindungen zusammengehaltenen Basenpaare der doppelsträngigen DNA haben, diskutiert worden. Eine erhöhte Löslichkeit der Kohlenwasserstoffe in DNA-Lösungen, Verschiebungen von UV-Absorptionsbanden der aromatischen Kohlenwasserstoffe zum Langwelligen und eine Erhöhung des „Schmelzpunktes" der DNA in derartigen Lösungen sind in diesem Sinne gedeutet worden[10,13,14,15]. Diese Effekte zeigen auch Proflavin-DNA-Lösungen.

Wenn Proflavin in seiner Wechselwirkung mit DNA und in seiner mutagenen Wirkung als Modell für den Wirkungsmechanismus krebserzeugender aromatischer Kohlenwasserstoffe und Amine zu betrachten ist, sollten sie Analogien zu entsprechenden Eigenschaften des Proflavins aufweisen, und es sollten sich krebserzeugend wirksame und unwirksame Aromaten voneinander unterscheiden. Diesen Fragen sind mehrere Arbeitsgruppen unseres Instituts in verschiedenen Richtungen nachgegangen.

1. BOYLAND und GREEN[10] und auch LIQUORI et al.[14] haben bei der Untersuchung der Eigenschaften von Kohlenwasserstoff-DNA-Lösungen (sie beschränkten sich im wesentlichen auf Pyren, 3,4-Benzpyren und 1,2,5,6-Dibenzanthracen) diese Lösungen dargestellt durch Verreiben der feinpulverisierten Kohlenwasserstoffe mit DNA-Lösungen. Dr. SONNENBICHLER hat eine andere Technik verwendet[13]: Wäßrige Lösungen von nativer Kalbsthymus-DNA wurden mit einer Lösung des Kohlenwasserstoffs in Dimethylformamid versetzt, anschließend wurde dialysiert, unter standardisierten Bedingungen mit Cyclohexan ausgeschüttelt und zentri-

fugiert. Ausgewertet wurden die UV-Spektren dieser Lösungen in bezug auf Bandenverschiebungen und Menge an gelöstem Kohlenwasserstoff, außerdem wurde auch der Schmelzpunkt der DNA bestimmt. Auch nach dieser Methode dargestellte Proflavin-DNA-Lösungen zeigen die Verschiebung der im Sichtbaren gelegenen Absorptionsbande des Proflavins um etwa 12 mμ zum Langwelligen, eine ausgeprägte Anreicherung des Proflavins bei der Gleich-

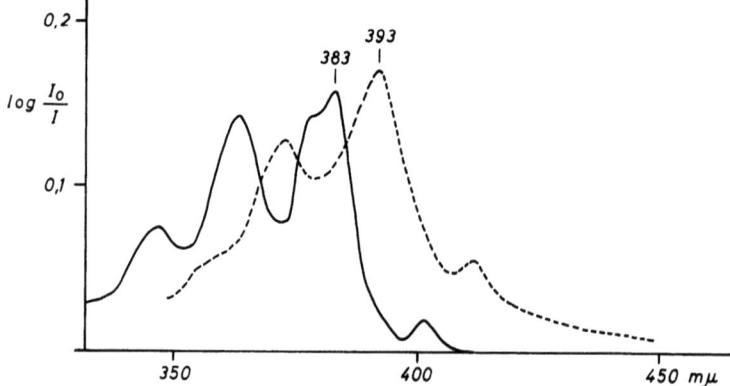

Abb. 2. UV-Absorption von 3,4-Benzpyren in Äthanol (———) und in DNA-Phosphatpuffer pH 6,8 (- - - -)

gewichtsdialyse (s. u. Tab. 2) und eine Erhöhung des Schmelzpunktes der DNA.

Mit dieser Methode wurden fünf krebserzeugend wirksame und drei unwirksame Kohlenwasserstoffe untersucht. Wie in den Versuchen von BOYLAND und GREEN wurde eine Verlagerung von Absorptionsbanden der Kohlenwasserstoffe zum Langwelligen gefunden (s. Kurven für 3,4-Benzpyren, Abb. 2), aber kein Zusammenhang zwischen der Größe des Effektes und der krebserzeugenden Wirksamkeit. Da die gleiche Verschiebung von Absorptionsbanden zum Langwelligen auch in entsprechenden Präparationen ohne DNA auftreten, z. B. beim 3,4-Benzpyren (Tab. 1)[17], möchten wir eher eine Anlagerung der Kohlenwasserstoffe an die DNA im Sinne eines stabilisierten Kolloids annehmen als einen spezifischen Einbau (vgl. dazu auch[18,19]). Aber auch eine derartige Anlagerung könnte mehr oder weniger spezifisch sein. Die Anreicherung von Kohlenwasserstoffen ist in den DNA-Lösungen gegenüber entsprechend bereiteten DNA-freien Lösungen um 1 bis 2 Zehnerpotenzen kleiner

Tabelle 1. *Absorptionsmaxima der langwelligen Banden von 3,4-Benzpyren (Schichtdicken 0,5 bis 10 cm)*

Gelöst in:	Absorptionsmaxima				
	mμ	mμ	mμ	mμ	mμ
Äthanol	347	363	383	403	—
0,7 mM Phosphatpuffer (zentrifugiert)	350	368	388	404	420
DNA/0,7 mM Phosphatpuffer					
„ (zentrifugiert)	355	373	393	404	—
„ (nicht zentrifugiert)	363	382	390	404	418
0,7 mM Phosphatpuffer					
(nicht zentrifugiert)	365	383	392	404	425

als beim Proflavin (s. Tab. 2)[17] und zeigt keinen eindeutigen Unterschied zwischen wirksamen und unwirksamen Kohlenwasserstoffen. Schließlich haben wir auch keine Erhöhung des Schmelzpunktes

Tabelle 2. *Anreicherung von aromatischen Kohlenwasserstoffen und Aminen in DNA-Lösungen*

Substanz	μMol* Liter in KHPO$_4$- Puffer (1)	μMol* Liter in DNA- Lösung (2)	‚gebundener' Anteil μMol Liter (2—1)	Anreicherung ($\frac{2}{1}$)
Coronen	1,54	2,10	0,56	1,4
Pyren	0,13	0,50	0,37	3,8
1,2-Benzpyren	0,35	0,90	0,55	2,6
1,2-Benzanthracen	0,03	0,50	0,47	17,0
1,2,5,6-Dibenzanthracen	1,2	6,7	5,5	5,5
3,4-Benzpyren	0,05	0,77	0,72	16,0
9,10-DMBA	0,05	0,13	0,08	2,6
3-Methylcholanthren	0,89	10,69	9,8	12,0
Acridinorange	0,07	16,07	16,0	230,0
Proflavin	0,08	22,18	22,1	280,0
2,7-Diaminofluoren	2,8	18,3	15,5	6,5
4-Dimethylaminoazobenzol	0,07	0,07	0	1,0

* Konzentration nach Gleichgewichtsdialyse und Extraktion mit Cyclohexan; pH 6,8.

der DNA in unseren Kohlenwasserstoff-DNA-Lösungen finden können[16].

Zusammenfassend ergibt sich aus unseren in-vitro-Untersuchungen kein Hinweis auf eine spezifische Wechselwirkung zwischen krebserzeugenden Kohlenwasserstoffen und DNA im Sinne der LERMANschen Hypothese. Eine Beziehung, deren Mechanismus man aber noch nicht kennt, ergibt sich dagegen aus in-vivo-Versuchen von P. BROOKES und P. D. LAWLEY[20].

2. Der Beweis einer Parallelität zwischen krebserzeugender und mutagener Wirkung steht bei den aromatischen Kohlenwasser-

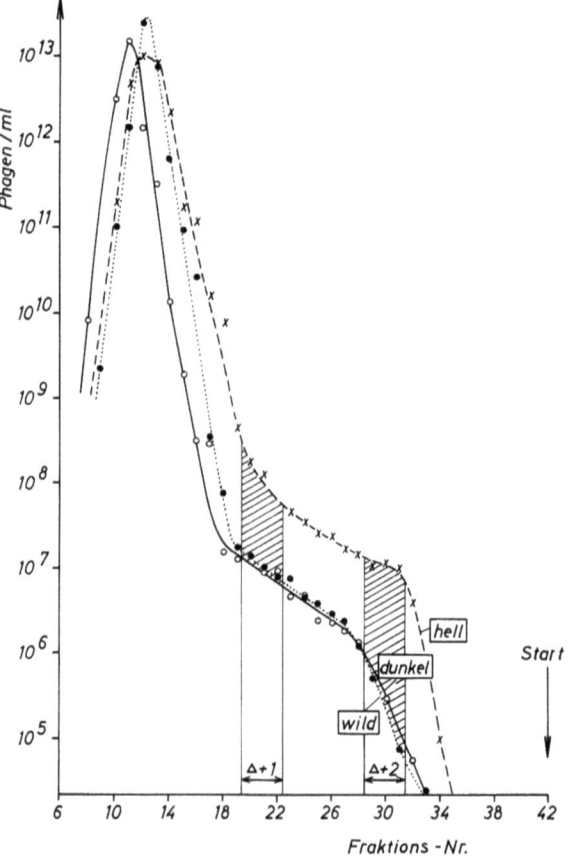

Abb. 3. Elektrophoretische Verteilungskurven für die Proflavin-Mutation, ohne (dunkel) und mit (hell) Photoinduktion gegenüber einer Vergleichskurve (wild); 4,5 γ/ml Proflavin. Eingetragen sind die Bereiche der Mutantenklassen △ +1 und △ +2[21]

stoffen und Aminen noch aus. Zur Untersuchung der Frage nach dem molekularen Wirkungsmechanismus haben G. BRAUNITZER und G. HOBOM[21] den Einfluß von Proflavin und 2,7-Diaminofluoren auf den E. coli-Bakteriophagen fd[22] geprüft. Dieser Phage, ein flexibles Stäbchen vom Partikelgewicht 11,6 Millionen, enthält eine einsträngige DNA vom Molgewicht 1,4 Millionen (etwa 4500 Nucleotide); die Untereinheit des Hüllproteins hat ein Molgewicht von 9000 bis 10000 entsprechend etwa 90 Aminosäuren.

Grundlage für die Erkennung von Mutanten und ihre Selektion sind Ladungsabweichungen im Hüllprotein. In der trägerfreien Elektrophorese von K. HANNIG[23] können Phagen mit unterschiedlicher Ladung im Hüllprotein voneinander getrennt werden[21].

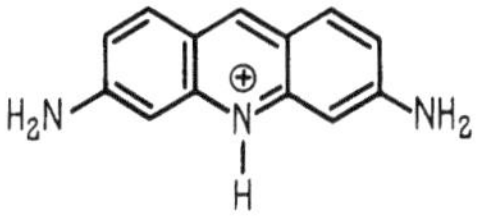

Proflavin mutagen

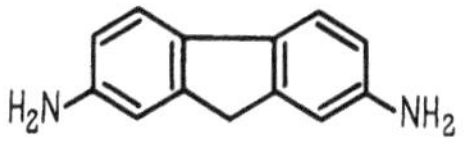

2,7-Diamino-fluoren carcinogen

Abb. 4

Berücksichtigt man nur Verschiebungen zur positiven Seite, so zeigt die Phagenfraktion einer Phagenaufzucht beim Wildtyp reproduzierbare elektrophoretische Verteilungskurven mit *einer* Hauptkomponente in bezug auf die Ladung und nur geringen Anteilen an Fraktionen mit den Ladungsabweichungen von $\Delta +1$ und $\Delta +2$ (s. Abb. 3). Wird die Phagenaufzucht in Gegenwart von Proflavin unter Ausschluß von Licht durchgeführt, so ist die elektrophoretische Verteilungskurve die gleiche wie beim Wildtyp (s. Abb. 3). Wird dagegen gleichzeitig mit Licht bestrahlt, welches von Proflavin absorbiert wird, so erfolgt als Ausdruck der mutagenen Wirkung eine Zunahme der $\Delta +1$- und $\Delta +2$-Hüllproteinphagen, bei $\Delta +2$ um etwa zwei Zehnerpotenzen (s. Abb. 3). Notwendiges Glied in der zur Proflavinmutation führenden Reaktionskette, scheint direkt oder indirekt das *angeregte* Proflavinmolekül zu sein.

Untersucht man in der gleichen Versuchsanordnung das dem Proflavin strukturell sehr ähnlich gebaute krebserzeugende 2,7-Diaminofluoren (s. Abb. 4), so ergibt sich auch hier eine etwa in der gleichen Größenordnung wie beim Proflavin in Gegenwart von

Licht liegende Zunahme der $\varDelta+1$- und $\varDelta+2$-Hüllproteinphagen (s. Abb. 5). Im Gegensatz zu Proflavin bedarf es beim 2,7-Diaminofluoren aber *nicht* der Mitwirkung von Licht. Darin besteht ein wesentlicher Unterschied in der mutagenen Wirkung dieser beiden

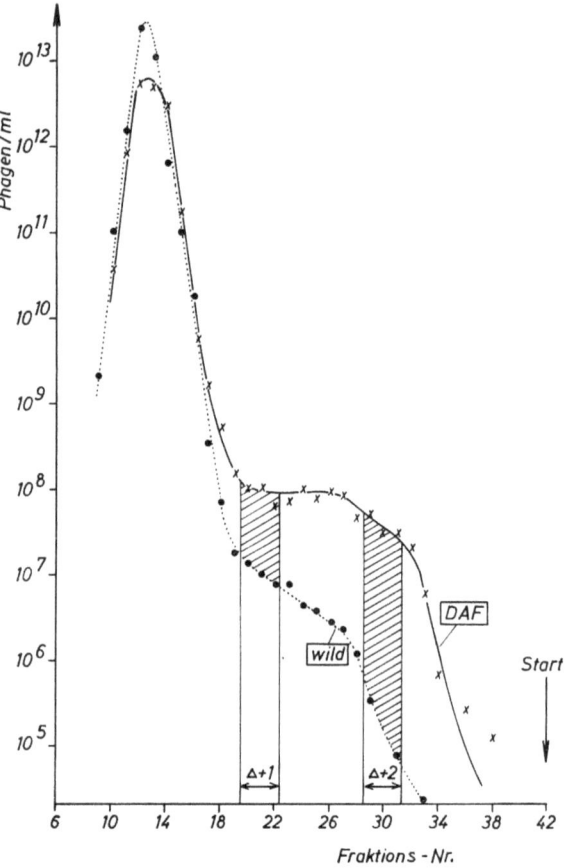

Abb. 5. Elektrophoretische Verteilung für 2,7-Diaminofluoren-Mutation im Dunkeln (DAF) gegenüber einer Vergleichskurve (wild); 22 γ/ml Diaminofluoren

Verbindungen. Weitere Untersuchungen müssen zeigen, ob der molekulare Wirkungsmechanismus in beiden Fällen identisch ist.

3. Schließlich soll noch kurz auf eine weitere Eigenschaft von Proflavin eingegangen werden. Proflavin hemmt ebenso wie Actinomyzin C_3 die von der DNA abhängigen RNA-Polymerase kataly-

sierte RNA-Synthese aus den Nucleotidtriphosphaten, wie von
W. ZILLIG u. Mitarb.[25] gezeigt wurde (s. Tab. 3). Ein Maß für die
Bildung der RNA ist der Einbau von ATP-^{14}C in säureunlösliche
RNA. Zwischen der Wirkungsweise von Actinomycin und Proflavin besteht aber ein Unterschied, und zwar hemmt Actinomyzin
diese Reaktion in Gegenwart vou einsträngiger und doppelsträngiger DNA, Proflavin dagegen nur in Gegenwart von doppelsträn-

Tabelle 3. *Einfluß auf die DNA-abhängige RNA-Synthese (Präger Thymin-DNA oder T_2-DNA)*

Substanz	Zusatz μMol/Ansatz	%Einbau
—	—	100,0
Actinomyzin C$_3$	0,05	0,8
Proflavin	0,05	45,0
2,7-Diaminofluoren	<0,1	99,7
2-Aminophenanthren	0,14	78,3
2-Aminophenanthren-N-(D)-ribosid	<0,1 (ges.)	100,8
3-Aminophenanthren	0,1	81,7
3-Dimethylaminophenanthren	<0,1	103,6
Benzidin	0,1	98,9

giger DNA, jedoch nicht bei einsträngiger DNA. Dieses Verhalten
des Proflavins könnte durch den intercalaren Einbau in die doppelsträngige DNA erklärt werden. In diesem Test wurde von W. ZILLIG
und E. FUCHS[26] auch eine Reihe krebserzeugender aromatischer
Amine geprüft, wobei als Lösungsvermittler 1 bis 2% Dimethylformamid verwendet wurde, welches in dieser Konzentration die
Reaktion nicht hemmt. Eine deutliche Hemmung des ATP-^{14}C-
Einbaus, aber wesentlich schwächer als diejenige von Proflavin,
zeigten nur 2-Amino- und 3-Aminophenanthren, unwirksam waren
dagegen 2,7-Diaminofluoren, 2-Acetaminofluoren, Benzidin, 4-Dimethylaminoazobenzol (Tab. 3). Aromatische Kohlenwasserstoffe zeigten, vielleicht auf Grund ihrer zu geringen Löslichkeit, keine Wirkung.

Zusammenfassend ergibt sich aus den besprochenen Untersuchungen, daß der für das Proflavin heute diskutierte Mechanismus
nicht ohne weiteres auf die Wechselwirkung zwischen Desoxyribonucleinsäuren und krebserzeugenden aromatischen Kohlenwasserstoffen und Aminen übertragen werden kann.

Literatur

[1] Zusammenfassung: MARTIN, CH. M.: Progr. exp. Tumor Res. **5**, 134 (1964).
[2] DE MAEYER, E., and J. DE MAEYER-GUIGNARD: Science **146**, 650 (1964).
[3] HSU, W.-T., J. W. MOOHR, and S. B. WEISS: Proc. nat. Acad. Sci. (Wash.) **53**, 517 (1965).
[4] JENSEN, E. V., E. FORD, and CH. HUGGINS: Proc. nat. Acad. Sci. (Wash.) **50**, 454 (1963).
[5] BRENNER, S., L. BARNETT, F. M. C. CRICK, and A. ORGEL: J. molec. Biol. **3**, 121 (1961).
[6] LERMAN, L. S.: J. molec. Biol. **3**, 18 (1961); J. cell. comp. Physiol. **64**, Suppl. 1,1 (1964).
[7] LIERSCH, M., u. G. HARTMANN: Biochem. Z. **43**, 16 (1965).
[8] BLAKE, A., and A. PEACOCKE: Nature (Lond.) **206**, 1009 (1965).
[9] DRUMMOND, D., V. SIMPSON-GILDEMEISTER, and A. PEACOCKE: Biopolymers **3**, 135 (1965).
[10] BOYLAND, E., and E. B. GREEN: Brit. J. Cancer **16**, 507 (1962).
[11] HUGGINS, CH., and N. C. YANG: Science **137**, 257 (1962).
[12] DANNENBERG, H.: Dtsch. med. Wschr. **88**, 605 (1963).
[13] BOYLAND, E., B. GREEN, and S.-L. LIU: Biochim. biophys. Acta (Amst.) **87**, 653 (1964).
[14] LIQUORI, A. M., B. DE LERMA, F. ASCOLI, C. BOTIE, and M. TRASCIETTI: J. molec. Biol. **5**, 521 (1962).
[15] BOYLAND, E., and B. GREEN: Biochem. J. **87**, 14 P (1963).
[16] DANNENBERG, H., u. J. SONNENBICHLER: Z. Krebsforsch. **67**, 127 (1965).
[17] SONNENBICHLER, J.: Unveröffentlicht.
[18] GIOVANELLA, B. C., L. E. MCKINNEY, and CH. HEIDELBERGER: J. molec. Biol. **8**, 20 (1964).
[19] BOYLAND, E., and B. GREEN: J. molec. Biol. **9**, 589 (1964).
[20] BROOKES, P., and P. D. LAWLEY: Nature (Lond.) **202**, 781 (1964).
— J. cell. comp. Physiol. **64**, Suppl. 1, 111 (1964).
[21] HOBOM, G.: Dissertation Universität München 1965.
[22] MARVIN, D. A., and H. HOFFMANN-BERLING: Nature (Lond.) **197**, 517 (1963). — Z. Naturforsch. **18** b, 884 (1963).
[23] HANNIG, K.: Z. analyt. Chem. **181**, 244 (1960).
[24] BRAUNITZER, G.: Ber. Bunsenges. Physik. Chem. **68**, 733 (1964).
[25] DOERFLER, W., W. ZILLIG, E. FUCHS und M. ALBERS: Hoppe Seylers Z. physiol. Chem. **330**, 96 (1962).
[26] FUCHS, E.: Dissertation Universität München 1965.

Die cocarcinogene Wirkung der Phorbolester

Von E. HECKER

Deutsches Krebsforschungszentrum, Biochemisches Institut, Heidelberg

Mit 4 Abbildungen

In der Absicht, die Carcinogenese in möglichst frühen Stadien biochemisch zu untersuchen, beschäftigen wir uns mit dem Zusammenwirken von carcinogenen aromatischen Kohlenwasserstoffen und Cocarcinogenen an der Mäusehaut. Entsprechende Experimente sind 1941 von BERENBLUM[1] angegeben und 1944 von MOTTRAM[2] präzisiert worden, wobei als Cocarcinogen das entzündlich wirksame Crotonöl — das Samenöl einer Euphorbiacee — Verwendung fand. Das Crotonöl erwies sich in der Folgezeit als bei weitem wirksamstes Cocarcinogen.

Die für das „BERENBLUM-Experiment" typische Anordnung mit 9,10-Dimethyl-1,2-benzanthracen (DMBA) als Carcinogen und Crotonöl als Cocarcinogen ist in Tab. 1 schematisch wiedergegeben. Im ersten Versuch wird eine Gruppe von Mäusen fortlaufend mit einer bestimmten Dosis D von DMBA behandelt. Nach einer Anzahl von Applikationen bilden sich die Tumoren. Die Dosis des carcinogenen Kohlenwasserstoffs wird so gewählt, daß eine einzige Dosis D nicht zu Tumoren

Tabelle 1. *Zusammenwirken von Carcinogen und Cocarcinogen im BERENBLUM-Experiment an der Mäusehaut*

Applikation	Tumoren
1. D D D . . . . . . D D D	+
2. D	—
3. O C C . . . . . . C C C	—
4. D C C . . . . . . C C C	+

D = 9,10-Dimethyl-1,2-benzanthracen (carcinogen)
C = Crotonöl (cocarcinogen)

führt (Versuch 2); die Einzeldosis D kann daher als unterschwellig bezeichnet werden. Auch die fortlaufende Applikation bestimmter Crotonöldosen C — wie in Versuch 3 — soll im Idealfall keine Tumoren erzeugen. Die Kombination der unterschwelligen Einzeldosis D mit der fortlaufenden Applikation der für sich unwirksamen Dosis C des Crotonöls hingegen führt, wie in Versuch 1, zur Bildung von Tumoren.

Die Interpretation dieser und ähnlicher Experimente führte BERENBLUM u. a. zur Zwei- oder Mehrstufenhypothese der Cancerisierung der Mäusehaut. Sie besagt, daß gewisse Zellen der Mäusehaut durch die unterschwellige Einzeldosis von carcinogenem Kohlenwasserstoff in potentielle Tumorzellen umgewandelt werden, die ohne weitere Behandlung latent bleiben. Diese erste Stufe wird als *Initialphase* der Cancerisierung bezeichnet. Die potentiellen Tumorzellen können aber in einer zweiten Stufe, der *Entwicklungsphase*, zu Tumoren realisiert werden, wenn auf die Haut Crotonöl aufgetragen wird. Der Ausbildung potentieller Tumorzellen in der Initialphase und ihrer Entwicklung zu sichtbaren Tumoren in der Entwicklungsphase gilt unser besonderes Interesse, weil dieses Modell der Carcinogenese im Gegensatz zu fast allen anderen Ansätzen die Möglichkeit bietet, für die Cancerisierung essentielle biochemische Veränderungen normaler Zellen zu einem sehr frühen Zeitpunkt auf molekular-biologischer Ebene zu erfassen[3,4]. Dies setzt allerdings voraus, daß die von BERENBLUM gegebene Interpretation des Ansatzes zu Recht besteht, was in der Literatur häufig angezweifelt wurde (z. B.[5]).

Im übrigen sind für die beabsichtigten biochemischen Untersuchungen zunächst die molekularen Voraussetzungen zu schaffen[3,6], denn die Wirkstoffe des Crotonöls und ihre chemische Natur waren bis vor einigen Jahren unbekannt. 1962—1965 ist es uns erstmals gelungen, diese Wirkstoffe zu isolieren, sie rein darzustellen und ihre wichtigsten strukturellen Merkmale zu identifizieren[3,6-10]. Bis jetzt konnten insgesamt elf biologisch hochaktive Substanzen rein erhalten werden (Tab. 2). Sie treten im Crotonöl in zwei chromatographisch durch den Rf-Wert (0,3 bzw. 0,4) unterscheidbaren Gruppen auf, die mit A und B bezeichnet werden und die sich durch multiplikative Verteilung in die reinen Komponenten A1—A4 sowie B1—B7 auftrennen lassen. Alle isolierten Wirkstoffe sind Diester desselben polyfunktionellen Grundalkohols „Phorbol" $C_{20}H_{28}O_6$ mit jeweils einer kurz- und einer langkettigen Fettsäure. Neben Essigsäure wurden (+)-S-2-Methylbuttersäure sowie Tiglinsäure (trans-2-Methylcrotonsäure) als kurzkettige Fettsäuren gefunden. Die geradzahligen, gesättigten und unverzweigten Fettsäuren von C_8-C_{16} kommen als langkettige Fettsäuren vor.

Phorbol ist ein tetracyclisches Diterpen (Tab. 2). Es enthält fünf Hydroxylgruppen, nämlich eine primär-allylische Hydroxylgruppe

Die cocarcinogene Wirkung der Phorbolester 107

(a) (Formel I), eine sekundäre Hydroxylgruppe (b) und eine tertiäre Hydroxylgruppe (c), die mit Pyridin/Acetanhydrid acetylierbar sind, sowie zwei tertiäre, unter denselben Bedingungen nicht acetylierbare Hydroxylgruppen (d) und (e). Die 6. Sauerstoffunktion (f) ist eine α,β-ungesättigte Carbonylgruppe. Auf Grund der Analyse von UV-, IR-, KMR- und Massenspektren und unter Heranziehung chemischer Reaktionen wurde von uns für Phorbol die Struktur I vorgeschlagen[11], die mit allen bis dahin erhobenen chemischen und

Tabelle 2. *Die elf aus Crotonöl isolierten Wirkstoffe und ihre chemische Struktur*

Substanz	Brutto-formel	Rf-Wert*	Säurereste identifiziert als	Grund-alkohol
A1	$C_{36}H_{56}O_8$		Essigsäure, Myristinsäure	
A2	$C_{32}H_{48}O_8$	0,3	Essigsäure, Caprinsäure	
A3	$C_{34}H_{52}O_8$		Essigsäure, Laurinsäure	
A4	$C_{38}H_{60}O_8$		Essigsäure, Palmitinsäure	$C_{20}H_{28}O_6$
				Phorbol
B1	$C_{37}H_{58}O_8$		(+)-S-2-Methylbuttersäure, Laurinsäure	(Diterpen)
B2	$C_{35}H_{51}O_8$		(+)-S-2-Methylbuttersäure, Caprinsäure	5 OH 1 C=O
B3	$C_{35}H_{52}O_8$	0,4	Tiglinsäure, Caprinsäure	2 C=C
B4	$C_{34}H_{52}O_8$		Essigsäure, Laurinsäure	
B5	$C_{33}H_{50}O_8$		(+)-S-2-Methylbuttersäure, Caprylsäure	
B6	$C_{33}H_{48}O_8$		Tiglinsäure, Caprylsäure	
B7	$C_{32}H_{48}O_8$		Essigsäure, Caprinsäure	

* im System CH_2Cl_2/Aceton = 3:1, Kammersättigung, Kieselgel Merck, HF 254

physikalischen Daten des Phorbols in Einklang stand. — Die Strukturvorschläge anderer Autoren[12] wurden nicht näher begründet und stehen im Widerspruch zu den von uns ermittelten NMR-Daten des Phorbols[13].

Neue chemische Befunde aus unserem Laboratorium sowie Messungen des Circulardichroismus bestimmter Phorbolderivate zeigen, daß unser Strukturvorschlag I noch zu modifizieren ist[10]. Der neue Strukturvorschlag enthält in den Partialstrukturen II, III und IV alle C-, H- und O-Atome; auch die von uns beschriebenen[11] sechs Sauerstoffunktionen des Phorbols sind unverändert. Der Unterschied zu Strukturvorschlag I geht im wesentlichen darauf zurück, daß wir inzwischen gefunden haben, daß Phorbol ein tertiäres und nicht ein sekundäres Acyloin ist, und daß seine reduzierenden Eigenschaften, die ein sekundäres Acyloin vermuten ließen[11,14], auf das tertiäre Cyclopropanhydroxyl zurückgehen. Durch Kombination der Partialformeln II—IV lassen sich für Phorbol mehrere

tetracyclische Strukturen angeben, zwischen denen aber auf Grund der bis jetzt vorliegenden Daten noch nicht endgültig entschieden werden kann.

In den isolierten Wirkstoffen sind die kurz- bzw. langkettigen Fettsäurereste mit den Hydroxylgruppen (b) und (c) des Phorbols verestert, während die Hydroxylgruppe (a) sowie die beiden tertiären Hydroxyle (d) und (e) frei vorliegen. Bei Vertauschung ungleichartiger Fettsäurereste an (b) und (c) sind daher Isomeren*paare* zu erwarten. Durch selektive Veresterung des Phorbols mit Essigsäure bzw. den entsprechenden langkettigen Fettsäuren konnten wir drei solche Isomerenpaare darstellen[15]. Fünf dieser sechs Phorbolester sind nach allen physikalischen und chemischen Kriterien identisch mit den im Crotonöl enthaltenen Wirkstoffen (vgl. Tab. 2)

Die cocarcinogene Wirkung der Phorbolester 109

A1, A2, A3 sowie B4 und B7[9,10,15]. Dies beweist, daß wir die Chemie der funktionellen Gruppen des Phorbols beherrschen, wenn auch die Struktur des Kohlenstoffskelets noch nicht in allen Einzelheiten festliegt. — Die in der Literatur beschriebene Synthese[16] eines Acetyl-phorbol-myristinsäureesters verläuft ungezielt und berücksichtigt das Isomerenproblem nicht.

Die beschriebenen Phorbolester sind auf partialsynthetischem Wege nunmehr leicht und in beliebigen Mengen zugänglich. Damit ist die Überprüfung der von verschiedenen Autoren (z. B.[5]) in Frage gestellten cocarcinogenen oder tumor-promovierenden (promoting) Wirkung dieser Substanzen möglich. Als biologisches Testsystem wird die Anordnung des BERENBLUM-Experiments in einer von uns entwickelten quantitativen Variante herangezogen.

In Tab. 3 sind einige Versuche zusammengefaßt, die der Ermittlung der unterschwelligen Dosis von DMBA bei dem verwendeten Mäusestamm dienen. Es werden Gruppen von 28 Mäusen im

Tabelle 3. *Einmalige Applikation von DMBA*
Stamm: NMRI, 14/14 Mäuse pro Gruppe, Lsgm. Aceton p. a.

DMBA (μMol)	Fortlaufend 2 × wöchentl	Latenzzeit (Wochen)	Tumoren/Gesamtzahl lebender Tiere			Signifikanz- niveau
			Stand Latenzzeit	Stand 24. Woche	Stand 48. Woche	
1	—	8	2/28	4/25	1/16	$0,05 < p < 0,1$
1	—	16	2/30	4/28	1/23	$0,05 < p < 0,1$
0,1	—	24	4/28	4/28	3/21	$0,1 < p < 0,2$
0,1	Aceton	> 60	—	0/22	0/18*	
0,01	Aceton	> 26	—	0/28	—	
0	Aceton	> 60	—	0/26	0/17**	

* 29.—41.Woche 2—3/22; ** 42., 43. Woche 1/20

Geschlechtsverhältnis 1:1 verwendet. Wie man Tab. 3 entnehmen kann, erzeugt eine einmalige Dosis von 1 μMol DMBA vereinzelte Tumoren mit einer Latenzzeit von 8 bzw. 16 Wochen bis zum Auftreten des ersten Tumors. Die Latenzzeit, mit der diese Tumoren auftreten, ist statistisch nicht befriedigend zu sichern. Wird die einmalige Dosis von DMBA auf 0,1 μMol reduziert, so ist die Latenzzeit verlängert. Bei derselben Dosis von DMBA mit nachfolgender zweimal wöchentlicher Applikation von 0,1 ml Aceton sind dagegen lediglich in der 29. bis 41. Woche zwei bis drei Tumorträger pro 22 überlebende Tiere zu beobachten. Die Verminderung der

DMBA-Dosis auf 0,01 μMol mit nachfolgender Acetonbehandlung ergibt 0/28 Tumorträger (bis zur 26. Woche, das Experiment ist noch im Gange). Es ist interessant zu vermerken, daß bei alleiniger Applikation von zweimal 0,1 ml Aceton/Woche in der 42. und 43. Woche 1/20 Tumorträger auftritt. Zusammenfassend läßt sich feststellen, daß für den Stamm NMRI eine Dosis von 1 μMol DMBA wohl noch als unterschwellig bezeichnet werden kann. Mit größerer Sicherheit als unterschwellig zu betrachten ist die Dosis

Tabelle 4. *Wirkung einer einmaligen Dosis von 0,1 μMol DMBA und 2 × 0,1ml 0,01% Wirkstofflösung pro Woche auf t_{50}*
Stamm: NMRI, 14/14 Mäuse pro Gruppe, Lsgm. Aceton

Wirkstoff	0,01% Wirkstoff			
	mit DMBA (Wochen)	Signifikanz-niveau	ohne DMBA (Wochen)	Signifikanzniveau
A1	8,0 ± 2,5	p < 0,001	24,5 ± 10,0	0,01 < p < 0,05
B1	10,0 ± 2,5	p < 0,001	30,0 ± 12,0	0,01 < p < 0,05
B2	8,5 ± 2,5	p < 0,001	30,0 ± 12,5	p < 0,001
B3	10,0 ± 1,8	p < 0,001	29,0 ± 12,5	p < 0,001
B4	11,0 ± 4,0	p = 0,001	24,5 ± 11,0	p < 0,001
B7	12,0 ± 3,5	p < 0,001	32,0 ± 4,0	p < 0,001

0,1 μMol DMBA bei nachfolgender Applikation von 0,1 ml Aceton. Dieser Versuch ist zugleich das exakte Kontrollexperiment für die kombinierte Applikation von Carcinogen und Cocarcinogen.

Daten für die kombinierte Wirkung von unterschwelliger einmaliger Dosis DMBA und einigen der aus Crotonöl isolierten reinen cocarcinogenen Wirkstoffen sind in Tab. 4 zusammengefaßt. Vergleicht man die Zeit, nach der 50% der Tiere mindestens einen Tumor tragen (mittlere Latenzzeit t_{50}), so erkennt man (Tab. 4, 2. Spalte), daß die einzelnen Wirkstoffe keine deutlichen Unterschiede zeigen. Als wirksamste Cocarcinogene erscheinen die Substanzen A1 und B2.

Wie aus dem Vergleich der Werte für t_{50} mit und ohne die Initialdosis DMBA hervorgeht (Tab. 4, 4. Spalte), treten auch bei Applikation der Crotonölwirkstoffe allein Tumoren auf. Die mittlere Latenzzeit liegt jedoch um den Faktor 2 bis 3 höher als bei Kombination mit der Initialdosis DMBA. Dies bedeutet, daß den reinen Wirkstoffen des Crotonöls in den angewendeten Dosierungen eine gewisse carcinogene Wirkung zukommt. Die Verhältnisse, besonders auch im Hinblick auf die Tumorrate bei alleiniger Applikation von

DMBA, lassen sich am besten in einem quantitativen BERENBLUM-Experiment übersehen (Abb. 1).

In einem 1. Versuch wurde eine Gruppe von 28 Mäusen mit einer einmaligen Dosis DMBA und — eine Woche danach — zweimal wöchentlich fortlaufend mit derselben Dosis DMBA behandelt. Dabei entwickelten sich Tumoren mit einer Latenzzeit t_{50} von 12 ± 3 Wochen bei einer applizierten Gesamtdosis D_{50} von $2,4\pm0,6$ µMol DMBA. Der Verlauf der Tumorbildung im Wahrscheinlichkeitsnetz

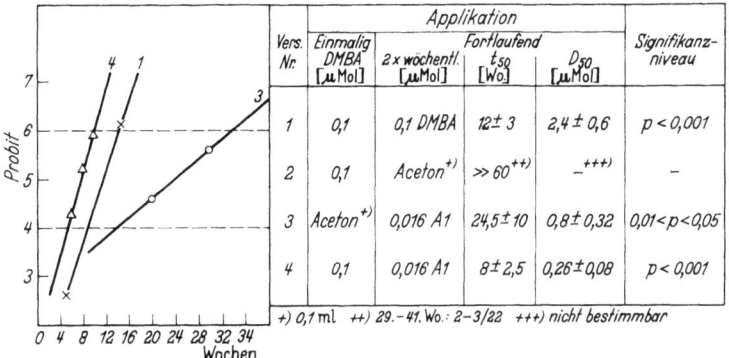

Abb. 1. Quantitatives BERENBLUM-Experiment mit 0,1 µMol DMBA und 0,016 µMol Wirkstoff A_1. Stamm NMRI; 14/14 Mäuse/Gruppe; Lsgm.: Aceton p. a.

ist linear (Dosiswirkungskurve 1), wenn auf der Ordinate die Prozente der Mäuse mit Tumoren in Probiteinheiten und auf der Abszisse die Zeit in Wochen aufgetragen wird. Der Versuch Nr. 2, mit einer Einzeldosis von 0,1 µMol DMBA und zweimal wöchentlicher Applikation von 0,1 ml Aceton, ist bereits in Tab. 3 enthalten und ergibt praktisch keine Tumoren. Die zweimal wöchentliche Applikation einer Einzeldosis von 0,016 µMol A1 führt mit einer Latenzzeit t_{50} von $24,5\pm10$ Wochen (vgl. auch Tab. 4) und einer Gesamtdosis D_{50} von $0,8\pm0,32$ µMol A1 zu Tumoren. Dem entspricht die Dosiswirkungskurve 3. Der flache Anstieg dieser Geraden kann als Ausdruck einer carcinogenen Aktivität des Wirkstoffs A1 gewertet werden, die wesentlich geringer ist als die des DMBA. Die Kombination von 0,1 µMol DMBA mit zweimal wöchentlicher Applikation von 0,016 µMol A1 ergibt mit einer Latenzzeit von $8\pm2,5$ Wochen (Tab. 4) und einer Gesamtdosis D_{50} von $0,26\pm0,08$ µMol entsprechend der Dosiswirkungskurve 4 eine höhere Tumorausbeute als die Behandlung mit Cocarcinogen allein. Der Anstieg der Dosis-wirkungskurve 4 ist mit dem Anstieg der Kurve 1 praktisch identisch.

Es ist bemerkenswert, daß trotz der niedrigen Einzeldosis des Wirkstoffs A1, nämlich ca. $^1/_6$ der DMBA-Einzeldosis in Versuch Nr. 1, die Latenzzeit des Versuchs Nr. 4 niedriger liegt als die des Versuchs Nr. 1. Auch die applizierte Gesamtdosis D_{50} dieses Versuchs

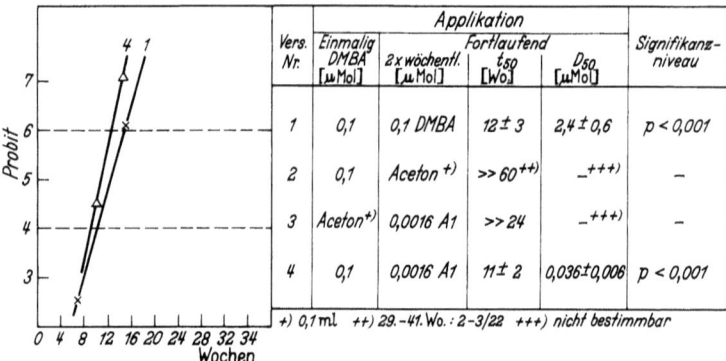

Abb. 2. Quantitatives BERENBLUM-Experiment mit 0,1 μMol DMBA und 0,0016 μMol Wirkstoff A_1. Stamm NMRI; 14/14 Mäuse/Gruppe; Lsgm.: Aceton p. a.

ist — trotz gleicher Tumorausbeute — um etwa den Faktor 10 niedriger als in Versuch Nr. 1. Dieser Unterschied in der Latenzzeit und in der Gesamtdosis zwischen den Versuchen 1 und 4 kann als Hinweis darauf gewertet werden, daß der Wirkstoff A1 neben der schwach carcinogenen auch eine definitive cocarcinogene — d. h. im Sinne von BERENBLUM tumor-promovierende — Qualität besitzt.

Die saubere Differenzierung zwischen diesen beiden Qualitäten scheint eine Frage der Dosierung zu sein, wie aus dem in Abb. 2 dargestellten Experiment hervorgeht. Dieser Versuch unterscheidet sich von dem vorausgehenden nur darin, daß die Einzeldosis des Wirkstoffs A1 mit 0,016 μMol um den Faktor 10 niedriger liegt. Wie aus Versuch 3 hervorgeht, zeigt Wirkstoff A1 in dieser Dosierung keine carcinogene Qualität mehr. Diese Aussage gilt zumindest für 25 Behandlungswochen, nach denen das Experiment abgebrochen wurde. Die in Versuch 4 bei Kombination mit 0,1 μMol DMBA gemessene Latenzzeit t_{50} ist mit der Latenzzeit in Versuch Nr. 1 praktisch identisch, obwohl die zweimal wöchentlich applizierte

Die cocarcinogene Wirkung der Phorbolester 113

Dosis von A1 nur $^1/_{60}$ der DMBA-Dosis in Versuch Nr. 1 ausmacht. Die Wirksamkeit von A1 ist beachtlich. Zum Vergleich sei daran erinnert, daß z. B. Östron im Allen-Doisy-Test an der Maus (vgl. [17]) mit 1 µg $\hat{=}$ 0,0037 µMol/Maus östrogene Wirksamkeit zeigt.

Die Ergebnisse der Experimente mit dem reinen Wirkstoff A1 sprechen dafür, daß den Wirkstoffen des Crotonöls neben der

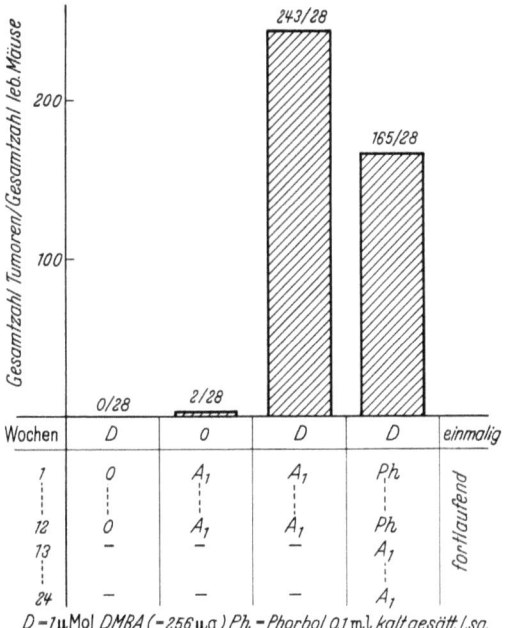

Abb. 3. Der Zeitfaktor im BERENBLUM-Experiment. 1 µMol DMBA und in 12 Wochen zweimal 0,016 µMol Wirkstoff A₁ wöchentlich. Stamm NMRI; 14/14 Mäuse; Lsgm.: Aceton p. a.

schwachen carcinogenen auch eine besondere tumor-promovierende Qualität definitiv zukommt. Die Erfüllung dieser für die beabsichtigten biochemischen Untersuchungen zum Mechanismus der Carcinogenese essentiellen Voraussetzung kann jedoch durch die beschriebenen Versuche noch nicht völlig zweifelsfrei bewiesen werden. Der von uns entwickelte quantitative Ansatz ist dazu jedoch prinzipiell geeignet und das Problem läßt sich mit einigen weiteren Versuchen, die in unserem Laboratorium bereits im Gange sind, bald endgültig entscheiden.

Im Zusammenhang mit unseren weiteren Untersuchungen ist die Wirkung des freien, unveresterten Phorbols von besonderem Interesse. Sie wird charakterisiert durch die in Abb. 3 zusammengefaßten Ergebnisse, die den Stand eines Berenblum-Experiments nach 12 Wochen darstellen. Wie man erkennt, tritt nach einer Initialdosis von 1 μMol DMBA innerhalb von 12 Wochen kein Tumor auf. Bei fortlaufender Applikation von 0,032 μMol Wirkstoff A1 (zweimal wöchentlich 0,016 μMol) werden nach 12 Wochen zwei Tumoren bei 28 überlebenden Mäusen festgestellt, während die kombinierte Anwendung von 1 μMol DMBA und 0,032 μMol A1 nach 12 Wochen zu 243 Tumoren bei 28 Tieren führt. Appliziert man jedoch nach der einmaligen Dosis von 1 μMol DMBA fortlaufend 0,1 ml einer gesättigten Lösung von Phorbol, so werden nach 12 Wochen keinerlei Tumoren gefunden. Unverestertes Phorbol wirkt also in dieser relativ hohen Dosierung im Gegensatz zu seinem c-Acetyl-b-myristoyl-Derivat (= Wirkstoff A1) nicht tumor-promovierend. Schließt man in der 13. Woche eine fortlaufende Applikation von zweimal wöchentlich 0,016 μMol A1 an (Abb. 3), so werden nach

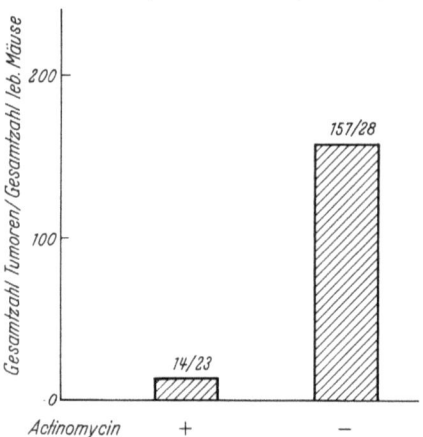

Abb. 4. Actinomycin im BERENBLUM-Experiment. Einmalig: 0,1 μMol DMBA und sechsmal 0,5 μg Actinomycin; in 12 Wochen: zweimal 0,02 μMol Wirkstoff A_1 wöchentlich. Stamm NMRI; 14/14 Mäuse; Lsgm.: Aceton p. a.

12 weiteren Behandlungswochen, also in der 24. Woche vom Zeitpunkt 0 an gerechnet, 165/28 Tumoren erzielt. Dies bedeutet, daß die durch eine einmalige Dosis von 1 μMol DMBA an der Mäusehaut hervorgerufene Veränderung trotz der zwischendurch erfolgten

zwölfwöchigen Phorbolbehandlung fast vollständig erhalten geblieben ist.

In Versuchen mit radioaktiv markiertem DMBA haben wir gefunden[18], daß die Radioaktivität 3 Tage nach der (oralen) Applikation aus der Mäusehaut fast vollständig verschwunden ist. Da ferner die mittlere Lebensdauer (Generationszeit) der Zellen des Stratum basale normaler Mäusehaut etwa 6 Tage beträgt[19], beweisen die Ergebnisse des Experiments in Abb. 3, daß die ,,potentiellen Tumorzellen'' über viele Zellgenerationen erhalten bleiben. Wir schließen aus diesem Versuch, daß die durch unterschwellige Dosen eines carcinogenen Kohlenwasserstoffes erzeugten potentiellen Tumorzellen gegenüber den normalen Zellen einen irgendwie veränderten Informationsgehalt besitzen (vgl. dazu [4]). Es ist in diesem Zusammenhang von Interesse, daß die Ausbeute an Tumoren stark vermindert wird (Abb. 4), wenn man gleichzeitig mit der unterschwelligen Dosis DMBA Actinomycin appliziert (vgl. auch [2]), das bekanntlich den Informationsfluß von der DNS zur m-RNS blockiert. Dieses Versuchsergebnis kann als Hinweis darauf verstanden werden, daß für die Initiierung von potentiellen Tumorzellen durch das Carcinogen möglicherweise die Synthese von m-RNS erforderlich ist.

Die vorliegenden Ergebnisse sind das Resultat intensiver Mit- und Zusammenarbeit folgender Mitarbeiter: H. BARTSCH, Dr. H. BRESCH, Dr. E. CLARKE, M. GSCHWENDT, Dr. E. HÄRLE, G. KREIBICH, Dr. H. KUBINYI, Dr. J. G. MEYER, D. PAUL, H. U. SCHAIRER, CH. V. SZCZEPANSKI, Dr. H. W. THIELMANN. Die statistische Bearbeitung der Tierversuche verdanken wir Herrn Dr. H. IMMICH, Institut für Dokumentation, Information und Statistik (Dir. Prof. Dr. G. WAGNER) am Deutschen Krebsforschungszentrum, Heidelberg.

Literatur

[1] BERENBLUM, I.: Cancer Res. 1, 44, 807 (1941).
[2] MOTTRAM, I. C.: J. Path. Bact. 56, 181, 391 (1944).
[3] HECKER, E.: Angew. Chem. 74, 722 (1962); Angew. Chem. internat. Ed. 1, 602 (1962); vgl. auch Mitteil. der Max-Planck-Gesellschaft. 1963, 41.
[4] — Plenarvortrag auf dem Deutschen Krebskongreß am 24. 2. 1966. Strahlentherapie (im Druck).
[5] NAKAHARA, W.: Progr. exp. Tumor Res. 2, 158 (1961); Med. Welt 61, 661 (1966).
[6] HECKER, E., H. BRESCH und J. G. MEYER: I. World Fat Congress, Hamburg 1964, Abstracts of Papers p. 176; vgl. Fette-Seifen-Anstrichmittel 67, 78 (1965).
[7] —, u. H. BRESCH: Z. Naturforsch. 20 b, 216 (1965).
[8] —, u. H. KUBINYI: Z. Krebsforsch. 67, 176 (1965).
[9] CLARKE, E., u. E. HECKER: Z. Krebsforsch. 67, 192 (1965).

[10] KUBINYI, H., u. E. HECKER: Analytikertagung Lindau 13. bis 16. April 1966; Z. analyt. Chem. **221**, 424 (1966).
[11] HECKER, E., H. KUBINYI, CH. v. SZCZEPANSKI, E. HÄRLE und H. BRESCH: Tetrahedron Letters 1965, 1837.
[12] ARROYO, E. R., and J. HOLCOMB: Chem. and Ind. **1965**, 350.
— — J. med. Chem. 8, 672 (1965).
[13] HECKER, E., H. KUBINYI, H. BRESCH, and CH. v. SZCZEPANSKI, J. med. Chem. **9**, 246 (1966).
[14] KAUFFMANN, TH., u. H. NEUMANN: Chem. Ber. **92**, 1715 (1959).
[15] HECKER, E., H. KUBINYI, H. U. SCHAIRER, CH. v. SZCZEPANSKI und H. BRESCH: Angew. Chem. **77**, 1076 (1965); Angew. Chem. Internat. Ed. **4**, 1072 (1965).
[16] DUUREN, B. L. VAN, and L. ORRIS: Cancer Res. **25**, 1871 (1965).
[17] HECKER, E., u. G. FARTHOFER-BOECKH, Biochem. Z. **338**, 628 (1963).
[18] Unveröffentlichte Versuche mit D. PAUL.
[19] DOERMER, P., H. TULINIUS und W. OEHLERT: Z. Krebsforsch. **66**, 11 (1964).
[20] GELBOIN, H. V., M. KLEIN, and R. R. BATES: Proc. nat. Acad. Sci. (Wash.) **53**, 1353 (1965).

Neue Reaktionen der carcinogenen Kohlenwasserstoffe 3,4-Benzpyren, 9,10-Dimethyl-1,2-benzanthracen und 20-Methylcholanthren [1]

Von M. WILK, W. BEZ und J. ROCHLITZ

Institut für Organische Chemie der Universität Frankfurt am Main

An aktiven Oberflächen (Kieselgel, Papier) adsorbiertes 3,4-Benzpyren unterliegt leicht einer monovalenten Oxydation, wobei über die Stufe des Radikalkations im Zuge einer elektrophilen Substitution (neben einigen Chinonen und höheren Oxydationsprodukten) im Schwerpunkt ein dimerer Kohlenwasserstoff, das 5,5'-Bis-benzpyrenyl gebildet wird. Als Oxydationsmittel kann im Gegensatz zu den Befunden von SZENT-GYÖRGYI[2] sowie von ALLISON and NASH[3] Joddampf verwendet werden. Die Reaktion verläuft aber auch mit $FeCl_3$ oder Luftsauerstoff als Oxydationsmittel. Durch Analyse, IR-, UV- und Massenspektren sowie durch Synthese, konnte die Struktur des dimeren Aromaten gesichert werden.

Beim DMBA und 20-Methylcholanthren bilden sich unter den gleichen Bedingungen hauptsächlich tetramere Kohlenwasserstoffe. Bei den untersuchten nicht carcinogenen Verbindungen dieser Reihe war die Bildung von Chinonen und anderen Reaktionsprodukten bevorzugt. Die entstandenen Dimere und Tetramere der carcinogenen Aromaten fallen in atropisomeren, racemischen Formen an.

Dieser neue Reaktionstyp des elektrophilen Angriffs der Radikalkationen aromatischer Kohlenwasserstoffe kann durchaus mit deren carcinogener Wirksamkeit in Zusammenhang stehen, zumal in der Zelle ausreichend „aktive Grenzflächen" vorhanden sind und die monovalente Oxydation mit Hilfe von freiem Sauerstoff bei Zimmertemperatur erfolgen kann. Für den Mechanismus der chemischen Carcinogenese würden sich daraus einige neue Gesichtspunkte ergeben:

1. Haben die größeren dimeren oder tetrameren Moleküle keine Möglichkeit mehr, die Zelle oder deren Strukturbereiche zu

verlassen, so wäre dieses „Käfigmodell" eine indirekte Bestätigung der *Druckreyschen Theorie*[1] über die streng additive Wirkung auch kleinster Dosen über lange Zeiträume. Diese Vorstellung setzt voraus, daß die Oligomeren

a) biologisch nicht abgebaut werden können und daß sie

b) bei Hautpinselungen keine carcinogene Aktivität zeigen dürfen, weil sie nach unseren Vorstellungen nicht mehr in die Zelle eindringen können. Die Möglichkeit a) eines biologischen Abbaus erscheint uns kaum realisierbar, da diese Substanzen im Gegensatz zu den „Monomeren" weder einer Photooxydation noch einem Angriff durch molekularen Sauerstoff (bzw. Jod, $FeCl_3$) unterliegen und auch mit konzentrierter Schwefelsäure nicht mehr die bekannten Farbreaktionen liefern. Die biologischen Teste b) zeigten in Übereinstimmung mit unseren Vorstellungen keine carcinogene Aktivität[5].

Als Hinweis für das mögliche Auftreten von dimerem Benzpyren in der Zelle können die Untersuchungen von GRAFFI[6] dienen. Er fand, daß etwa ein Tag nach der Pinselung in den Hautzellen der Maus die blaue Fluorescenz des gelösten Benzpyrens im Cytoplasma und an den Mitochondrien abnimmt und sich ein fast geschlossener, grüngelb fluorescierender Ring um den Zellkern ausbildet. Es kann sich hierbei durchaus um das in gleicher Farbe fluorescierende Bisbenzpyrenyl handeln. Allerdings ist nicht auszuschließen, daß dieser Fluorescenzumschlag von kristallinem 3,4-Benzpyren herrührt, welches sich in Kernnähe ansammelt[7].

2. Bei der Oligomerisierung der carcinogenen Aromaten (Kohlenwasserstoffe, Heterocyclen, Amine und Azofarbstoffe) entstehen Atropisomere. Damit gewinnt der Hinweis von LETTRÉ[8] weitere Bedeutung, wonach die meisten carcinogenen Kohlenwasserstoffe „planare Asymmetrien" aufweisen. Die von ihm genannten Ausnahmefälle symmetrischer Verbindungen mit carcinogener Wirksamkeit führen im Zuge der Oligomerisierung ebenfalls zu asymmetrischen Atropisomeren, die in optische Antipoden spaltbar sein müssen.

3. Reagieren die primär gebildeten Radikalkationen carcinogener Kohlenwasserstoffe (Heterocyclen usw.) mit Zellstrukturen beliebiger (auch planarer) Asymmetrie, so müssen ebenfalls Atropiso-

meriefälle auftreten, die eine tiefgreifende Änderung der Zellstruktureinheiten zur Folge haben.

Literatur

[1] Originalarbeit erscheint in *Tetrahedron*.
[2] SZENT-GYÖRGYI, A.: Proc. nat. Acad. Sci. (Wash.) **46**, 1444 (1960).
[3] ALLISON, A. C., and T. NASH: Nature (Lond.) **197**, 758 (1963).
[4] DRUCKREY, H., H. HAMPERL und BROCK: Z. Krebsforsch. **50**, 431 (1940).
[5] Die Teste wurden freundlicherweise von E. L. WYNDER, Sloan-Kettering-Institute for Cancer Research, New York, durchgeführt.
[6] GRAFFI, A.: Z. Krebsforsch. **52**, 165 (1942).
[7] WILK, M., u. H. SCHWAB: Angew. Chem. **75**, 1128 (1963).
[8] LETTRÉ, H.: Z. physiol. Chem. **280**, 28 (1944).

Diskussion

DANNENBERG (München): Herr Dr. MAGEE, Sie haben gezeigt, daß im Organismus Nucleinsäuren durch Dimethylnitrosamin am Guanin methyliert werden. 7-Methylguanin tritt aber auch normalerweise im Harn als Abbauprodukt von Nucleinsäuren auf. 1. Frage: Hat man nach Verabfolgung anderer Nitrosamine (DRUCKREY hat sehr viele derartige Verbindungen dargestellt) auch eine Alkylierung von Nucleinsäuren beobachtet und hat man an Stelle von 7-Methylguanin andere entsprechende 7-Alkylguanine gefunden? 2. Frage: Ist bekannt, ob bei Tierstämmen, z. B. Mäusestämmen mit einer hohen Rate an Spontantumoren, 7-Methylguanin vermehrt ausgeschieden wird im Vergleich zu Tierstämmen ohne Spontantumoren? Oder ist bereits untersucht worden, ob im Harn von Krebskranken 7-Methylguanin vermehrt ausgeschieden wird?

MAGEE (Carshalton): Answer 1: We are working on this problem, but we have not yet got any definite results.
Answer 2: BOREK and his colleagues [MANDEL, L. R., P. R. SRINIVASAN, and E. BOREK: Nature **209**, 586 (1966)] in New York have recently published a paper on the origin of the normal — or if you like — on the spontaneous 7-methylguanine by the rat and they have shown that it derives from methionine; it is labelled, when radioactive methionine is injected, and the excretion is increased in tumor bearing animals. I believe also that there is an increased excretion in some leukemic patients [PARK. R. W., J. F. HOLLAND, and A. JENKINS: Cancer Res. **22**, 469 (1962)]. BOREK certainly made the point, that the tumor in the animal which he studied was quite small but there was a twofold increase in the amount of 7-methylguanine excreted in the urine in a tumor bearing animal.

HEIDELBERGER: I like to make a comment to Dr. MAGEE's elegant and admirably restrained paper. It seems to be tacitly assumed that chemical carcinogenesis converts normal cells into tumor cells. I think this is the basic assumption that most of us have been working under. This is one possible biological mechanism. A second one was quickly alluded to by Dr. MAGEE and I think dismissed for very good reasons, that is, that carcinogenic compounds can activate latent viruses. Now there is a third possibility which is also quite different and which cannot be dismissed. This was really first put forward clearly by PREHN and it is the clonal selection theory of carcinogenesis, a completely different biological mechanism, in which he suggests that a carcinogen selects for and permits the multiplication of preexisting cancer cells, that is that cancer cells are already present. This is a biological mechanism which has to be recognized seriously. I believe that it can only be answered by proper experiments on carcinogenesis in vitro, perhaps Dr. SACHS will say something about this tomorrow. It cannot be dismissed as a possibility at the present time.

HECKER: Well, Dr. MAGEE, the comparison of mutagenic activity and carcinogenic activity confronts us with a principle problem: very often the mutagenic activity is assayed in microorganisms whereas carcinogenic activity refers mostly to mammals. Would you like to make a comment as to whether the mutagenic activities you have measured are assayed in yeast or whatever microorganism it may be or in mammals? If the latter is correct what method of mutagenic assay in mammals did you use?

MAGEE: Well, I think, that advice was very good, because when giving my main presentation, I did not make clear that I have not done the experiments with the nitrosamines. These experiments have been done by several groups of people from the point of view of comparing carcinogenic activity with mutagenic activity. Most of this work has been done in Germany. Dr. LUISE PASTERNAK was the first person to demonstrate the mutagenic action of dimethylnitrosamine. This was done in *Drosophila*. From the same laboratory I think it was GEISSLER, if I remember rightly, who showed that dimethylnitrosamine was not mutagenic in microorganisms. This, we feel, is consistent with the presence of a demethylating enzyme in *Drosophila* and its absence in the microorganism. On the other hand the nitrosamide carcinogens, for example Nitrosomethylharnstoff (nitrosomethylurea) is mutagenic in *Drosophila* and in a wide range of microorganisms. Dr. MARQUARDT and his group at Freiburg have done a large series of experiments on this, and as far as I have seen it in print N-Nitroso-N'-nitromethylguanidine is the most effective mutagenic agent in microorganism systems [ADELBERG, E. A., M. MANDEL, and G. C. C. CHEN: Biochem. biophys. Res. Comm. 18, 788 (1965)]. So the answer is, that I think all the conventional systems for testing mutagens have been used by different groups in different countries. And broadly speaking, the nitrosamines only act in the fruit fly and not in the simple organisms, but the amides, for example nitrosomethylurea or nitrosomethylurethane act in all of these systems.

HECKER: One short comment: Of course there is still a big difference between *Drosophila* and mammals as far as metabolism is concerned. And in your carcinogenesis I suppose metabolism is definitely included before the carcinogenic agent is produced.

MAGEE: Yes, that is certainly true, but if we consider insects there is evidence that several insecticides are in fact demethylated by enzyme systems which might be able to demethylate the nitrosamines as well. It is known, that insects, I am speaking generally, are capable of performing these demethylation reactions. But as far as I know the nitrosamines have not been extensively studied on mutagenicity in mammals because one needs very many animals. It is a pity, but these things will perhaps be done in future.

KERSTEN (Münster/Westf.): Herr Professor DANNENBERG hat gezeigt, daß keine Relation besteht zwischen der carcinogenen Wirkung der geprüften Substanzen und ihrem Effekt auf die gemessenen physikalisch-chemischen Charakteristika der DNA. Auch wir fanden, daß Dimethylbenzanthracen in verschiedenen Solventien keinen Einfluß hat auf das Schmelzverhalten von DNA^2.

Wir haben vergleichende Untersuchungen durchgeführt zur Bindung von verschiedenen Gruppen von Antibiotica an Nucleinsäuren und ihrer Wirkung auf das Bakterienwachstum und den Nucleinsäure- und Proteinstoffwechsel[1]. Dabei zeigte sich, daß keine einfache Beziehung besteht zwischen der Veränderung physikalisch-chemischer Charakteristika der DNA durch diese Substanzen und ihrem Einfluß auf DNA-abhängige enzymatische Prozesse oder ihrem hemmenden Einfluß auf das Wachstum von Bakterien oder Tumorzellen[3,4].

Alle untersuchten Substanzen binden sich an DNA, was sich am sichersten dadurch zeigen läßt, daß sie bis zu einer gewissen Konzentration an DNA gebunden sedimentieren.

Die Substanzen (Anthracycline und Acridine), die den stärksten Effekt auf die Schmelztemperatur, die Viskosität und den Sedimentationskoeffizienten der DNA haben, sind weniger stark wirksam im biologischen System als z. B. Actinomycine oder Chromomycin und die diesen chemisch verwandten Substanzen, die geringen oder keinen Einfluß haben auf diese Eigenschaften. Am ehesten besteht noch eine Beziehung zwischen der Veränderung der Dichte der DNA in CsCl durch die Antibiotica und ihrer biologischen Wirkung.

Aber auch diese Beziehung ist nicht sehr streng. Die Verminderung der Dichte der DNA könnte am ehesten durch eine Veränderung der Hydratation erklärt werden. In vitro sind die DNA-abhängigen enzymatischen Prozesse unterschiedlich empfindlich gegenüber den Antibiotica. So sind die DNA-abhängige RNA-Synthese und die Methylierung von DNA recht empfindlich, während die DNA-abhängige DNA-Synthese mittels des KORNBERG-Enzyms und der Abbau der DNA durch Endo- und Exonucleasen weniger gehemmt werden.

Erst weitere Untersuchungen und die Aufklärung der Art und des Ortes der Bindungen von verschiedenen Substanzen an DNA kann den Zusammenhang zwischen der Chemie und der biologischen Wirkung erbringen.

Literatur

[1] KERSTEN, W., u. H. KERSTEN: Die Bindung von Daunomycin, Cinerubin und Chromomycin A_3 an Nucleinsäuren. Biochem. Z. **341**, 174 (1965).

[2] KERSTEN, W., H. KERSTEN, and W. SZYBALSKI: Physicochemical Properties of Complexes between Deoxyribonucleic Acid and Antibiotics which Affect Ribonucleic Acid Synthesis (Actinomycin, Daunomycin, Cinerubin, Nogalamycin, Chromomycin, Methramycin and Olivomycin). Biochemistry 5, 236 (1966).
[3] HARTMANN, G., H. GOLLER, K. KOSCHEL, W. KERSTEN und H. KERSTEN: Hemmung der DNA-abhängigen RNA- und DNA-Synthese durch Antibiotica. Biochem. Z. 341, 126 (1964).
[4] KOSCHEL, K., G. HARTMANN, W. KERSTEN und H. KERSTEN: Die Wirkung des Chromomycins und einiger Anthracyclinantibiotica auf die DNA-abhängige Nucleinsäuresynthese. Biochem. Z. 344, 76 (1966).

DANNENBERG: Herr Dr. KERSTEN, Untersuchungen, nach denen Sie fragen, haben wir nicht gemacht. Uns kam es in den Untersuchungen nur darauf an, zu zeigen, daß Parallelen zu dem bestehen, was man für den Wirkungsmechanismus des Proflavins hält, also in bezug auf den Einbau (Intercalation).

HEIDELBERGER: I would like to comment briefly Professor DANNENBERG's paper. You refer to the work of BOYLAND and GREEN on the solubilization of hydrocarbons in solutions of DNA. In our group we showed quite conclusively that the experiments of BOYLAND were wrong. Nevertheless since that time he has published three more papers ignoring our objections! And I am very glad to see that Professor DANNENBERG did not get the correlation between the reaction *in vitro* of the hydrocarbons and the DNA. On the other hand there is very definite evidence that there are *in vivo* interactions of carcinogenic hydrocarbons and DNA. We first demonstrated this in 1959. And since then BROOKES and LAWLEY have extended and confirmed this work with several other hydrocarbons. Rather than quarrel about it, I am glad to say that Dr. BROOKES is spending a year in our laboratory and we are investigating this phenomenon. It seem squite clear, as Dr. MAGEE has pointed out with the nitrosamines, that hydrocarbons also react more or less in a parallel fashion with DNA, RNA and protein. And there seems to be a rough correlation between carcinogenic activity of the compounds and such binding. Recently BROOKES and I have studied the binding of hydrocarbons to DNA, RNA and proteins of cells grown in culture. We made the rather striking finding that in normal cells there is a very high order of binding, for example the specific activity of the DNA is about 200 in arbitrary units. If these cells are transformed into tumor cells by *polyoma virus* this binding goes down to about 15. So there is a very striking loss of binding in tumor cells. This low binding to DNA is also found in tumor cells transformed in vitro by hydrocarbons, it is found in long established cell lines as HeLa-cells and L-cells in mice. And this goes strikingly parallel to the toxicity which these compounds exert on normal cells and not on tumor cells. We are now currently investigating the chemistry of the binding of the hydrocarbons with DNA in a cellular system to see how it compares with the *in vivo* system in mouse skin.

Diskussion 123

LAWLEY (London): In reply to Dr. HEIDELBERGER and in defence of Professor BOYLAND, I would quote the report by Dr. L. S. LERMAN, who prepared complexes of DNA and hydrocarbons by a technique involving the prior solubilization of the hydrocarbon by caffeine; the solution was put on a column of DEAE-Sephadex and then the DNA passed down this column. By this method DNA was shown to bind quite considerable quantities of certain hydrocarbons, pyrene being most readily solubilized. Application of the techniques as used for proflavin-DNA complexes led to the conclusion that these complexes are of the same type as with proflavin. There is also the publication by BALL, MCCARTER and SMITH who claimed similar results to those of BOYLAND and GREEN. So I would say that there is pretty good evidence that one can form intercalation complexes between hydrocarbons and DNA.

HEIDELBERGER: I will not go into too many details. Let me just say to Dr. LAWLEY, that his dear friend and colleague Dr. BROOKES in my laboratory has tried on several occasions to repeat the LERMAN-column experiment which you mentioned. We have not been able to do so. It seems that the column absorbs the hydrocarbon much better than either the caffeine or the DNA.

EMMERICH (Berlin): Wir haben zur Untersuchung der Wechselwirkung zwischen cancerogenen Aromaten und Zellbestandteilen die cancerogenen Aromaten an Larven von *Drosophila* verfüttert und feststellen können, daß diese Verbindungen in den Riesenchromosomen der Speicheldrüsen im Verhältnis 1:2 gegenüber dem Cytoplasma angereichert sind. Diese Anreicherung, die nur in den Speicheldrüsen und nicht in den anderen Geweben, z. B. im Fettkörper, feststellbar ist, ist vermutlich darauf zurückzuführen, daß die anderen Insektengewebe sehr lipidreich sind. Zum zweiten sind diese Cancerogene in den Riesenchromosomen relativ leicht auswaschbar mit apolaren Solventien; schon mit Alkohol werden sie zu einem großen Teil aus den Chromosomen entfernt. Der nicht-cancerogene Aromat Anthracen wird wesentlich schwächer in die Chromosomen eingebaut, so daß in diesem Fall kein gesteigerter Einbau gegenüber dem Cytoplasma festzustellen ist. Dieses Phänomen kann aber genau so gut auch auf einen erhöhten Umsatz des Anthracens im Stoffwechsel der Speicheldrüse von *Drosophila* zurückzuführen sein.

HECKER: A short comment: Referring to a new paper from OEHLERT and his coworkers [Z. Krebsforsch. 68, 14 (1966)] they were able to demonstrate by autoradiographic techniques that protein — as well as RNA — and DNA-binding is occurring after treatment of mouse skin with carcinogenic hydrocarbons. This is the same effect as mentioned by Dr. HEIDELBERGER a few minutes ago and which was seen with other techniques. This indicates that binding as such to these three kinds of macromolecules is not specific. Therefore search for specific binding sites is important rather than measuring grossbinding to all the three types of macromolecules.

124   Diskussion

STAUDINGER(Gießen): Herr HECKER, haben Sie irgendwelche Vorstellungen über die Wirkungsweise der Phorbolester als Cocarcinogene? Sie nehmen an, daß die erste Applikation des eigentlichen Carcinogens bereits das definitive Ereignis ist und daß die Cocarcinogene die phaenotypische Realisierung beschleunigen. Ist eigentlich der Versuch gemacht worden, die Reihenfolge umzukehren, also erst die Cocarcinogene zu geben und dann das Carcinogen? Haben Sie eine Erklärung für die cocarcinogene Wirkung in Formulierungen des Zellstoffwechsels?

HECKER: Das umgekehrte Experiment ist schon von BERENBLUM und später von anderen gemacht worden und fällt praktisch negativ aus.

Die Reindarstellung und die Chemie der Phorbolester ist — wie ich in der Einleitung zu meinem Referat bemerkt habe — keineswegs das letzte Ziel unserer Bemühungen, wohl aber eine wichtige Voraussetzung für die von Ihnen angesprochene Bearbeitung der biochemischen Wirkungsweise dieser Ester.

Natürlich sind wir bereits mit der biochemischen Analyse des Wirkungsmechanismus der Cocarcinogene und der initialen Dosen von carcinogenen Kohlenwasserstoffen beschäftigt. Ein relativ einfacher Ansatzpunkt dafür wäre gegeben, wenn die Phorbolester eine gewisse Wachstumsstimulierung verursachen würden, die wir im Tierexperiment als hyperplasinogene Wirkung der Phorbolester gut kennen. Es könnte sein, daß die allgemeine Wachstumsstimulierung, die als Folge einer Entzündung auftritt, entscheidende Bedeutung hat. Sie könnte z. B. außer einer unspezifischen auch eine für potentielle Tumorzellen spezifische Komponente enthalten. In diesem Zusammenhang untersuchen wir DNS-, RNS- und Proteinsynthese unter der Einwirkung von Phorbolestern. Aber es ist noch zu früh, hier über die Resultate im einzelnen zu berichten.

GRAFFI (Berlin-Buch): Sie haben uns durch Ihre schönen Versuche davon überzeugt, daß die Phorbolester als die tumorrealisierenden Komponenten des Crotonöls gleichzeitig auch eine echte cancerogene Wirkung haben. Ich möchte jedoch davor warnen, nunmehr eine cocancerogene Wirkung und damit den Zweiphasenmechanismus der Cancerogenese überhaupt in Zweifel zu stellen. Es muß nämlich daran erinnert werden, daß der cocancerogene Effekt an der Haut auch durch mechanische und thermische Reize und die daran sich anschließenden Regenerationsprozesse ausgelöst werden kann, also durch Einwirkungen nicht-chemischer Natur, wie dieses DEELMAN bereits vor über 40 Jahren an der Mäusehaut und ROUS und Mitarbeiter am Kaninchenohr nachweisen konnten. Im letzteren Fall wurden mittels eines Korkbohrers kleine runde Hautfelder ausgestanzt, in deren Bereich es im Zuge der Hautregeneration nur dann regelmäßig zur Tumorbildung kam, wenn das betreffende Hautfeld unterschwellig mit cancerogenem Kohlenwasserstoff vorbehandelt worden war.

Zweitens möchte ich Sie fragen, was Sie histologisch an der Mäusehaut nach der Applikation von Actinomycin D festgestellt haben. Es besteht nämlich die Möglichkeit, daß allein durch Zelluntergang in der Epidermis und den Haarbälgen im Gefolge einer übermäßigen, rein toxischen Actino-

Diskussion

mycinwirkung ein beträchtlicher Anteil der cancerogenen Initialwirkung, die man sich ja als einen irreversiblen cellulären Prozeß vorzustellen hat, ausgelöscht wurde.

HECKER: Es würde mir leid tun, Herr GRAFFI, wenn bei Ihnen und vielleicht auch beim Auditorium der Eindruck entstanden sein sollte, als ob ich der carcinogenen Wirkung der Phorbolester Bedeutung beimessen würde; im Gegenteil, es ist eine Dosierungsfrage, ob man die carcinogene und/oder die cocarcinogene Wirkung erfassen kann. Wie ich gezeigt habe, läßt sich durch entsprechend niedrige Dosierung der Phorbolester sehr wohl erreichen, daß die "tumor promoting activity" dieser Ester zur Wirkung kommt, ohne daß der schwache carcinogene Effekt zum Tragen kommt. Auch BOUTWELL hat [Progr. exp. Tumor Res. (Basel) **4**, 207 (1964)] einen qualitativen Unterschied zwischen initialen Dosen von carcinogenen Kohlenwasserstoffen und Crotonöl demonstriert, indem er zeigen konnte, daß die tumorpromovierende Wirkung des Crotonöls reversibel ist, während die initiale Wirkung carcinogener Kohlenwasserstoffe bekanntlich irreversibel ist.

Zu Ihrer Frage nach der Wirkung von Actinomycin allein, kann ich auf die Experimente von GELBOIN u. Mitarb. [Proc. Nat. Acad. Sci. (Wash.) **53**, 1353 (1965)] hinweisen, die im letzten Jahr gezeigt haben, daß relativ hohe Dosen von Actinomycin zu Schäden an der Mäusehaut führen. Sie heilen aber innerhalb von 2 Wochen ab. Wir haben uns außerdem bemüht, die Dosierungen so zu wählen (3mal 0,5 μg pro Maus vor und 3mal 0,5 μg nach der Applikation des DMBA), daß nur geringfügige Schädigungen auftraten.

SACHS (Rehovoth, Israel): I would just like to make a comment regarding terminology which I trust Prof. HECKER will agree with. I prefer at the present stage to use the term promotor, as used by Prof. BERENBLUM, rather than cocarcinogen for a substance such as croton oil, in view of the possibility that this substance may only be able to promote the development of single tumor cells into visible tumors. The occasional tumors found *in vivo* after treatment even with the purified fractions, could be due to the croton oil promoting the development of tumors from single tumor cells that have arisen spontaneously. It would seem worth while further to explore the possibilty that croton oil may be useful in measuring the spontaneous rate of transformation of normal cells into tumor cells in different strains of mice.

Mechanism of Action of Alkylating Agents: Comparisons with other Cytotoxic, Mutagenic and Carcinogenic Agents

By P. D. LAWLEY

Chester Beatty Research Institute, Institute of Cancer Research, Royal Cancer Hospital, London S.W. 3, Great Britain

With 5 Figures

The general theory of mutagenesis based on Watson-Crick model for DNA structure and replication[1,2] appears to be well established. The principal mutagenic mechanisms envisaged are the transition[1] and exchange error[2] types. Furthermore alkylation of DNA can lead to mutation[3], and the induction of transition mutations by such reaction appears to be the best established mechanism for studies with micro-organisms, although it may well not be the sole mechanism[4]. Knowledge of the reactions of chemical carcinogens with DNA is also increasing (cf. review, ref. 5) and permits some reasonable speculations to be made concerning the possible relevance of such studies to the molecular mechanism of carcinogenesis.

It has been pointed out[3] that alkylation of DNA at the most reactive site, viz. N-7 of guanine, would be expected to change the hydrogen-bonding potentiality of the nucleoside, either due to ionisation or tautomeric change. Similarly 5-bromodeoxyuridine[7] has a greater tendency to ionise at neutral pH than thymidine. Another case where a typical reagent inducing transition mutations can yield a product with a lower acidic pK_a value than the normal nucleoside is presented by hydroxylamine. This can replace the amino group of the cytosine moiety in deoxycytidine by hydroxylamino; this substitution[8] is simultaneous with, not subsequent to, reaction of hydroxylamine at the 4.5-double bond. It should be noted that the pK_a of this product of 10.2, although lower than that of cytosine by about 2 units, is not such as to cause ionisation at neutral pH. However the changed electronic distribution which this pK difference reflects may cause a tautomeric shift in the guanine-N^6-hydroxycytosine base pair-to be favoured, analogous to that suggested[9] for the 7-alkylguanine cytosine pair.

There is still doubt, however, as to the nature of the reaction of hydroxylamine with the cytosine ring in DNA which causes the replication error. BROWN and PHILLIPS[10] consider that their studies on the modification of the template activity of poly-C by hydroxylamine indicate the 4,5-double bond to be the essential reactive site, the product being 4-hydroxylamino-4,5-dihydrocytosine. On the other hand, JANION and SHUGAR[11] point out that 5-substitution of the cytosine ring, as in phage T2-DNA, blocks reaction at this bond, and the substitution at N^6 becomes the sole reaction; nevertheless T2 phage is susceptible to hydroxylamine mutagenesis. Also the observations of FREESE[12] that denatured transforming DNA is more readily mutated by hydroxylamine than is the native form suggest that the essential reaction occurs at a group, such as the N^6-amino group, involved in hydrogen-bonding in native DNA.

Possible correlations between these chemical mutagens and chemical carcinogens follow from the findings that aromatic hydrocarbons, such as benz/a/anthracene and dibenz/a,h/anthracene, are metabolised via epoxide intermediates[13], while certain carcinogenic aromatic amines, *e.g.* acetylaminofluorene, are metabolised to arylhydroxylamines, *e.g.* N-hydroxy-2-acetylaminofluorene. Both types of metabolite could react as carbonium ions, *i.e.* in the same way as the alkylating agents. Thus while an arylhydroxylamine might be expected to react with nucleic acid in the same way as hydroxylamine itself, the evidence available shows, on the contrary, that guanine residues in nucleic acids are the site of reaction, not cytosine[15]. The reagent used was 2-fluorenylhydroxylamine, and the product was suggested to result from attack on the C-8 atom of guanine; however there remains the possibility that carbonium ion attack at N-7 of guanine had occurred. It is known that products consistent with the latter mechanism result from reaction of arylhydroxylamines with the S-atom of cysteine[16] and methionine[17]. It has further been suggested that the product of arylation of the methionine S-atom in cellular protein by 4-dimethylaminoazobenzene *in vivo* could act as a methylating agent.

Apart from considerations of the chemical nature of reactions with DNA which could cause mutations, attention has also been drawn to the question of subsequent biochemical processes involved in mutation fixation[18]. An important factor here is the assumed existence of enzymic mechanisms for repair of chemically induced lesions in DNA.

As discussed, for example, by MULLER[19], it is reasonable to argue that repair mechanisms for radiation-induced lesions have developed during evolution because organisms possessing them would have selective advantage. (However the search for evidence of a radioprotective mechanism in *Drosophila* did not yield a positive result[19]). It has further been suggested[20] that the precise chemical nature of the lesion in DNA may to some extent be irrelevant to

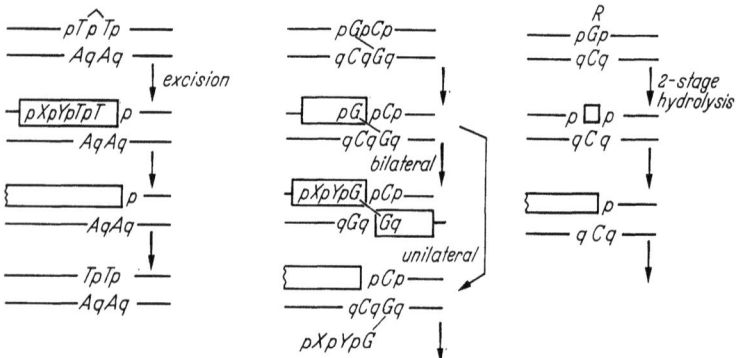

Fig. 1. Hypothetical mechanisms for excision and repair of lesions in DNA, following SETLOW (ref. 21). a) thymine dimer; b) cross-linking by a difunctional alkylating agent; c) single strand break induced by monofunctional alkylating agent

its ability to stimulate repair. For example, the cross-resistance of an *E. coli* strain to u.v. and difunctional alkylating agents has been attributed to the possibility that the first stage of repair, *viz.* excision of the lesion from DNA[21], can be effected by the same enzyme acting on either the u.v.-induced thymine dimer lesion, or the cross-link induced by the alkylating agent (Fig. 1 a, b).

Evidence has also been obtained that a lesion induced by ionising radiation, *viz.* the single-strand break in DNA, which also, appears to be the essential cytotoxic lesion due to methylation[22], can be repaired without the first stage of excision[23] (Fig. 1 c).

However it should be noted that whereas the principal u.v.-induced lesion, the thymine dimer, is formed in a single strand of DNA, the characteristic product of alkylation of DNA by difunctional agents, the di-guanin-7-yl derivative, is formed by crosslinking of the two DNA strands. Furthermore, excision of the diguaninyl derivative was observed[21], together with a smaller loss of the mono-guaninyl product, from DNA of *E. coli* B/r after cells had been treated with the difunctional alkylating agent mustard gas, di-(2-chloroethyl) sulphide.

It appears necessary therefore to invoke two stages of excision, as in the bilateral excision of Fig. 1 b, for removal of the lesion induced by a difunctional alkylating agent. Following unilateral excision, the presence of a bulky substituent consisting of the excised oligonucleotide still connected by the alkyl chain to the macromolecule, might suffice to stimulate a further excision. This consideration draws attention to the probable importance of the nature of the substituents attached to DNA by covalent reactions of mutagenic or carcinogenic compounds. There is evidence[25] that monofunctional alkylating agents capable of attaching by a single covalent bond a group which can interact strongly with DNA by non-covalent binding possess cytotoxic potency markedly greater than that of the aliphatic monofunctional agents. Thus the high cytotoxic potency of nitrogen mustards derived from acridine dyes, e.g. 2-methoxy-9-[3-(ethyl-2-chloroethyl)amino-propylamino] acridine, was attributed to the affinity of the aminoacridine moiety for binding to the bases of DNA.

We have been interested in the effects of acridine dyes from a related point of view, *viz.* the sensitisation of *E. coli* to the cytotoxic action of sulphur mustards. One objective was to investigate quantitatively the dose -response relationship between cytotoxic action, measured by decrease in colony-forming ability, and extent of alkylation of cellular DNA. It was thought that the sensitive mutant *E. coli* B_{s-1} (HILL), which is deficient in repair enzymes, might be expected to be reproductively inactivated by a single lesion in its DNA, this target being a single macromolecule. The lesion was expected to be a single cross-link in this DNA molecule.

However, it was found by LOVELESS[23] that recovery of this sensitive strain after alkylation with mustard gas can occur, although more slowly than for strain B/r. It seemed possible therefore that the mean lethal dose of this alkylating agent could be reduced by sensitising agents, and this in fact was found (Tab. 1).

The mean lethal doses are here defined as those reducing survival of colony forming ability to 37%, i.e. the kinetics of inactivation are assumed to be of the single-hit type. In fact the curves of (log survival) *versus* dose which were obtained approximated to the linear form required by this assumption, although at lower survival

Table 1. Effect of sulphur mustards, mustard gas, H, di-(2-chlorcethyl) sulphide, or hemi-H, 2-chloroethyl 2-hydroxyethyl sulphide, on colony forming ability of *E. coli*, strains B/r or B_{s-1}.
Cells grown to 7×10^8/ml in M9 medium were harvested, resuspended in M9 buffer at 5×10^9/ml., and treated with ^{35}S-labelled mustards (specific radioactivity 100 mc/mmole) for 15 min at 37°, then diluted and plated on nutrient agar. Colony formation was assayed after 16 hours. Iodoacetamide (3 mM) was added 45 min before addition of mustard; acriflavine (5 μg/ml) or caffeine 2 mg/ml) were dissolved in plating media.
Results were expressed as D_{37}, the dose required to reduce colony forming to 37% of controls without mustard, by interpolation from linear plots of log. survival *versus* dose, over a range of survival down to 10^{-2}. Extent of alkylation of DNA was determined as described in ref.[24] by assay of radioactivity in DNA isolated from treated cells.
Values for HeLa cells were provided by Drs. A. R. CRATHORN and J. J. ROBERTS.

	D_{37} values			
	H, mustard gas		hemi-H	
	μg/ml	μmole/mole DNA-P	μg/ml	μmole/mole DNA-P
B/r	5.8	35	> 160	> 800
B/r + iodoacetamide	2.0	24		
B/r + caffeine	3.0	18		
B/r + acriflavine	0.7	4		
B_{s-1}	0.8	5	10	50
B_{s-1} + iodoacetamide	0.12	1.4		
B_{s-1} + caffeine	0.36	2		
B_{s-1} + acriflavine	0.27	1.6		
HeLa cell	0.1	1.5	1.3	approx. 20

values than about 10^{-3}, the effectiveness of the alkylating agent decreases[27]. A possible explanation is that there are phenotypic differences between cells in their sensitivity, since colonies derived from survivors of alkylation did not show enhanced resistance when re-treated.

The mean lethal doses can in turn be expressed in terms of the extent of alkylation of cellular constituents, from knowledge of the relationship between these quantities and the doses expressed as concentration of the agent in the treatment media; this was found to be a simple proportionality. The extents of alkylation of protein by both the difunctional sulphur mustard, di-(2-chloroethyl) sulphide, mustard gas, H, and the corresponding monofunctional mustard, of equal chemical reactivity with half-life of reaction about 1 min at 37°, 2-chloroethyl 2-hydroxyethyl sulphide, hemi-H, were somewhat greater than those of nucleic acids (Tab. 2). The

Table 2. Extent of reaction of cellular constituents in *E. coli* B_{s-1} cells treated with alkylating agents; ^{35}S-labelled H or hemi-H were reacted for 15 min using concentrations up to 200 µg/ml, iodoacetamide-[2-^{14}C] for 1 or 21 hours, concentration 0.14 mM. Results are expressed as µg alkylating agent bound per gram cellular constituent at a unit dose of 1 µg/ml, derived for H and semi-H from linear plots of extent of binding *versus* dose. TCA-insoluble refers to material insoluble on cold 5% (w/v) trichloracetic acid; protein was obtained by methanol precipitation of phenol extracts of lysed cells.

	DNA	RNA	Protein	TCA-insoluble
H, mustard gas	2.8	2.8	5.1	5.3
hemi-H	2.0	—	2.6	2.8
iodoacetamide, 1 hour	0.02	0.14	24	19
iodoacetamide, 21 hours	0.04	0.08	33	21

time of treatment of the bacterial suspensions was 15 min at 37° in M9 buffer, dilution for plating then being carried out immediately.

The mean lethal dose of H to strain B_{s-1} was estimated in this way to be 5 µmole H/mole DNA-P. In order to convert this value to the number of cross-links induced in the cellular DNA, we assume that the proportion of di-(guanin-7-ylethyl) sulphide in the DNA alkylation products estimated chromatographically (Fig. 2) gives the number of molecules of the alkylating agent involved in cross-links, *i.e.* 20%. For the cellular DNA content we take a value of 2.7×10^7 DNA-P, as determined by us, in reasonable agreement with other values for *E. coli* B strains[23]. There remains the question of the number of DNA molecules which this quantity represents; for this we have assumed an average of 3, following Witkin's report[2] of a minimal value of 2 nuclear bodies in stationary cultures, and 4 in growing cultures. Thus the D_{37} value becomes 45 mole H/nuclear DNA content of 9×10^6 DNA-P, or 9 cross-links in the nuclear DNA. The lowest values found for D_{37} using the sensitising agents were:— 1.4 μmole H/mole DNA-P for cells

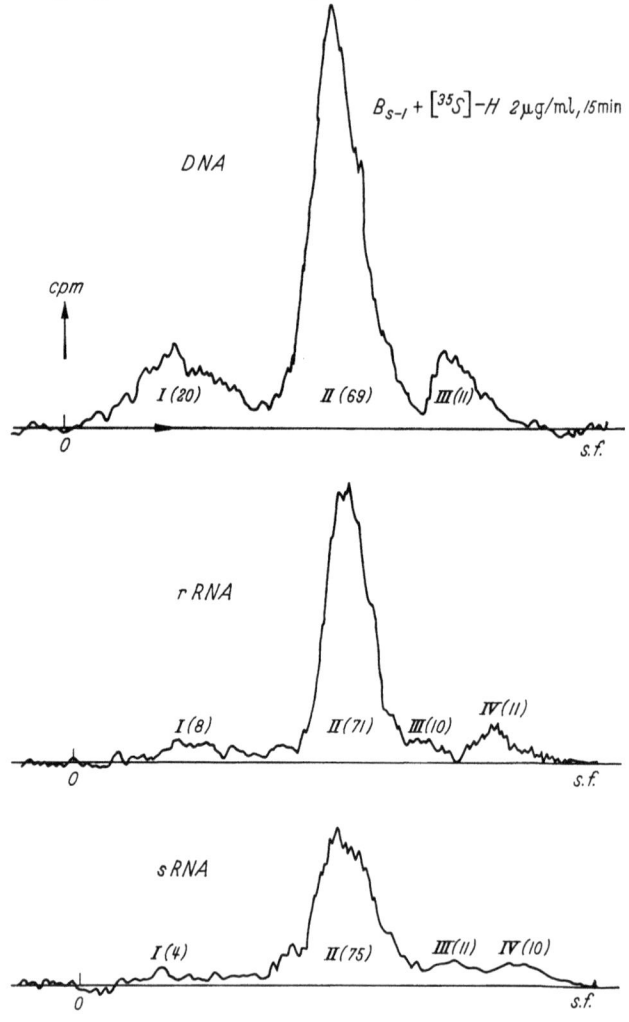

Fig. 2. Analysis of products of in-vivo-alkylation of DNA and RNA of *E. coli* $B_{s\text{-}1}$ by ^{35}S-labelled mustard gas. Isolated nucleic acids were hydrolysed with N-HCl at 100° and chromatographed on paper using isopropanol-HCl-H_2O, 68:16.4:15.6, by volume as solvent. (I) di-(guanin-7-ylethyl) sulphide; (II) 7-(2-hydroxyethylthioethyl)guanine; (III) 3-(2-hydroxyethylthioethyl)adenine (DNA); or 1-(2-hydroxyethylthioethyl)adenine (RNA); (IV) 1-(2-hydroxyethylthioethyl)cytidylic acid. Figures in brackets indicate molar proportions of products

incubated with 3 mM-iodoacetamide for 45 min before treatment; and 1.6 µmole H/mole DNA-P for cells treated in the usual way but plated on nutrient agar containing 5 µg/ml acriflavine. Thus a minimum observed value of between 2 and 3 cross-links in the nuclear DNA was obtained for the mean lethal dose; it may be noted that a similar value was obtained[30] for the number of cross-links in DNA of phage T2 or T4 required to inactivate plaque

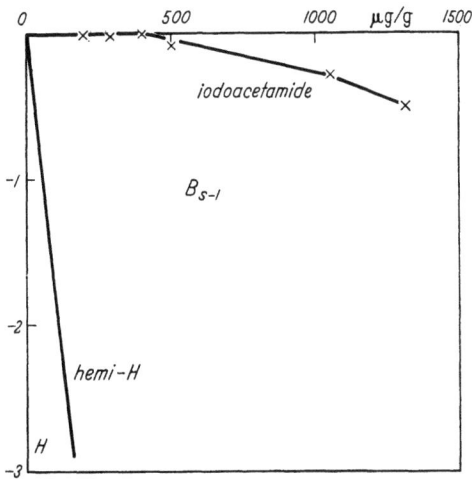

Fig. 3. Survival of colony forming ability of $E.$ $coli$ $B_{s\text{-}1}$ after alkylation with ^{35}S-labelled mustard gas (H), 2-chloroethyl 2-hydroxyethyl[^{35}S]sulphide (hemi-H), or ^{14}C-labelled iodo-acetamide; dose expressed as µg alkylating agent reacted per gram cellular material insoluble in cold 5% trichloracetic acid

formation. Possible reasons for the deviation from the theoretical value of unity are that complete inhibition of repair mechanisms had not been obtained; that some of the di-guaninyl derivative did not derive from interstrand cross-links; and the multiplicity of targets presented by the existence of more than a single nuclear body per cell.

The action of iodoacetamide which was observed confirmed that this compound is an atypical alkylating agent in comparison with the typical monfunctional agent hemi-H. It was found that iodo-acetamide-[2-^{14}C] reacted extensively with cellular proteins, the initial rate corresponding to a first-order reaction of half-life 30 min, but this rate fell off at higher extents of reaction, possibly because the most reactive sites became saturated. As shown in Fig. 3

extensive reaction of iodoacetamide with cellular protein did not cause loss of colony forming ability of $E.$ $coli$ B_{s-1}, and the contrast with the sulphur mustards will be evident from Figs. 3 and 4. Analysis of the amino-acids in HCl-hydrolysates of protein obtained from phenol extracts of lysed cells reacted *in vivo* with ^{14}C-labelled iodoacetamide showed that 95% of the bound radioactivity was

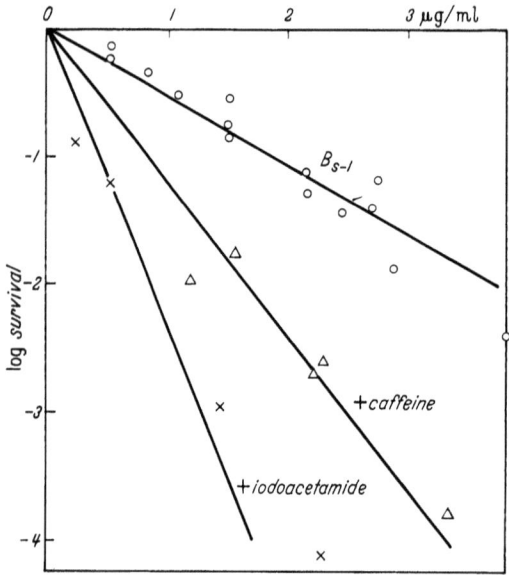

Fig. 4. Survival of colony forming ability of E. coli B_{s-1} after alkylation with ^{35}S-labelled mustard gas, for 15 min at 37°. ○ plated on nutrient agar; △ plated on nutrient agar containing 2 mg/ml caffeine; × incubated with iodoacetamide (0.84 mM) for 45 min at 37° before treatment with mustard

recovered as ^{14}C-labelled carboxymethylcysteine, whereas almost no radioactivity was found in isolated RNA or DNA (Tab. 2). Since the mustards were able to react almost equally as well with nucleic acids as with protein, it appears that protein alkylation is not responsible for the characteristic cytotoxic action of the mustards.

The results also showed that the difunctional mustard was about ten times as effective as the monofunctional mustard in the reproductive inactivation of strain B_{s-1}; this provides confirmative evidence for the view that cross-linking is the essential characteristic inactivating reaction of the difunctional agents. It will also be noted, from Fig. 2, that the proportion of the di-guaninyl derivative found in RNA isolated from H-treated cells is markedly less than that found in DNA, confirming previous findings from *in vitro*-reactions that the double-stranded DNA structure favours formation of this derivative. It may further be recalled that the concept of cross-linkage of DNA as an effective lesion has also been supported by observations that the di-guaninyl derivative is preferentially excised from DNA of the resistant strain B/r; that the primary biochemical effect of DNA alkylation is inhibition of DNA synthesis, but not RNA or protein synthesis[24]; that this inhibition is temporary with resistant cells[21,23,31]; and that recovery in DNA synthesis is subsequent to excision, which presumably therefore restores template activity[21,31]. Analogous findings have recently been made from studies of the action of mustard gas on cultured HeLa cells[32], although in this case both the products of mono- and difunctional alkylation were excised without the marked preferential excision of the di-guaninyl derivative found for *E. coli* B/r.

The sensitising action of iodoacetamide appears to be due to two factors; firstly, an enhanced extent of alkylation of cellular DNA resulted from pretreatment of cells with iodoacetamide, but this could not account for all the sensitising action; secondly iodoacetamide was indicated to inhibit the excision stage of repair[24], possibly by inactivation of the enzyme concerned. The range of concentration causing this inhibition is also that effective for sensitisation.

A reasonable assumption to explain the sensitising action of acriflavine, following WITKIN[33], is that it binds to cellular DNA and thus interferes with the repair process. Thus acridine dyes are known[34] to inhibit recombination, which is indicated to involve breakage and reunion of DNA molecules[35]. LERMAN has proposed[34] that the essential mode of binding involved is intercalation of dye molecules between base-pairs in double-stranded DNA molecules. An alternative possibility is that dye molecules bind to single-stranded DNA, which is a postulated intermediate in repair[21] and recombination[35,36] processes. With regard to this possibility, it may be noted that denatured DNA binds more aminoacridine than native DNA, more especially at higher ionic strengths[37,38].

Thus much remains to be learned not only about the lesions induced in DNA by mutagenic agents and radiomimetic cytotoxic agents, which are indicated to act by inhibiting DNA replication or causing lethal mutations, but also about the ways in which these lesions can be repaired, and this repair in turn inhibited. Nevertheless it seems useful even at this stage to attempt to formulate a generalised scheme, as in Fig. 5, to represent the reactions of mutagenic chemicals *in vivo*, and to speculate on the way in which the repair mechanisms might be expected to affect

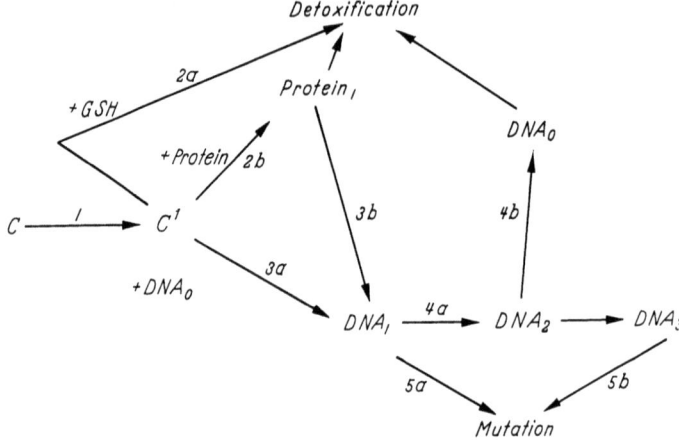

Fig. 5. A scheme for the reactions of a chemical mutagen or carcinogen *in vivo*. (1) The chemical mutagen or carcinogen (C) is converted to a reactive molecular species (C') either, (a) by chemical reaction, *e.g.* alkylating agent → carbonium ion, or (b) by metabolism, *e.g.* aromatic hydrocarbon → epoxide, aromatic amine → N-hydroxyamine → carbonium ion, dialkylnitrosamine → carbonium ion. (2) The reactive species combines with (a) glutathione, or some other cellular constituent, leading to its detoxification, or (b) with protein, leading to detoxification, or to generation of another reactive species (cf. ref. 17). (3) The reactive species combines with DNA, (a) directly, or (b) a further reactive species from protein reaction combines with DNA. (4) Enzymatic repair of altered DNA, (a) first stage, excision of altered part of DNA, (b) subsequent stages, synthesis to replace excised portion, followed by recombination. (5) Failure of repair, (a) enzymatic repair not activated, (a') repair blocked at first stage, abnormal DNA_1 replicates, (b) subsequent stage of repair blocked, *e.g.* exchange error occurs during recombination, and an abnormal DNA different from DNA_1, DNA_3, replicates. The mutation may be viable, *i.e.* give rise to a mutant cell line, or lethal, *i.e.* the abnormal cell does not divide

the expression of these reactions as mutations capable of yielding viable mutant cell lines, or as lethal mutations preventing cell division. Furthermore, if tumour initiation is assumed to involve somatic mutation, such a scheme would have relevance to the problem of the molecular mechanism of carcinogenesis.

There is, as mentioned earlier, an increasing body of evidence that chemical carcinogens, other than those classified as alkylating agents simply by virtue of their chemical reactivity, are metabolised to reactive molecular species which can give rise to carbonium ions, *viz.* polynuclear aromatic hydrocarbons and aromatic amines. Another, and the best established case, is presented by the carcinogenic dialkylnitrosamines[39,40], which have been shown to alkylate DNA *in vivo* in the tissues in which they can induce tumours, these alkylation reactions being mediated by products of oxidative metabolism.

Whereas there is evidence that aromatic hydrocarbons react with DNA *in vivo*, the nature of the products remains undetermined. However, it has been shown that an epoxide derived from naphthalene, 1,2,3,4-tetrahydronaphthalene-1,2-epoxide reacts with DNA or deoxyguanylic acid in the manner characteristic of other alkylating agents[41]. However, it should be noted that metabolism through epoxides could not be detected with 7,12-dimethylbenz/a/anthracene, although metabolites resulting from oxidation of the methyl groups were found[42]. It can also be suggested that the reactive metabolic species from aromatic hydrocarbons may be radicals, as formed on u.v. irradiation[43,44], rather than epoxides.

Quantitative data relating the extent of reaction of carcinogens with DNA *in vivo* with the carcinogenic response must also be admitted to be small in extent at present. However the general impression from the available data[39,40] is that relatively high extents of alkylation of nucleic acids in tissues susceptible to the carcinogenic action of dialkylnitrosamines result from administration of these carcinogens, of the order of 100 μmole/mole DNA-P. Whereas with the aromatic carcinogens the observed extents of reaction are lower. For example, with the most reactive hydrocarbon of those examined[4], 7,12-dimethylbenz/a/anthracene, the relatively high dose of 1 μmole applied to mouse skin gave an extent of reaction of 40 μmole/mole DNA-P. It has been reported[45] that in a sensitive strain of mice a dose of 0.01 μmole of this hydrocarbon could cause carcinogenesis. Following a carcinogenic dose of 4-dimethylaminoazobenzene, an extent of reaction of 8 μmole/mole DNA-P with DNA of a susceptible tissue, the liver of hooded rat, was found[46].

It seems possible therefore that, adopting the hypothesis that reaction of a carcinogen with DNA is the cause of tumour initiation, aromatic carcinogens are more effective than the classical aliphatic alkylating agents in this respect. To account for this, perhaps two types of response to chemical modification of DNA should be distinguished. Firstly, that the initial reaction, to give DNA_1, as in Fig. 5, should either inactivate the cell reproductively, due to failure or saturation of the repair mechanism, or lead to a modified DNA capable of replication, with the chance that mutation may result. Secondly, that a block in repair at a subsequent stage could yield another type of altered DNA, denoted DNA_3, also potentiating mutation.

It might then be envisaged that reaction of DNA with molecules containing special structural features, such as aromatic groups, could favour the second type of alteration in DNA. Whereas with aliphatic agents this secondary block might not be so probable, although, as with iodoacetamide, inactivation of repair enzymes might achieve such a result; this would appear however to require a relatively high extent of protein alkylation. In order to achieve such an extent of alkylation of protein, extensive alkylation of DNA capable of saturating the available repair enzymes and thus preventing DNA replication would be expected, since most alkylating agents react equally well with DNA or protein.

Some justification for the consideration that aromatic groups bound to DNA may confer an enhanced potential for inactivation of the template over that of aliphatic groups is provided by the example of the sensitising action of acriflavine. If this is attributed to its ability to bind to DNA, a role for the non-covalent binding of aromatic carcinogens to DNA, which has been shown to resemble that of acridine dyes[47,48,49], can be envisaged. It should be noted, however, that acridine dyes appear to act as mutagens where recombination, or breakage and reunion, of DNA occurs[50], so that it remains necessary to retain the need for covalent binding of the aromatic carcinogens in order to activate such a mechanism.

The nature of the non-covalent binding postulated for inhibition of repair remains uncertain; it could conceivably be intercalation, or some other type of binding to single-stranded DNA might be involved. Here it may be noted that denaturation of DNA favours the binding of stacked complexes of dye molecules; and that MARMASSE has reported[51] the binding of small micelles of the carcinogen 4-dimethylaminoazobenzene to RNA *in vitro*.

With reference to the two types of altered DNA envisaged in the scheme of Fig. 5, it might be expected that DNA_1 would give rise to mutations of the transition type, while the erroneous repair resulting in DNA_3 might be expected to potentiate mutations of the exchange error type. There is evidence that this type of mutation is more frequent in spontaneous mutations, e.g. it was found that spontaneous mutations were more frequent in meiotically rather than in mitotically reproducing yeast[52]. Assuming therefore that spontaneous errors in repair can occur in somatic cells, these might give rise to spontaneous mutations constituting spontaneous tumour initiation. Furthermore viral induction of cancer could be included in this scheme, since it has been reported that incorporation of viral genomes, or replicas thereof, into cellular DNA, occurs during carcinogenesis by polyoma[53] or Rous virus[54].

It should of course be stressed that the concept of somatic mutation as tumour initiation remains unproved, and the possible mechanism for a subsequent promotion stage in carcinogenesis has not been considered, although this might reasonably be supposed to involve stimulation of cell division enabling growth of the initiated cell line[45]. Nevertheless it would appear that studies of the effects of chemical mutagens, carcinogens and cytotoxic agents on DNA continue to possess relevance to the question of the molecular mechanism of carcinogenesis.

I thank Miss E. J. SHAW for valuable technical assistance in the studies of the action of alkylating agents on *E. coli* reported in this paper, and Miss P. SIMSON for data on protein analysis.

This investigation has been supported by grants to the Chester Beatty Research Institute (Institute of Cancer Research: Royal Cancer Hospital) from the Medical Research Council and the British Empire Cancer Campaign for Research, and by the Public Health Service Research Grant No. CA-03188-08 from the National Cancer Institute, U. S. Public Health Service.

References

[1] WATSON, J. D., and F. H. C. CRICK: Nature (Lond.) **171**, 964 (1953).
[2] CRICK, F. H. C., L. BARNETT, S. BRENNER, and J. WATTS-TOBIN, Nature (Lond.) **192**, 1227 (1961).
[3] LOVELESS, A.: Proc. roy. Soc. **B 150**, 497 (1959).
[4] BROOKES, P., and P. D. LAWLEY: J. cell. comp. Physio. **64**, Suppl. 1, 111 (1964).

[5] LAWLEY, P. D.: Progr. Nucleic Acid Res. and Mol. Biol. 5 (1966) (In press).
[6] —, and P. BROOKES: Naturel (Lond.) 192, 1081 (1961).
[7] — — J. molec. Biol. 4, 216 (1962).
[8] — Unpublished.
[9] NAGATA, C., A. IMAMURA, H. SAITO, and K. FUKUI: Gann. 54, 109 (1963).
[10] BROWN, D. M., and J. H. PHILLIPS: J. molec. Biol. 11, 663 (1965).
[11] JANION, C., and D. SHUGAR: Acta biochim. pol. 12, 337 (1965).
[12] FREESE, E.: Proc. XI Internat. Congr. Genetics 2, 297 (1963).
[13] BOYLAND, E., and P. SIMS: Biochem. J. 97, 7 (1965).
[14] MILLER, E. C., J. A. MILLER, and H. A. HARTMANN: Cancer Res. 21, 815 (1961).
[15] KRIEK, E.: Biochem. biophys. Res. Commun. 20, 560 (1965).
[16] BOYLAND, E.: Z. Krebsforsch. 65, 378 (1963).
[17] SCRIBNER, J. D., J. A. MILLER, and E. C. MILLER: Biochem. biophys. Res. Commun. 20, 560 (1965).
[18] AUERBACH, C.: Proc. XI Internat. Congr. Genetics 2, 275 (1963).
[19] MULLER, H. J.: Proc. XI Internat. Congr. Genetics 2, 265 (1963).
[20] HAYNES, R. H., and P. C. HANAWALT: Biochem. biophys. Res. Commun. 19, 462 (1965).
[21] SETLOW, R. B.: J. cell. comp. Physiol. 64, Suppl. 1, 51 (1964).
[22] STRAUSS, B. S., R. WAHL-SYNEK, H. REITER, and T. SEARASHI: Proc. Symp. on Mutation, Prague 1965. (In press).
[23] BOYCE, R. P., u. P. HOWARD-FLANDERS: Z. Vererbungslehre 95, 345 (1964).
[24] LAWLEY, P. D., and P. BROOKES: Nature (Lond.) 206, 480 (1965).
[25] PRESTON, R. K., R. M. PECK, E. R. BREUNINGER, A. J. MILLER, and H. J. CREECH: J. med. Chem. 7, 471 (1964).
[26] LOVELESS, A.: Genetic and Allied Effects of Alkylating Agents, p. 163. London: Butterworths 1966.
[27] LOVELESS, A., J. WHEATLEY, and P. COOK: Nature (Lond.) 205, 980 (1965).
[28] HAROLD, F. M., and Z. Z. ZIPORIN: Biochim. biophys. Acta (Amst.) 28, 482 (1958).
[29] WITKIN, E.: Cold Spring Harbor Symp. Quant. Biol. 16, 357 (1951).
[30] LAWLEY, P. D., and P. BROOKES: Biochem. J. 89, 138 (1963).
[31] PAPIRMEISTER, B., and C. DAVIDSON: Biochem. biophys. Res. Commun. 17, 608 (1964).
[32] CRATHORN, A. R., and J. J. ROBERTS: Nature (Lond.). 211, 150 (1966).
[33] WITKIN, E. M.: J. cell. comp. Physiol. 58, Suppl. 1, 135 (1958).
[34] LERMAN, L. S.: J. cell. comp. Physiol. 64, Suppl. 1, 1 (1964).
[35] MESELSON, M.: J. molec. Biol. 9, 734 (1964).
[36] WHITEHOUSE, H. L. K.: Nature (Lond.) 199, 1034 (1963).
[37] LAWLEY, P. D.: Biochim. biophys. Acta (Amst.) 22, 451 (1956).
[38] DRUMMOND, D. S., V. F. W. SIMPSON-GILDEMEISTER, and A. R. PEACOCKE: Biopolymers 33, 135 (1965).
[39] CRADDOCK, V. M., and P. N. MAGEE: Biochem. J. 89, 32 (1963).
[40] — — Biochim. biophys. Acta (Amst.) 95, 677 (1965).
[41] BROOKES, P., and P. D. LAWLEY: Unpublished.
[42] BOYLAND, E., and P. SIMS: Biochem. J. 95, 780 (1965).
[43] T'SO, P. O. P., and P. LU: Proc. nat. Acad. Sci. (Wash.) 51, 272 (1964).

[44] RICE, J. M.: J. Amer. chem. Soc. **86**, 1444 (1964).
[45] BOUTWELL, R. K.: Progr. exp. Tumor Res. **4**, 207 (1964).
[46] ROBERTS, J. J., and G. P. WARWICK: Int. J. Cancer **1**, 179 (1966).
[47] BOYLAND, E., and B. GREEN: Brit. J. Cancer **16**, 507 (1962).
[48] LERMAN, L. S.: Proc. 5th Nat. Cancer Conference, p. 39. Philadelphia: Lippincott 1964.
[49] BALL, J. K., J. A. MCCARTER, and M. F. SMITH: Biochim. biophys. Acta (Amst.) **103**, 275 (1965).
[50] MAGNI, G. E., R. C. VON BORSTEL, and S. SORA: Mutation Res. **1**, 227 (1964).
[51] MARMASSE, C.: Nature (Lond.) **202**, 1010 (1964).
[52] MAGNI, G. E.: J. cell. comp. Physiol. **64**, Suppl. 1, **165** (1964).
[53] AXELROD, D., K. HABEL, and E. T. BOLTON: Science **146**, 1466 (1964).
[54] TEMIN, H. M.: Proc. nat. Acad. Sci. (Wash.) **52**, 323 (1964).

DPNase-Induktion, RNSase-Erhöhung und Hemmung der DNS-Synthese durch cytostatische Agenzien

Von H. HILZ

Physiologisch-Chemisches Institut der Universität Hamburg

Die Einwirkung cytostatischer Agenzien in Konzentrationen, welche die Zellteilung inhibieren, führt zu einer einheitlichen Reaktion der behandelten Zellen hinsichtlich folgender drei Parameter:

1. Eine indirekte oder direkte Inhibierung der DNS-Synthese,
2. Eine Induktion der DPNase (E.C. 3.2.2.5),
3. Eine Zunahme der freien RNSase-Aktivität (E.C. 2.7.7.16).

Ad 1. Die *Inhibierung der DNS-Synthese* ist nach den meisten Cytostatika (in teilungshemmender Konzentration) indirekt, wie die Produktion von *polyploiden* Riesenzellen nach Röntgenstrahlen, Trenimon (Triäthyleniminobenzochinon), Actinomycin, Colchicin und Cortisol zeigt. Der umgekehrte Fall, direkte Hemmung der DNS-Synthese, ist bei Fluoruracilbehandlung realisiert.

Ad 2. Ein dosisabhängiger *Anstieg der DPNase-Aktivität* wird nach Behandlung von Ehrlich-Ascitestumorzellen mit den erwähnten Cytostatika einschließlich Fluoruracil beobachtet. Die Aktivitätszunahme ist mit ziemlicher Sicherheit eine Induktion (Neusynthese des Enzyms). Sie betrifft fast ausschließlich die mikrosomale DPNase, während das im Nucleolus gebundene Isoenzym nur wenig erhöht ist.

Da nach niedriger Dosierung mit Röntgenstrahlen oder Trenimon der DPN-Gehalt der Zellen nicht absinkt, sondern übernormal wird, da weiterhin unter den gleichen Bedingungen die DPN-Synthese aus Nicotinsäure im gleichen prozentualen Ausmaß erhöht ist wie die DPNase-Aktivität, läßt sich die Zunahme der DPNase als Ausdruck eines erhöhten DPN turnovers deuten.

Ad 3. Die *Zunahme der Ribonuclease* betrifft hauptsächlich die freie Form des Enzyms, obleich auch die latente RNSase ansteigt.

Mit zunehmender Dosis an cytostatischem Agenz kommt es jedoch zu einer immer stärkeren Verschiebung zugunsten der freien Form.

Diese erhöhte Enzymaktivität steht vermutlich im direkten Zusammenhang mit dem Abbau einer nuclearen RNS-Fraktion nach Actinomycin oder Röntgenstrahlen (HÖLZEL, KOCH und TRAMS).

Diese dreifache Reaktion der Tumorzellen auf niedrigdosierte, teilungshemmende Cytostatikakonzentrationen — Hemmung der DNS-Synthese, Induktion der DPNase und Zunahme (Induktion?) der freien RNSase — zeigt eine weitgehend lineare Abhängigkeit vom Grad der Proliferationshemmung. Sie wurde bisher beobachtet nach Röntgenstrahlen, Actinomycin, 5-Fluoruracil, Colchicin, Trenimon und, wie GREEN und BODANSKI fanden, auch nach N-Lost.

Aus diesen Beobachtungen lassen sich folgende Schlüsse ziehen:

1. Keines der von uns verwendeten Cytostatika (Röntgenstrahlen, Trenimon, Actinomyzin, Colchicin, Fluoruracil, Cortisol und Jodacetamid) führt bei teilungshemmender Dosierung zu einer generellen und irreparablen Schädigung der DNS-Matrix. Replikation wie Transkription und Translation müssen im Prinzip funktionieren: Einerseits entstehen in fast allen Fällen Zellen mit übernormalem DNS-Gehalt, zum anderen ist auch die Induktion von Enzymen noch realisiert. Dasselbe ergibt sich auch bei synchronisierten Hefezellen, die nach 50000 r zwar eine totale Teilungshemmung aufweisen, aber noch zu einer fast normalen Synthese von Enzymen des Embden-Meyerhof-Weges fähig sind.

2. Obgleich die verwendeten Cytostatika sich in ihrem primären Wirkungsmechanismus unterscheiden, ist ihre Wirkung in den niedrigen Konzentrationsbereichen immer die einer Proliferationshemmung. Es sollte daher ein besonders sensitives „target" existieren, vielleicht ein Teilungsoperator, vielleicht eine bestimmte Membran oder auch nur ein bestimmtes System von Rückkoppelungsmechanismen, jedenfalls aber ein gemeinsamer Denominator, der die außerordentliche Empfindlichkeit der Zellteilung erklärt. Dieser Denominator ist offensichtlich eine Funktion der proliferierenden Zellen, da sich in Geweben mit geringer Teilungsrate (z. B. Leber, Muskel, Blut) durch Cytostatika keine Zunahme von DPNase oder RNSase induzieren läßt.

Literatur

[1] RÜTER, J., H. VACHEK, M. OLDEKOP, J. WÜPPEN und H. HILZ: Biochem. Z. **344**, 153 (1966).
[2] ERBE, W., J. PREISS, R. SEIFERT, and H. HILZ: Biochem. biophys. Res. Commun. **23**, 392 (1966).
[3] GREEN, S., and O. BODANSKY: J. biol. Chem. **239**, 2613 (1964); **240**, 2574 (1965).
[4] ECKSTEIN, H., V. PADUCH und H. HILZ: Biochem. Z. Im Druck.
[5] HÖLZEL, F.: Pers. Mitteilung.
[6] KOCH, G., u. G. TRAMS: Pers. Mitteilung.

Einfluß alkylierender Cytostatika auf biochemische Parameter und die Transplantierbarkeit von Tumorzellen

Von E. Liss

Nuklearmedizinische Abteilung der Medizinischen Kliniken der Freien Universität Berlin

Mit 1 Abbildung

Unsere früheren in-vitro- und in-vivo-Versuche hatten ergeben, daß die DNS-Synthese von Ehrlich-Ascitestumorzellen wesentlich empfindlicher gegen alkylierende Cytostatika ist als DPN-Gehalt, Glykolyse und Proteinsynthese. Weitere Untersuchungen haben jedoch gezeigt, daß auch die Hemmung der DNS-Synthese durch

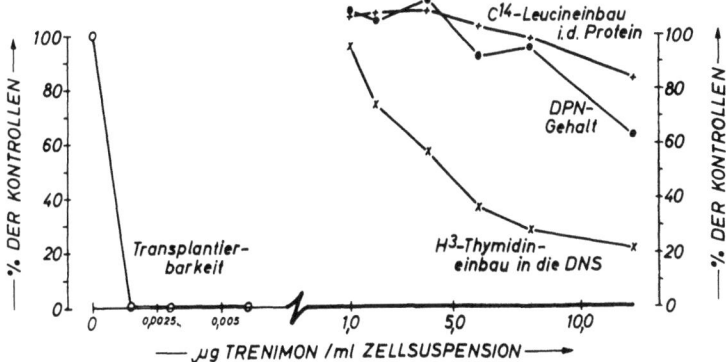

Abb. Biochemische Parameter und Transplantierbarkeit von Yoshida-Ascites-Sarkomzellen nach in-vitro-Behandlung mit Trenimon (alle Angaben in % der unbehandelten Kontrollen)

Trenimon nicht der für die cytostatische Wirkung ausschlaggebende biochemische Parameter sein dürfte, da sie erst bei einer Konzentration deutlich wird, die zehnmal höher ist als die für die vollständige Aufhebung der Transplantierbarkeit der Zellen ausreichende Konzentration. Bei neueren Versuchen mit Zellen des Yoshida-Sarkoms wurde das Fehlen eines Zusammenhanges zwischen cytostatischer Wirkung des Trenimons und bisher bekannten biochemischen Veränderungen noch deutlicher (Abb.). Die Hemmung der DNS-Synthese auf 73% der Kontrollen erforderte mehr als das

1000fache der Konzentration, die zur Aufhebung der Transplantierbarkeit ausreichte. Auch in diesem Fall waren jedoch die Proteinsynthese und der DPN-Gehalt unempfindlicher als die DNS-Synthese. Die bei der Untersuchung der Wirkung des Trenimons erhobenen Befunde sind nicht auf die Wirkung von Lostverbindungen wie Mitomen und Endoxan übertragbar. Bei diesen Verbindungen bleibt die Möglichkeit eines Zusammenhanges zwischen Hemmung der DNS-Synthese und cytostatischer Wirkung noch bestehen.

Neben einer Hemmung der DNS-Synthese als Folge der Beeinflussung von Fermenten wird häufig die direkte Alkylierung der DNS diskutiert. Bei Untersuchungen mit Endoxan (in der Lostgruppe H^3-markiert) haben wir gezeigt, daß nach in-vivo-Verabreichung einer cytostatisch wirksamen Dosis weniger als 1 Endoxanmolekül pro 100000 DNS-Nucleotide an die DNS von Ehrlich-Ascitestumorzellen gebunden war. Dem Einwand, daß eine an sich stärkere Alkylierung der DNS nicht nachgewiesen wurde, weil Alkylgruppen oder alkylierte Basen in der Zelle oder bei der Aufarbeitung abgespalten wurden, läßt sich experimentell schwer begegnen. Am Beispiel des Trenimons läßt sich jedoch sicher beweisen, daß die bei Anwendung therapeutischer Dosen eintretende Alkylierung der DNS äußerst gering sein muß. Bei in-vitro-Inkubationen der Zellen des bereits erwähnten sehr empfindlichen Yoshida-Sarkoms wurde die kleinste Trenimondosis ermittelt, die noch zu einer deutlichen Verringerung des Wachstums der Zellen nach der Reimplantation führte. Beim Vergleich der insgesamt vorhandenen Trenimonmenge mit der in den inkubierten Zellen vorhandenen DNS-Menge ergab sich, daß von vornherein höchstens 1 Trenimonmolekül pro 80000 DNS-Nucleotide zur Verfügung stand. Die tatsächliche Alkylierungsrate der DNS ist zweifellos wesentlich geringer, weil neben der DNS noch andere Zellbestandteile — vor allem die Proteine — um das alkylierende Agens konkurrieren. Es taucht daher die Frage auf, ob die cytostatische Wirkung einer alkylierenden Verbindung als Folge einer Alkylierung von weit weniger als einem von 100000 DNS-Nucleotiden verstanden werden könnte, zumal manches dafür spricht, daß die Alkylierung der DNS-Nucleotide einer zufälligen Verteilung gehorcht. Auf jeden Fall kann man darauf verzichten, nach physikalisch-chemischen Veränderungen der DNS zu suchen. Sie sind bei den Dosen, die von therapeutischem Interesse sind, sicher nicht aufzufinden.

Untersuchungen über die Wirkung von Trenimon (2,3,5-Trisäthyleniminobenzochinon-1,4) auf die DNS von Ehrlich-Ascites-Tumorzellen

Von H. GRUNICKE

Biochemisches Institut der Universität Freiburg

Untersuchungen über den Einfluß von Trenimon auf die Starteraktivität der DNS im RNS-Polymerasesystem ergaben, daß Trenimonkonzentrationen von 10^{-4} bis 10^{-3} M notwendig sind, um eine deutliche Erniedrigung der Starteraktivität und als Folge davon eine Hemmung der RNS-Synthese zu bewirken[1]. Befunde von LISS und PALME[2] zeigen jedoch, daß in intakten Zellen eine Hemmung

Tabelle. *Einfluß von Trenimon auf die Extrahierbarkeit der DNS aus Ehrlich-Ascites-Tumorzellen*
DNS-Isolierung nach COLTER et al.[3]. DNS-Bestimmung nach BURTON[4]. Protein wurde mit Hilfe der Biuretmethode bestimmt. Vers. 1: Behandlung der Tumorzellen in vivo. Vers. 2: Behandlung der Tumorzellen in vitro. 10% Zellsuspension in Krebs-Ringer-Phosphat-Puffer pH 7,4; 5 mg/ml Glucose enthaltend. Inkubation 1 Std bei 37°C.

	DNS-Gehalt im Rohextrakt	DNS-Ausbeute	DNS im Protein	
	mg	mg	mg	µg
	ml Zellen	ml Zellen	ml Zellen	mg Protein
Vers. 1: Kontrolle	7,0	4,4	0,7	6,0
+ Trenimon (1,76×10⁻⁶ Mol/kg)	6,8	1,4	3,1	21,7
Vers. 2: Kontrolle		4,2		8,3
+ Trenimon 4,4×10⁻⁷ M		2,9		18,5
+ Trenimon 4,4×10⁻⁶ M		1,4		36,3

der RNS-Synthese bereits bei 100fach geringeren Trenimonkonzentrationen meßbar ist. Es ist daher anzunehmen, daß die Hemmung der RNS-Synthese in intakten Zellen nicht allein auf einer

Alkylierung der DNS beruht, sondern daß hier ein zusätzlicher Mechanismus wirksam ist.

Isoliert man die DNS aus trenimonbehandelten Ascitezellen mit Hilfe der Phenolmethode nach COLTER et al.[3], so zeigt sich, daß aus den trenimonbehandelten Zellen regelmäßig weniger DNS isoliert werden kann als aus den unbehandelten Kontrollen, obwohl die DNS-Gehalte von behandelten und unbehandelten Zellen gleich groß sind (vgl. Tabelle). Als Ursache für die geringere Extrahierbarkeit ergab sich, daß bei der Isolierung der DNS aus den trenimonbehandelten Zellen in Abhängigkeit von der Trenimonkonzentration ein beträchtlicher Anteil der DNS in dem Proteinrückstand zwischen wäßriger und phenolischer Phase zu finden ist. Eine Gelbildung aus reiner DNS konnte als Ursache des Effektes ausgeschlossen werden. Durch Behandlung mit konzentrierter NaCl-Lösung läßt sich die DNS nicht vom Protein ablösen.

Als Erklärung für diese Befunde nehmen wir an, daß unter der Wirkung von Trenimon cross-links zwischen DNS und Protein auftreten. Vorläufige Versuche ergaben, daß es sich hierbei nicht um Verknüpfungen zwischen DNS und Histon handelt. In Analogie zu Befunden von STEELE[5] und GOLDSTEIN und RUTMAN[6] über die Wirkung bifunktioneller S- und N-Lost-Verbindungen vermuten wir, daß Trenimon cross-links zwischen DNS und saurem Kernprotein bewirkt. Angesichts des Ausmaßes dieses Effektes — bei einer Trenimondosis von $1{,}7 \times 10^{-6}$ Mol/kg finden sich etwa 70% des DNS-Gehaltes der Zelle in dem oben beschriebenen Proteinrückstand — ist es nicht unwahrscheinlich, daß den DNS/Protein — cross-links wesentliche Bedeutung für das Auftreten des cytostatischen Effektes zukommt.

Literatur

[1] GRUNICKE, H., F. ZIMMERMANN, and H. HOLZER: Biochem. biophys. Res. Commun. **18**, 319 (1965).
[2] LISS, E., u. G. PALME: Z. Krebsforsch. **66**, 196 (1964).
[3] COLTER, J. S., R. A. BROWN, and K. A. O. ELLEM: Biochem. biophys. Acta (Amst.) **55**, 31 (1962).
[4] BURTON, K.: Biochem. J. **62**, 315 (1956).
[5] STEELE, W. J.: Proc. Amer. Ass. Cancer Res. **3**, 364 (1962).
[6] GOLDSTEIN, N. O., and R. J. RUTMAN: Cancer Res. **24**, 1363 (1964).

Some New Points of View Concerning the Mode of Action of Biological Alkylating Agents

By I. P. HORVÁTH, and L. INSTITÓRIS

Chinoin, Laboratory for Cancer Research, Budapest

On basis of our experimental observations we would like to suggest some points of view concerning the mode of action of dibromomannitol and dibromodulcitol which represent a new type of the myelotropic alkylating agents. We found that in vivo they are more stable than other biological alkylating agents. Half of the compound decomposes in four to six hours, and organic bromine compounds were found as the main metabolites. Therefore, it is difficult to explain their cytostatic effects merely by their alkylating action.

Considering that the dibromohexitol is present in the cell for a relatively long time, it can be adsorbed on some critical points, and it may disturb the electron distribution pattern of macromolecular systems. In subsequent alkylation or hydrolysis hydrobromic acid may be formed, inducing temporary changes in the local hydrogen ion concentration.

In this connection we should recall the hydrochloric acid hypothesis of MARSHALL, — although it was found untenable in the case of the powerful alkylating mustards. However, in the case of agents with a much weaker alkylating activity, the role of an intracellular acidification should be revaluated. Then, protons and anions formed on a critical point of the cell may increase or replace the effects of alkylation. The balance of ionic atmospheres may be upset. Membranes and lamellae may be damaged. Protons liberated in the vicinity of DNA can protonate the lone electron pairs of nucleotide bases, inducing thereby anomalous base pairing and depurination even without a direct alkylation of the nucleic acid.

For similar reasons, the unexpected similarity in the neutrophile-depressing actions of Myleran and dimethyl-myleran may be due perhaps to the effects of the methanesulfonic acid which is liberated by both compounds in different reactions.

Diskussion

NEUBERT (Berlin-Dahlem): Nachdem wir eine RNS-Polymerase in isolierten Mitochondrien nachgewiesen haben, untersuchten wir den Einfluß von Cyclophosphamid (Endoxan) auf die DNS-abhängige RNS-Polymerase-Aktivität von Tumormitochondrien.
Die Untersuchungen wurden gemeinsam mit Herrn Dr. H. HELGE durchgeführt[1,2]. Folgende Versuchsanordnung wurde gewählt: Tumortragende Ratten erhielten eine einmalige Dosis von 20 mg/kg Cyclophosphamid — das ist etwa $^1/_5$ bis $^1/_{10}$ der LD_{50} — und 24 Std später wurde der Einbau von C^{14}- oder H^3-UTP in die RNS der durch Gradientenzentrifugation gereinigten, isolierten Zellkerne und Mitochondrien bestimmt.

Der UTP-Einbau in die *Mitochondrien*-RNS von Yoshida-Sarkom, Walker-Carcinom und Shay-Chloro-Leukom (alles sehr Cyclophosphamid-empfindliche Tumoren) war unter diesen Versuchsbedingungen hochgradig vermindert, der Einbau in die mitochondriale RNS des DS-Carcinosarkoms (recht Cyclophosphamid-unempfindlich) nur mäßig stark gehemmt. Beim Morris-Hepatom 5123 D konnte, wie bei Lebergewebe unter dem Einfluß von Cyclophosphamid keine Verminderung der RNS-Polymerase-Aktivität in den Mitochondrien nachgewiesen werden.

Bemerkenswerterweise war unter den gewählten Versuchsbedingungen der UTP-Einbau weder in die RNS isolierter Leber-*Zellkerne* noch in die von Tumor-*Zellkernen* gehemmt (mit Ausnahme der Kerne des sehr Cyclophosphamid-empfindlichen Yoshida-Sarkoms). Bei einer Erhöhung der Cyclophosphamiddosis auf 60 mg/kg wurde dann neben der Hemmung der mitochondrialen RNS-Polymerase auch der UTP-Einbau in die isolierten Zellkerne des Shay-Chloro-Leukoms vermindert gefunden. Auch bei diesen hohen Dosen des alkylierenden Agens war jedoch die RNS-Polymerase-Aktivität der isolierten Zellkerne aus Leber und Morris-Hepatomen unverändert.

Nach Gabe von H^3-Cyclophosphamid (20 mg/kg) haben wir die DNS aus Zellkernen und Mitochondrien isoliert und die spezifische Aktivität bestimmt. In diesen Untersuchungen zeigte die Mitochondrien-DNS eine wesentlich höhere Markierung (Alkylierung) als die Zellkern-DNS. Allerdings war die Markierung der Leber-DNS eher höher als die der Tumor-DNS!

Es soll noch erwähnt werden, daß nach Untersuchungen von Herrn Dr. H. COPER in unserem Institut die *DPN-Konzentration* in den Shay-Chloro-Tumoren 24 Std nach einer Dosis von 20 mg/kg Endoxan höchstens um 10 bis 20% vermindert ist.

Die Untersuchungen decken eine besonders ausgeprägte Empfindlichkeit der mitochondrialen RNS-Synthesereaktion gegenüber alkylierenden Verbindungen auf. Obgleich wir Anhaltspunkte dafür haben, daß wenigstens ein Teil der mitochondrialen RNS „messenger"-Eigenschaften besitzt[3], bleibt die Bedeutung der mitochondrialen Nucleinsäuren so lange unklar, bis aufgeklärt ist, welche und wieviele der mitochondrialen Proteine von diesen Zellpartikeln in eigener Regie gebildet werden können.

Mit den hier vorgestellten Befunden soll daher auch noch keine Erklärung des Wirkungsmechanismus alkylierender Substanzen gegeben werden. Es erscheint jedoch angebracht, bei den Untersuchungen über die Wirkung von Pharmaka nicht nur die Zellkern-DNS im Auge zu haben, sondern auch die Mitochondrien-DNS als möglichen Angriffspunkt zu erwägen.

Literatur

[1] HELGE, H., D. NEUBERT, R. BASS und N. BROCK: Naunyn-Schmiedebergs Arch. exp. Path. Pharmak. **253**, 44 (1966).
[2] NEUBERT, D.: Naunyn-Schmiedebergs Arch. exp. Path. Pharmak. **253**, 152 (1966).
[3] —, and H. HELGE: 2. FEBS-Meeting, Wien e 84 (1965). 3. FEBS-Meeting, Warschau G 27 (1966).

BÜCHER (München): Ich darf darauf hinweisen, daß die DNS aus Mitochondrien kürzlich dargestellt worden und nach elektronenmikroskopischen Untersuchungen der Gruppe von BORST in Amsterdam ein ringförmiges Molekül ist [E. F. J. VAN BRUGGEN et al.: Biochem. biophys. Acta (Amst.) **119**, 437 (1966)].

LAWLEY: With regard to effects of X-rays on RNA and protein synthesis, these may not be due to the direct action of X-rays on the DNA template, although I think there is strong evidence this is the cause of the temporary inhibition of DNA synthesis. The repair enzymes break down DNA and thus magnify the lesion; this has been shown in bacteria by POLLARD and others and therefore we may expect that this would suffice to affect RNA synthesis. However, the studies of my colleagues Dr. CRATHORN and Dr. ROBERTS on the action of sulfur mustards on cultured cells established that the only biochemical effect which immediately follows alkylation with a dose of about four mean lethal doses is a temporary inhibition of DNA synthesis. But this does not rule out the possibility of an interference with RNA and protein synthesis following the primary damage to the DNA template.

Passing on to the quantitative aspect of alkylation of DNA which I mentioned, I am not surprised that a very low extent of alkylation by the ethyleneimine derivatives can cause inhibition of cell division, because for sulfur mustard CRATHORN and ROBERTS showed that one alkylation in about 500000 DNA nucleotides sufficed to inhibit the ability of cultured cells of mouse lymphoma or of HeLa to divide. This, of course, seems a very low extent of alkylation expressed in this way. But if you recall that this corresponds to something of the order of one thousand crosslinks in the total nuclear DNA of a mammalian cell, then this amount is much higher than the value of two crosslinks in the whole nuclear DNA of bacteria of a sensitive strain, which I showed will suffice to prevent these cells from dividing. Therefore the mammalian cell is more resistant than the bacterial cell, if we consider the dose as being the number of alkylations in the total DNA. In this regard, we envisage that all the DNA must replicate to give a viable cell. In fact, the lethal dose of mustard does not prevent doubling of the amount of DNA, but DNA synthesis is slowed down. Therefore I would envisage, from the evidence obtained recently by my colleagues CRATHORN and ROBERTS, that the mammalian cell resembles the resistant bacterial cell in its response to mustard. First there is a temporary block in DNA synthesis which can be shown by a technique of pulse labelling. Then we have the excision of the mustard groups from DNA, as shown by a technique of double labelling of DNA first with tritiated thymidine and then with ^{35}S-labelled mustard. They were able to show that the mustard label was removed from the DNA more rapidly than the thymidine prelabel, the latter decrease being due to DNA synthesis.

Concerning DNA-protein crosslinking by ethyleneimines, I would mention that these are compounds with low reactivity, in contrast to sulfur mustard and can survive in neutral solutions for very long time in potential reactive form. Therefore, if we react one group of an ethyleneimine with DNA it is quite possible that the other groups will not have reacted with water. This is in contrast to the case of the sulfur mustards, and, therefore it is not surprising with the latter we do not observe DNA-protein-crosslinkage to any great extent. With ethyleneimines it is possible that one "arm" of the alkylating agent reacts with DNA, then after disruption of the cells groups in the proteins are liberated which can react with another "arm" of the reagent. Therefore you are not sure that the protein-DNA-crosslinkage occurred during the primary reaction of the ethyleneimine with the cell.

With regard to Dr. HORVARTH's very interesting contribution we have certainly considered this question of liberation of acid in connection with the action of the relatively unreactive agent methyl-methanesulfonate. The results of Drs. A. LOVELESS and B. S. STRAUSS on the action of methyl-methanesulfonate on bacteriophage and on transforming DNA show that methyl-methanesulfonate is more effective in the inactivation of DNA while the reagent is present in the treatment medium. The rate of degradation of DNA under these conditions appears to be greater than it is if the reagent is removed by dilution and then one follows the subsequent inactivation of the DNA. The only explanation we could think of was that the acid liberated by the alkylating agent while it was present in the treatment medium could enhance the rate of hydrolysis of DNA.

HEIDELBERGER (Madison/Wisconsin): Dr. LAWLEY very properly emphasized the effects of carcinogens on DNA. I would like to emphasize once again the point that Dr. MAGEE made yesterday, that compounds also react with RNA and with protein. I would like to point out that in the case of carcinogenic hydrocarbons, the azo dyes and acetylaminofluorene there are very impressive correlations between the binding of these compounds to a specific protein fraction and their carcinogenic activity. This correlation is quite good. If carcinogenesis with chemicals results from interaction with DNA one would expect a somatic mutation as indicated. On the basis of this it would be very improbable, that tumor cells so induced could revert back to normal cells. There is a growing body of evidence in the literature now that in many well controlled experiments there is actually a reversibility of carcinogenesis back to normal. And the frequency is much higher than one could expect for back mutations. So this is just a plea to keep an open mind. It will be very difficult to establish conclusively the importance to carcinogenesis of the interactions of these compounds with DNA, RNA, protein and perhaps fat.

SCHOLTISSEK (Gießen): Bei der Beurteilung der Wirkung mutagener und carcinogener Agentien auf die Nucleinsäure- und Proteinsynthese muß jeweils geprüft werden, inwieweit diese Stoffe die Penetrationsfähigkeit der intakten Zelle beeinflussen. Es ist bekannt, daß solche Agentien zwar *in vitro* mit gewissen Makromolekülen reagieren können, *in vivo* aber nicht an den entsprechenden Wirkungsort der Zelle zu gelangen brauchen. Weiterhin muß immer damit gerechnet werden, daß die bei den Experimenten verwendeten Isotope infolge der Vorbehandlung der Zellen mit den genannten Agentien nicht in diese penetrieren. Folgendes Beispiel möge das veranschaulichen: Sowohl Proflavin als auch Acridinorange verhindern den Einbau von ^{14}C-Uridin in die RNS von Hühnerfibroblasten. Dabei verhindert Proflavin tatsächlich die RNS-Synthese, während Acridinorange nur die Markierung der säurelöslichen Vorläufer der RNS hemmt. Das Bayerpräparat A 139, ein Äthyleniminochinon, verhindert in relativ hohen Dosen ebenfalls die Markierung der RNS mit ^{14}C-Uridin. Dieser Effekt kann zu einem großen Teil auch auf einen Einfluß auf die Markierung des säurelöslichen Pools zurückgeführt werden. — Mit diesem Präparat gelang es uns außerdem, lipidhaltige Viren in vitro zu inaktivieren, ohne ihre an Proteine gebundenen biologischen Aktivitäten (Antigene, Enzyme) zu zerstören. Für diese Inaktivierung ist ein pH-Wert ≤ 7 notwendig.

HÖLZEL (Hamburg): Arbeitet man mit p-Benzochinon, also dem Kern des eben erwähnten Trenimonmoleküls, so ergibt sich in Ehrlich-Ascites-Tumorzellen — parallel zur Hemmung der Transplantierbarkeit — ebenso eine Hemmung von DNA- oder RNA-Synthese, die unabhängig vom Energiestoffwechsel ist, wie mit Trenimon.

Verfolgt man mit ^{14}C-markiertem p-Benzochinon den Verbleib der Radioaktivität in den Tumorzellen, so findet man eine Anreicherung des Benzochinons in den Zellkernen. Nach saurer Extraktion der Histone findet sich nach 2stündiger Inkubation 80 bis 90% der ^{14}C-Radioaktivität der Zellkerne in der Fraktion der sauren Kernproteine. DNA und RNA der vergleichbaren Ausgangszellmenge enthalten nur geringe Anteile Benzochinon. Die Anreicherung des p-Benzochinons im sauren Kernprotein ist mindestens 30- oder 100fach gegenüber DNA oder RNA.

Man sollte diese Fraktion der sauren Kernproteine im Zusammenhang mit den beschriebenen Hemmeffekten der DNA- oder RNA-Synthese vielleicht etwas mehr beachten.

HECKER (Heidelberg): Ich möchte im Zusammenhang mit dem, was Herr SCHOLTISSEK sagte, auf ein Experiment von BERENBLUM u. Mitarb. aufmerksam machen [Biochem. Pharmacol. **13**, 263 (1964)], die vor 2 Jahren gezeigt haben, daß Mutagene wie Acridinorange und Proflavin oder 5-Brombzw. 5-Joddesoxyuridin, deren Wechselwirkung mit DNS man partiell zu verstehen glaubt, nicht als Initiatoren der Carcinogenese an der Maus wirken. Diese Substanzen sind danach für Studien zum Mechanismus der Carcinogenese nicht geeignet.

154   Diskussion

KERSTEN (Münster): The carcinostatic antibiotic Mitomycin causes "crosslinks" in DNA and also alkylates DNA as has first been shown by IYER and SZYBALSKI. These damages in DNA can be repaired as has been observed by HOWARD-FLANDERS et al. We have found, that in a strain of E. coli, which is capable to repair DNA, the activity of an endo-DNAase is increased upon treatment with Mitomycin. We suggest, that this endo-DNAase is involved in the repair-mechanism. I would like to ask Dr. LAWLEY the following question: If tumor cells as you suggest are not able to repair damages in the DNA it might be possible, that tumor cells do not contain enzymes, which are involved in repair or that their activities are somehow blocked. Is there any experimental evidence as yet that tumor cells have no or very low endo-DNAase-activity?

LISS (Berlin): Ich möchte kurz zu dem Einwand von Herrn SCHOLTISSEK Stellung nehmen. Es ist prinzipiell wichtig, daß man bei Untersuchungen mit radioaktiven Vorstufen der DNS oder bei ähnlichen Versuchen prüft, ob nicht die Penetration beeinflußt und eine Hemmung von Synthesevorgängen nur vorgetäuscht wird. Im Falle des Trenimons haben wir diese Frage selbst geprüft. Hier kommen Penetrationseffekte nicht in Betracht; es handelt sich tatsächlich um eine Hemmung von Vorgängen, die mit der DNS-Synthese zusammenhängen.

KERSTEN: Wir können die Ergebnisse von Herrn Dr. HILZ bezüglich der Wirkung von alkylierenden Cytostatica auf freie Ribonuclease(n) auch für das Mitomycin bestätigen. Wir haben in B. subtilis und auch in AscitesTumorzellen nach Mitomycineinwirkung eine 80- bis 100%ige Zunahme der Aktivität freier RNAase(n) gefunden. Es besteht eine sehr gute Korrelation zwischen der cytostatischen Wirkung und der Steigerung der RNAase(n)-aktivität.

MAGEE (Carshalton): I would just like to add a small comment to Dr. LAWLEY's remarks on the possible importance of quantitative aspects of alkylation in carcinogenesis and to draw your attention to an experiment carried out by RIOPELLE and TASMIN in the University of Montreal [RIOPELLE, T. L., and G. TASMIN: Rev. canad. Biol. **22**, 365 (1963)], who have induced kidney tumors in the rat with dimethylnitrosamin of the type that I have demonstrated yesterday in my lecture. They found that if they fed this compound at a level of 0.5 mg twice daily to rats for seven days, they obtained a hundred per cent incidence of kidney tumors at seven month. But if they reduced the dose to 0.2 mg twice daily and continued this for one month they obtained no kidney tumors at all. This suggests that alkylation may be relevant to carcinogenesis. I did not say alkylation of what, alkylation of something, some cell component. The quantitative degree of alkylation may be extremely critical here.

GRUNICKE (Freiburg i. Br.): Dr. LAWLEY mentioned that DNA-protein crosslinks in the presence of alkylating agents are more likely to occur after the breakage of the cell.

DNA and nuclear proteins are closely associated within the cell. Homogenization in the presence of concentrated sodium chloride solutions, however, which is done in the COLTER procedure used in our studies, leads to a dissociation of DNA and protein. Therefore, I would say that the alkylating agent has a much greater chance to crosslink DNA and protein under in vivo conditions than after the disruption of the cell.

LAWLEY: The last point is well taken. I was thinking more of experiments in which cells are disrupted by phenol extraction of protein which I think would certainly liberate reactive groups by denaturation of proteins. However, you may well be right that this protein-DNA crosslinking can occur *in vivo*.

Regarding Dr. KERSTEN's question concerned with mitomycin, I should point out that crosslinks involving the N-7 atom of guanine moieties are not heat-stable; SZYBALSKI found that those formed by mitomycin are, *i.e.* they resemble nitrous acid-induced crosslinks in this respect. The nature of the reaction involved in mitomycin crosslinkage seems to me therefore to be unestablished.

With regard to the endo-DNAase I agree that this might well be concerned with the repair system. This has been observed in bacterial cells following irradiation. However, in mammalian cells the involvement of this type enzyme in a repair mechanism has yet to be established.

Fluorinated Pyrimidines, Biochemically and Clinically Useful Antimetabolites *

By CH. HEIDELBERGER

McArdle Laboratory for Cancer Research, The Medical School, University of
Wisconsin, Madison, Wisconsin, USA

With 29 Figures

This story, like all others dealing with life, begins with DNA, whose logically and aesthetically satisfying structure is well known. Since one of the most characteristic attributes, but by no means the only one, of cancer cells is their rapid growth or cell division,

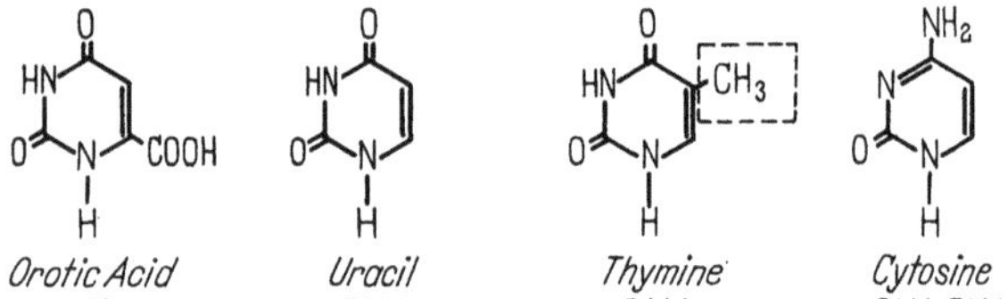

Orotic Acid Uracil Thymine Cytosine
NO RNA DNA DNA,RNA

Fig. 1. Naturally occurring pyrimidines

and since DNA must be replicated before cells divide it would seem self-evident that if one could inhibit DNA synthesis by some drug, the growth of the tumor would stop. Unfortunately for the cancer chemotherapeutists, there are also normal cells in the body, particularly in the bone-marrow and gastrointestinal tract, where the rate of cell division is as high or higher than that of most tumors. Hence drugs designed to inhibit DNA synthesis would be expected to be toxic to the bone-marrow and gastrointestinal tract, unless they were completely selective to tumor cells. It is somewhat embarrassing that at our present state of knowledge it is not yet possible to design a compound of sufficient selectivity of action against

---

* Other recent reviews on the same topic: HEIDELBERGER, C.: In Chemotherapy of Cancer, Pl. A. PLATTNER, ed., Elsevier, 1964, p. 88; HEIDELBERGER, C.: Progress in Nuclei Acid Research and Molecular Biology 4, 1 (1965). Supported in part by grant CA 7175, National Cancer Institute, National Institutes of Health, US Public Health Service.

tumors as to be non-toxic to the host, except possibly in the case of hormones.

The structures of the biologically important pyrimidines are shown in Fig. 1. Orotic acid is not found in the nucleic acids, but is ordinarily the primary precursor of the nucleic acid pyrimidines. Cytosine is found both in DNA and RNA, uracil is found only in RNA, and thymine, only in DNA. In order for thymine to be made, the methyl group must be attached to the ring of uracil at the

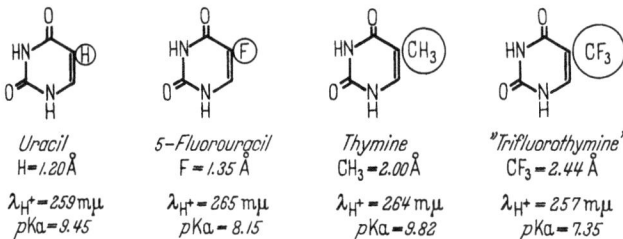

Uracil
H = 1.20 Å
λ_{H^+} = 259 mμ
pKa = 9.45

5-Fluorouracil
F = 1.35 Å
λ_{H^+} = 265 mμ
pKa = 8.15

Thymine
CH_3 = 2.00 Å
λ_{H^+} = 264 mμ
pKa = 9.82

"Trifluorothymine"
CF_3 = 2.44 Å
λ_{H^+} = 257 mμ
pKa = 7.35

Fig. 2. Structures and physical properties of pyrimidines and their fluorinated analogs

5-carbon. In 1954 RUTMAN et al. found that in a carcinogen-induced primary rat hepatoma, uracil was utilized to a significant extent as the precursor of the tumor nucleic acid pyrimidines, but not for those of the liver. This excited our curiosity, and we investigated the specificity and generality of this phenomenon in several trans-

$$FCH_2COOC_2H_5 + HCOOC_2H_5 \xrightarrow{K} KOCH = CHFCOOC_2H_5$$

Fig. 3. Synthesis of 5-fluorouracil

planted tumors of rats and mice. We compared the utilization of labeled uracil and orotic acid for DNA and RNA pyrimidine biosynthesis in the tumors and in intestinal mucosa, a rapidly dividing normal tissue, and found that in all tumors there was a considerably greater uptake of uracil than of orotic acid, which occurred to a lesser extent in the intestines (HEIDELBERGER et al. 1957b). Hence, with the idea of obtaining a potential drug with some selectivity of action against tumors, we wished to synthesize an antimetabolite that would resemble uracil as closely as possible, but with one significant change in the molecule. We decided to substitute a fluorine atom for a hydrogen atom in the uracil molecule, since such a change often produces profound biological effects (cf. the

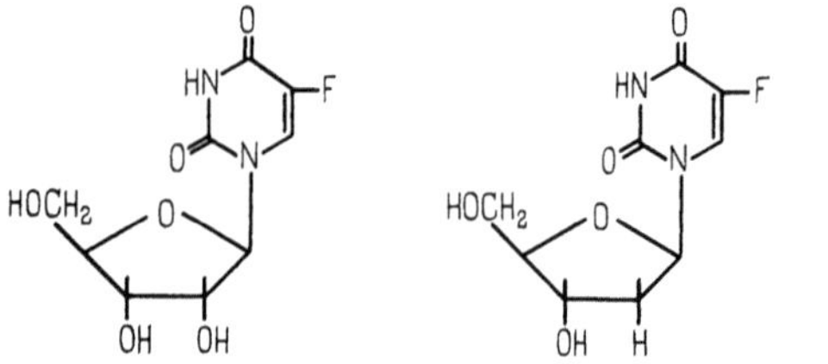

Fig. 4. Structures of fluorinated pyrimidine nucleosides

toxicity of fluoroacetate). There are only two positions in the uracil molecule where this can be done, at carbons 5 and 6. We chose the 5-position because it seemed more amenable to synthesis (in fact

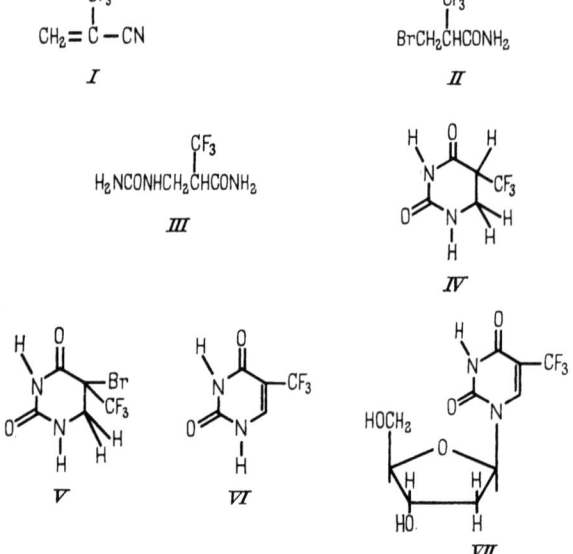

Fig. 5. Synthesis of "trifluorothymine"

6-fluorouracil was only synthesized in 1964, and turned out to be quite unstable, WEMPEN and FOX, 1964), and because the fluorine atom would be located on the same carbon to which the methyl group must be attached in order for thymine to be synthesized. Thus we anticipated, since the fluorine atom is almost the same size as the hydrogen atom, that 5-fluorouracil should be incorporated into RNA in place of uracil, and that it should block the biosynthesis of thymine. Consequently, 5-fluorouracil was prepared, and has the structure shown in Fig. 2. It was also decided to synthesize "trifluorothymine", which we expected would be incorporated into DNA in place of thymine, as are 5-bromouracil and 5-iodouracil.

The synthesis of 5-fluorouracil (FU) was accomplished according to the scheme shown in Fig. 3, which is used in the factory at the

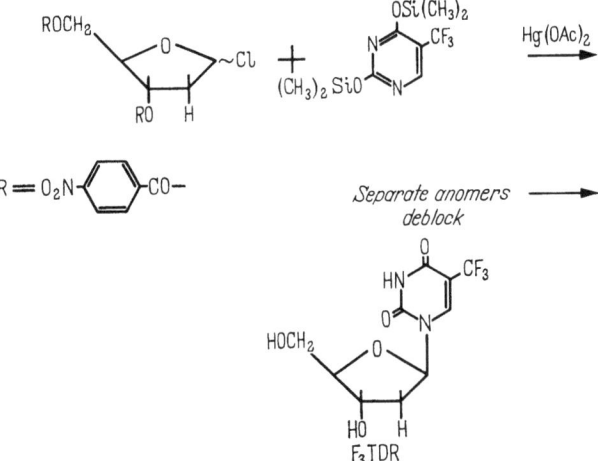

Fig. 6. Chemical synthesis of "trifluorothymidine"

present time (HEIDELBERGER et al., 1957a, DUSCHINSKY et al., 1957). In order to continue in a logical fashion the synthesis of the antimetabolites with only one structural change in the molecule, 5-fluorouridine (FUR) and 5-fluoro-2'-deoxyuridine (FUDR) were then prepared; their structures are shown in Fig. 4.

The synthesis of "trifluorothymine" (F_3T) was carried out at Wisconsin as shown in Fig. 5 (HEIDELBERGER et al., 1964b). However, since mammalian cells lack the enzyme to convert thymine into thymidine, we were not interested in F_5T, but immediately converted it into "trifluorothymidine" (F_3TDR), at first enzymatically, and later a chemical synthesis (shown in Fig. 6) was developed independently (RYAN et al, 1966).

Returning to 5-fluorouracil, it was soon demonstrated to have considerable tumor inhibitory activity against Sarcoma 180, as shown in Fig. 7, and against a wide variety of other transplanted tumors (HEIDELBERGER et al., 1958a). It was also found that FUR was more toxic, and in general less effective as a tumor inhibitor than FU, whereas FUDR was less toxic and more effective, and hence represents a chemotherapeutic improvement (HEIDELBERGER at al., 1958b). Following these findings, preclinical pharmacology was carried out, and clinical trials were initiated at the University of Wisconsin Hospitals (CURRERI et al., 1958). Much of the earlier clinical work has been reviewed (HEIDELBERGER and ANSFIELD, 1963).

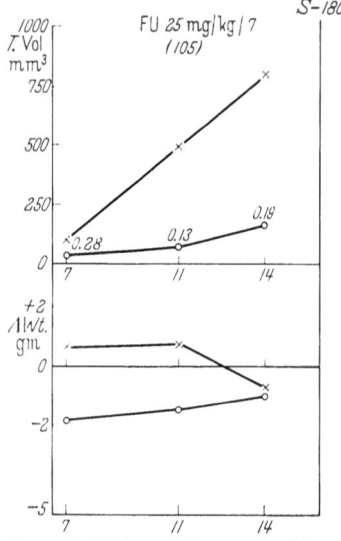

Fig. 7. Inhibition of the growth of Sarcoma-180 by 5-fluorouracil. T/C = volume of the treated tumors divided by the volume of the control tumors

Once the dose level for FU, and later for FUDR, became established in far advanced cancer patients, it became evident that

Toxicity	FU	FUDR
	811 patients (%)	128 patients (%)
Diarrhea	64	71
Stomatitis	48	42
Nausea and vomiting	30	41
Alopecia	20	2
Dermatitis	8	16
Pharyngitis and esophagitis	6	2
Epistaxis	3	0.8
Leukopenia		
Lowest wbc. between 2000—3000	22	18
Lowest wbc. under 2000	32	23

Fig. 8. Clinical toxicity manifestations of FU and FUDR

considerable toxicity was produced as shown in Fig. 8. As anticipated for compounds that would be expected to inhibit DNA biosynthesis, the toxicity is primarily to the bone-marrow and gastrointestinal tract. Nevertheless, with skill and experience, the toxicity can be managed in patients that are not in the terminal stages of the disease.

For various reasons, the initial clinical trials at Wisconsin were carried out in patients suffering from solid tumors, rather than

1. Measurable decrease in size of lesion.
2. Subjective improvement.
3. Leveling off or reversal in downward weight curve.
4. Improvement in performance status, resumes work, increased physical activity.
5. All of above must be maintained at least two months.

Fig. 9. Criteria of clinical improvement on solid tumors

those with leukemias that are easier to evaluate. Therefore, it was necessary to define criteria of objective response, as shown in Fig. 9. It was soon found that the highest percentage of objective responses was obtained with carcinomas of the breast and gastrointestinal

Study	Rapid Intravenous Injection			
	Objective Response	Gastrointestinal Cancer	Female Genital Cancer	Breast Cancer
Total cases reviewed	630/2375 27%	308/1041 30%	48/164 29%	151/472 32%
ANSFIELD et al. J. Amer. med. Ass. 181, 295 (1962)	91/428 21%	22/133 17%	10/46 22%	52/158 33%
Total cases minus ANSFIELD et al.	539/1947 28%	286/908 32%	38/118 32%	99/314 31%

Fig. 10. Clinical response of 3 tumors to 5-fluorouracil

tract. At first other clinicians were unable to confirm these positive results of the Wisconsin group, who under the most charitable circumstances were considered to be over-enthusiastic about the drug. However, with experience others obtained comparable results. That the Wisconsin group's experience (ANSFIELD et al., 1962) was actually reported in a conservative fashion is shown in Fig. 10, which demonstrates that the percentage of objective responses reported by others is higher than that reported by ANSFIELD et al., (1962) (HEIDELBERGER and ANSFIELD, 1963). Although a number of patients have survived for up to 7 years, none are cured, all have residual disease, and must be treated with monthly courses of FU therapy to prevent reactivation of their disease. Moreover, the eventual onset of drug-resistance is tragically inevitable.

Since temporary objective regressions of the tumors of patients are produced at the expense of some toxicity, it is legitimate to enquire whether patients actually derive any clinical benefit from this treatment. This was examined in a series of patients with disseminated breast carcinoma at the University of Wisconsin Hospitals, and the results are given in Fig. 11. The controls were

Group	Survival time	No. patients	No. surviving 48 months or more
Controls	13	127	2
5-FU	24	144	15
Non-Responders	20	86	4
Responders	29	58	11

Fig. 11. Mean survivals of patients with disseminated carcinoma of the breast

patients that were given all conventional forms of therapy, surgical, irradiation, ablative, and hormonal, and their mean survivals from the time of diagnosis of dissemination until death was 13 months, with only 2 surviving more than 48 months. There were compared with the first 144 patients to receive FU chemotherapy in addition to the modalities described for the controls. Their mean survival was 24 months with 12 living more than 48 months, which indicates that on the average a year of life was gained as a consequence of the chemotherapy (SAMP and ANSFIELD, 1966 in press). Furthermore, there was an increase of survival time even in those patients treated with FU in whom no objective response could be documented. Thus it would appear that patients with advanced breast carcinomas do derive significant clinical benefit from FU chemotherapy. Recently a new dosage schedule for FU treatment has been reported, which produces less toxicity but equally good clinical responses (ANSFIELD, 1964). It has also been shown in patients with inoperable bronchogenic carcinoma that a significant prolongation of survival can be obtained with a suitable combination of fluorouracil and radiotherapy (GOLLIN et al., 1962). Continuous intraarterial administration of FU to patients with inoperable carcinomas of the head and neck has produced favorable results (JOHNSON et al., 1965).

Some clinical pharmacological studies were carried out in cancer patients with FUDR-2-C^{14}, and it was found, as shown in Fig. 12,

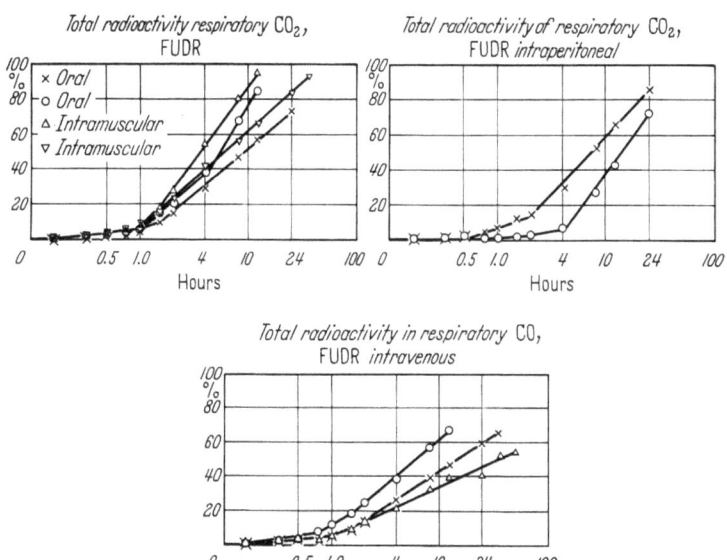

Fig. 12. Total excretion of radioactivity in the respiratory carbon dioxide in human cancer patients given FUDR-2-C^{14} by different routes of administration

that there was a rapid excretion of radioactivity in the respiratory carbon dioxide, amounting to about 90% of the administered dose by the oral route and 70% intravenously (MUKHERJEE et al., 1963a). This confirmed the rapid breakdown found earlier in mice, in which the degradative pathway of fluorouracil catabolism was shown to go from FU, dihydro-FU, α-fluoro-β-ureidopropionic acid, α-fluoro-β-guanidopropionic acid, urea, carbon dioxide and ammonia, and α-fluoro-β-alanine (CHAUDHURI et al., 1959; MUKHERJEE and HEIDELBERGER, 1960). The breakdown of FUDR to FU is carried out by nucleoside phosphorylase, a ubiquitous enzyme that diminishes the therapeutic effectiveness of FUDR (BIRNIE et al., 1963). This enzyme is inhibited by deoxyglucosylthymine (LANGEN and ETZOLD 1963).

Turning now to biochemistry, we first examined the incorporation of labeled FU into the nucleic acids of the Ehrlich ascites carcinoma. No radioactivity was found in the purified DNA, but as shown in Fig. 13, after alkaline hydrolysis of the RNA and ion-exchange chromatography, all the radioactivity was found in a single peak that was identified as 2′, 3′-fluorouridylic acid, proving that there

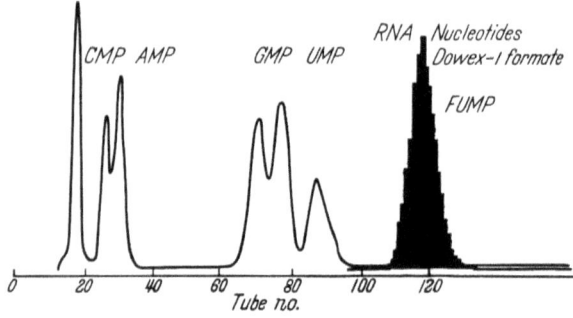

Fig. 13. Incorporation of FU-2-C^{14} into RNA of the Ehrlich ascites carcinoma. The graph shows the column chromatography of the alkaline hydrolysis of the RNA. The open lines denote the optical density at 260 mμ; the solid peak represents radioactivity

was incorporation into the cellular RNA (CHAUDHURI et al., 1958). The incorporation of FU into RNA was confirmed in other systems: in *E. coli* (HOROWITZ and CHARGAFF, 1959), in tobacco mosaic virus (GORDON and STAEHELIN, 1959), and in poliovirus (MUNYON and SALZMAN, 1962).

It is interesting to enquire about the consequences of the incorporation of fluorouracil into RNA. Although the coat proteins of poliovirus, (MUNYON and SALZMAN, 1962) and of tobacco mosaic virus (HOULABEK, 1962; SUTIC and DJORDJEVIC, 1962) containing up to 50% of the uracil replaced by FU, are normal in amino acid composition and immunochemical properties, nevertheless mutations have been observed in poliovirus (COOPER, 1964) and tobacco mosaic viruses (KRAMER et al., 1964) grown in the presence of FU. In coding experiments, it was found that poly-FU leads to the formation only of polyphenylalanine, and with less ambiguity than that produced by poly-U (WAHBA et al., 1963; GRUNBERG-MANAGO and MICHELSON, 1964).

5-Fluorouracil is also a powerful inhibitor of the induction of β-galactosidase in E. coli (HOROWITZ et al., 1960), and it was shown that a serologically related, but enzymatically incompetent, protein was produced (BUSSARD et al., 1960). Further studies demonstrated very clearly that these effects were a consequence of the incorporation of FU into messenger RNA (NAKADA and MAGASANIK, 1964).

As a result of an attempt to understand these diverse findings, a hypothesis was put forth (HEIDELBERGER, 1965) that the incorporation of FU in place of U in messenger RNA leads to no errors

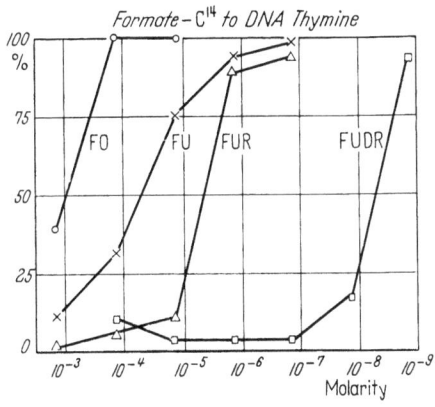

Fig. 14. Effects of fluorinated pyrimidines on the incorporation of formate-C^{14} into DNA thymine in suspensions of Ehrlich ascites carcinoma cells

in the process of translation into the amino acid sequence of proteins. However, due to the increased acidity of FU as compared to U, it was considered that FU might occasionally act as if it were cytosine, and base-pair with guanine during transcription or viral replication. This would have the consequence of having FU incorporated into mRNA in place of C, and read as if it were U, which can explain all the above facts. An attempt was made in our laboratory to demonstrate this, using a synthetic DNA of alternating sequence, purified RNA polymerase, α-P^{32}-FUTP and UTP, and carrying out nearest neighbor analyses. However, with the available purity of the enzyme it was not possible to show that errors of transcription occurred more frequently with FUTP than with UTP during mRNA synthesis (BUJARD and HEIDELBERGER, 1966).

The prediction that fluorouracil should prevent the attachment of the methyl group to the ring of uracil and hence inhibit thymine synthesis was tested in the experiment shown in Fig. 14. Here, Ehrlich ascites carcinoma cells were incubated with labeled formate, a specific precursor of the methyl group of thymine, in the presence of various concentrations of fluorinated pyrimidines. The DNA thymine was isolated chromatographically and its specific activity was determined. It is evident that FU does, in fact, inhibit this reaction, and that FUDR is a very effective inhibitor of this

Analog 5-Substituted	Input Multiplicity	Conc. mM.	Fr_s r-Frequency Spontaneous $\times 10^{-4}$	Fr_1 r-Frequency Induced $\times 10^{-3}$	Fr_1/Fr_s	Titer Increase		Ratio
						Control	Treated	
BrU	0.08	2.78	2.5	0.84	3	2220	2160	1
F_3T	0.08	2.78	2.5	44.9	18	2220	4	564
F_3T	1.3	2.78	3.6	1.9	5	95	1.7	56
BrUDR	0.08	2.78	2.5	13.1	52	2220	388	6
F_3TDR	0.08	2.78	2.5	14.9	60	2220	635	4
F_3TDR	0.81	2.7	2.6	8.1	31	273	5	55

Fig. 15. The induction of mutations at the r_{11} locus of bacteriophage T4 by halogenated pyrimidines and their nucleosides

metabolic pathway (BOSCH et al., 1958). It was further shown that this DNA inhibition, which will be considered in considerable detail below, is responsible for the inhibition of tumor growth, rather than the incorporation of FU into RNA (HEIDELBERGER et al., 1960). Furthermore, it was shown cytologically that Ehrlich ascites cells treated with FU *in vivo* had the normal amount of DNA but increased amounts of RNA and protein per cell, which is an indication of unbalanced growth or "thymineless death" (LINDNER, 1959).

I will now consider the newer analogs, "trifluorothymine" (F_3T) and "trifluorothymidine" (F_3TDR). When only small quantities were available in the laboratory it was felt that the best indication of possible incorporation of these compounds would be mutagenic activity in bacteriophage. Accordingly, they were tested in bacteriophage T_4 under comparable conditions to bromouracil (BrU) and bromodeoxyuridine (BrUDR), that are known to be mutagenic and incorporated into DNA in this phage (BENZER and FREESE, 1958). It can be seen from Fig. 15 that F_3T is more mutagenic than BrU, and that F_3TDR has about the same mutagenic activity as

BrUDR (GOTTSCHLING and HEIDELBERGER, 1963). Further proof of the incorporation into DNA came from the experiment in which the DNA was isolated from purified phages that had been grown in the presence of labeled F_3TDR and was subjected to cesium

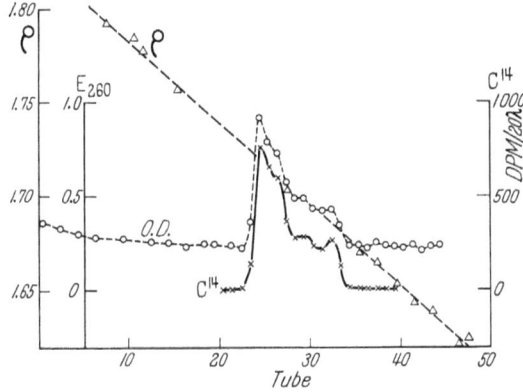

Fig. 16. Cesium chloride density gradient centrifugation of bacteriophage T4 DNA isolated after growth in a medium containing F_3TDR-2-C^{14}. (The two smaller peaks arose from inadvertant mixing during the collection of drops)

chloride density gradient centrifugation. The correspondence between radioactivity and optical density (Fig. 16) demonstrate the incorporation of the analog into DNA; in this experiment the extent of replacement of thymine was 11 percent (GOTTSCHLING and HEIDELBERGER, 1963). It had been found during the chemical

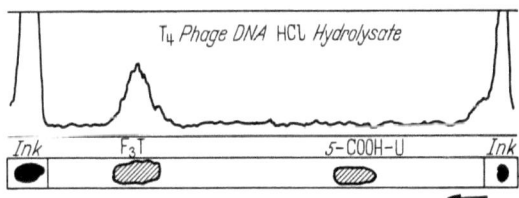

Fig. 17. Acid hydrolysis and paper chromatography of bacteriophage T4 DNA containing labeled F_3T

syntheses that the CF_3 group in both analogs was very easily hydrolyzed under mild alkaline conditions to a carboxyl group (HEIDELBERGER et al., 1964). Therefore, it became necessary to determine whether the compound incorporated into DNA retained the trifluoromethyl group. That this was so was demonstrated in the experiment shown in Fig. 17, in which the only radioactive product of acid hydrolysis of the phage DNA was F_3T (GOTTSCHLING and HEIDELBERGER, 1963). Since the proton on nitrogen-3 in F_3T and F_3TDR is two units more acidic than the corresponding proton of thymine and thymidine (Fig. 3), this change might be expected to alter the physical properties of the DNA. That this is so was shown in the experiment diagrammed in Fig. 18, where the melting temperature of the DNA containing 11% F_3T is lower than the normal bacteriophage T_4 DNA (GOTTSCHLING and HEIDELBERGER, 1963). Whether this can be attributed primarily to an alteration in hydrogen-bonding or stacking interactions is not known at present.

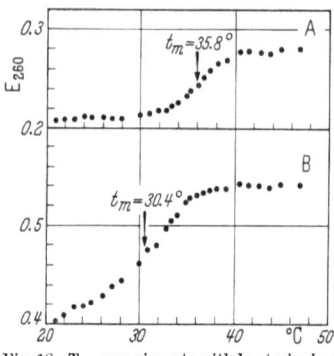

Fig. 18. T_m experiments with bacteriophage T_4DNA: A control, B containing F_3T

Another consequence of the incorporation of F_3TDR into the DNA of mammalian cells grown in culture is the sensitization to the lethal effects of x-rays, as shown in Fig. 19, in which the replacement of thymine was 5% (SZYBALSKI et al., 1963). It was found in the same study, that the toxicity of F_3TDR to these cells is reversed by thymidine.

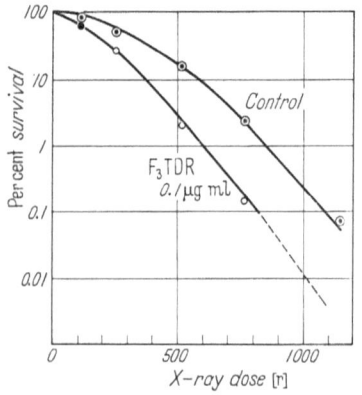

Fig. 19. Sensitization of human bone-marrow cells in culture to the lethal effects of X-radiation as a consequence of incorporation of F_3TDR into the DNA

When F_3T and F_3TDR were tested for their effects on the incorporation of labeled formate into DNA thymine in the Ehrlich ascites carcinoma cell system, considerable inhibition was obtained as shown in Fig. 20 (HEIDELBERGER et al., 1964a). In fact, F_3TDR is almost as effective as FUDR in this regard.

It is now necessary to consider in some detail the enzyme, thymidylate synthetase, which catalyzes the biosynthesis of thymidylate as shown in Fig. 21. This is a methylenetetrahydrofolate

Fluorinated Pyrimidines, Biochemically and Clinically Antimetabolites 169

coenzyme-catalyzed reaction as was shown by the elegant research of FRIEDKIN and his colleagus (cf. FRIEDKIN and ROBERTS, 1956). It was found in phage-infected *E. coli* that the nucleotide of FUDR, FUDRP, was a powerful inhibitor of this enzyme (COHEN et al., 1958). The kinetics of this inhibition was investigated in extracts of Ehrlich ascites carcinoma cells and was found to be competitive, using an isotopic method of assay (HARTMANN and HEIDELBERGER, 1961). With the availability of an optical method of assay of this enzyme (WAHBA and FRIEDKIN, 1961), an intensive study of the mechanism of Ehrlich ascites thymidylate synthetase was carried out (REYES and HEIDELBERGER, 1965). It was found by kinetic analysis that the sequence of events takes place as indicated in Fig. 22. The coenzyme attaches first to the enzyme, followed by the substrate, the product is released first, and then the dihydrofolate.

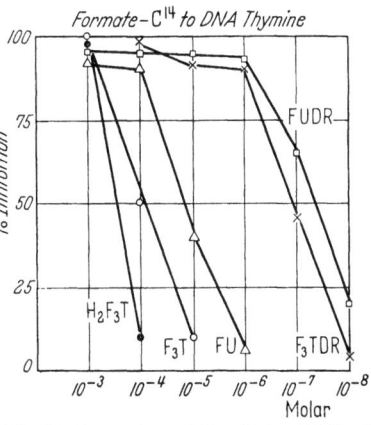

Fig. 20. Comparison of the effects of F_3T and F_3TDR and FU and FUDR on the incorporation of formate-C^{14} into DNA thymine in suspensions of Ehrlich ascites carcinoma cells

When the kinetics of the inhibition of thymidylate synthetase by FUDRP was reinvestigated, the inhibition was again found to be competitive whether or not preincubation with the inhibitor was carried out, as shown in Fig. 23 (REYES and HEIDELBERGER, 1965). Similarly, the inhibition by F_3TDRP was competitive without preincubation. However, when the enzyme was preincubated with F_3TDRP (Fig. 24) the kinetics changed to non-competitive (REYES and HEIDELBERGER, 1965). This suggested that as a result of preincubation F_3TDRP reacted irreversibly with the enzyme, and this was demonstrated by the dialysis experiment shown in Fig. 25. The alkali-lability of the CF_3 group, suggests that the compound may acylate a free amino group at the active site of the enzyme, and this is currently under investigation.

A summary of the biochemistry of the two series of compounds is given in Fig. 26. In the fluorouracil series the ring is degraded, the compound is converted to ribonucleotides and incorporated into RNA and not DNA, FUDRP is produced, which inhibits thymidylate synthetase powerfully and competitively. In the F_3T series, the

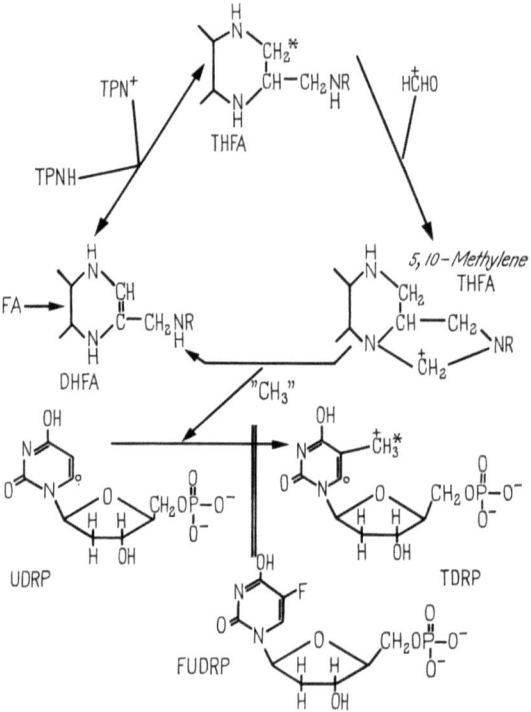

Fig. 21. Thymidylate synthetase scheme

ring is not degraded, the only catabolic product is 5-carboxyuracil (HEIDELBERGER et al., 1965), it is converted into deoxyribonucleo-

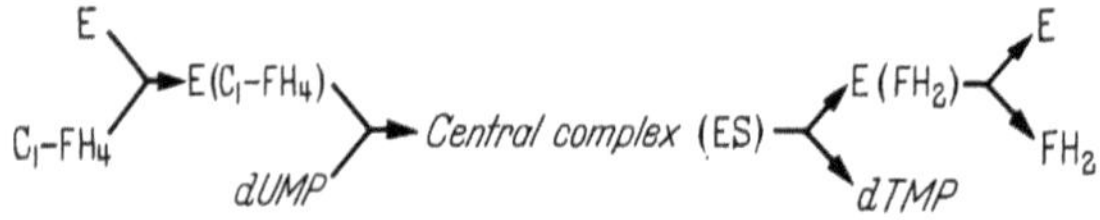

$$C_1\text{-}FH_4 \xrightarrow{E} E(C_1\text{-}FH_4) \xrightarrow{dUMP} \textit{Central complex } (ES) \xrightarrow{E(FH_2)} \xrightarrow{dTMP} \begin{matrix} E \\ FH_2 \end{matrix}$$

Fig. 22. The kinetic mechanisms of action of thymidylate synthetase

side phosphates and is incorporated into DNA and not RNA; F_3TDRP also inhibits thymidylate synthetase, but by a different mechanism than FUDRP. The possible inhibition of other enzymes by F_3TDR is under investigation.

The tumor-inhibitory properties of F_3TDR have been investigated. Against Sarcoma-180 and the Ehrlich ascites carcinoma

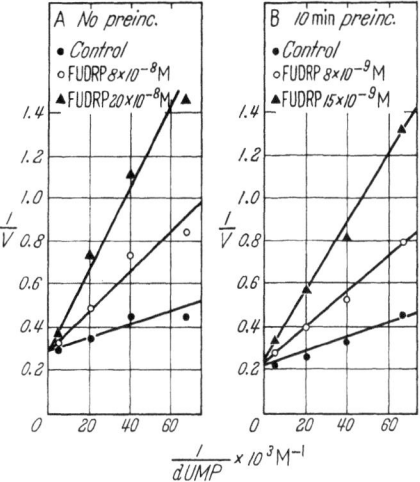

Fig. 23. Lineweaver-Burk plot of the kinetics of the inhibition of thymidylate synthetase by FUDRP, with and without preincubation

F_3TDR is less effective than FUDR; the two analogs show equal activity against L-1201 leukemia. However, against Adenocarci-

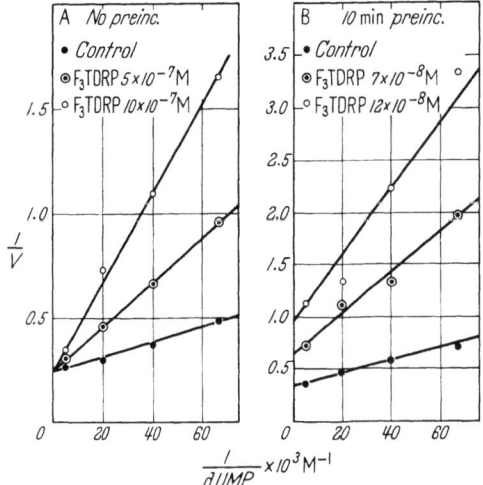

Fig. 24. Lineweaver-Burk plot of the kinetics of the inhibition of thymidylate synthetase by F_3TDRP, with and without preincubation

noma 755 F_3TDR has a better therapeutic index than FUDR, as shown in Fig. 27 (HEIDELBERGER and ANDERSON, 1964). For this reason and its biochemical interests, a large amount of F_3TDR

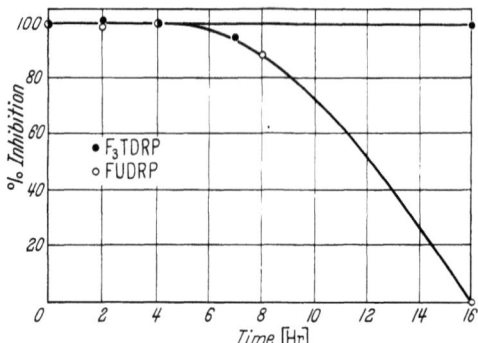

Fig. 25. Effect of dialysis on the inhibition of thymidylate synthetase after preincubation with FUDRP and F_3TDRP

has been prepared for clinical trial as a tumor-inhibitory compound. No results are as yet available.

Because of the clinical activity of iododeoxyuridine (IUDR) in treating herpes simplex infections of the eye, it was decided to test

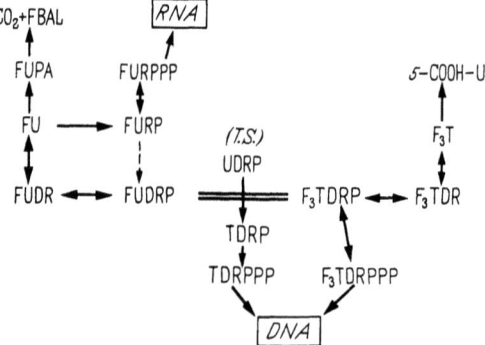

Fig. 26. Summary of the metabolic and biochemical effects of the fluorouracil and trifluorothymine series

F_3TDR by local application to rabbit's eyes infected with herpes simplex virus. As can be seen from Fig. 28, F_3TDR is 10 to 100 times as effective in this test system as IUDR and cytosine arabinoside. Furthermore, it is effective against an IUDR-resistant strain of the virus (KAUFMAN and HEIDELBERGER, 1964). We are currently investigating the mechanism of the antiviral activity of F_3TDR.

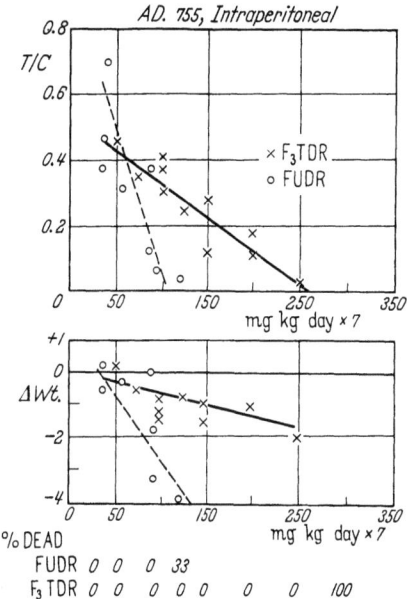

Fig. 27. Comparison of the inhibition of Adenocarcinoma 755 by FUDR and F_3TDR. For the definition of T/C, see Fig. 7

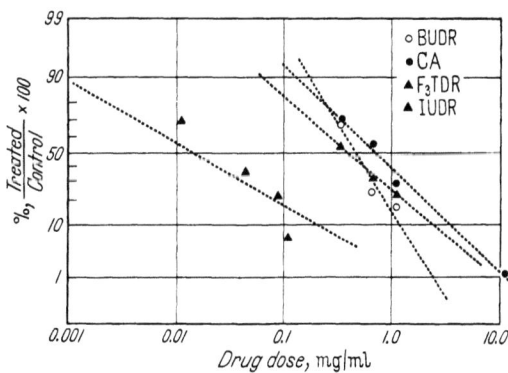

Fig. 28. The inhibition of herpes simplex keratitis in the rabbit's eye by F_3TDR, BUDR, IUDR, and cytosine arabinoside

It is evident that the present failure to achieve cures of cancer (except with choriocarcinoma of the female and possibly Burkitt's lymphoma) by means of chemotherapy is a consequence of the fact that we presently have no drugs available that are completely selective against tumors and not rapidly dividing normal tissues. Therefore, it seemed worthwhile to determine whether there was any selective of action of FUDR against carcinomas of the colon, where it is sometimes effective, as compared with normal intestinal mucosa. Accordingly, labeled FUDR was administered to patients prior to surgery, and samples were obtained from the carcinomas

Patient	Tissue	Time (hours)	Percent of sample in			
			UREA	FU and FUDR	FUPA	Nucleotides
Mrs. L. S.	Carcinoma of Colon	2.0	39	0	29	16
,,	Normal Intestinal Mucosa	,,	46	16	34	6.4
Mr. E. H.	Carcinoma of Colon	2.17	34	43	2.7	16.5
,,	Normal Intestinal Mucosa	,,	0	42	53	3.8
Mrs. H. S.	Carcinoma of Colon	6.0	36	7.2	52	10.7
,,	Normal Intestinal Mucosa	,,	14	14	34	5.9

Fig. 29. Distribution of radioactivity in tissues of human cancer patients following intravenous injection of $FUDR-2-C^{14}$, 30 mg/kg

of the colon and the normal intestinal mucosa. Although the total amount of radioactivity in the two tissues varied, when the radioactivity was fractionated the results shown in Fig. 29 were obtained. In every case, there was considerably more radioactivity present in the active, nucleotide form of the drug in the tumor than in the normal tissue, and less was present as degradation products. This, I believe, constitutes a demonstration of some selectivity of action against the tumor in the human cancer patient, and represents some cause for cautious optimism about the possibility of designing more selective compounds that would, hopefully, be more effective in the treatment of human cancer. We are currently attempting to do so.

References

ANSFIELD, F. J.: J. Amer. med. Ass. **190**, 686 (1964).
—, J. M. SCHROEDER, and A. R. CURRERI: J. Amer. med. Ass. **181**, 295 (1962).
BENZER, S., and E. FREESE: Proc. nat. Acad. Sci. (Wash.) **44**, 112 (1958).
BIRNIE, G. D., H. KROEGER, and C. HEIDELBERGER: Biochemistry **2**, 566 (1963).
BOSCH, L., E. HARBERS, and C. HEIDELBERGER: Cancer Res. **18**, 335 (1958).
BUJARD, H., and C. HEIDELBERGER: Biochemistry. In press 1966.

BUSSARD, A., S. NAONO, F. GROS, and J. MONOD: C. R. Acad. Sci. (Paris) **250**, 4049 (1960).
CHAUDHURI, N. K., B. J. MONTAG, and C. HEIDELBERGER: Cancer Res. **18**, 318 (1958).
—, K. L. MUKHERJEE, and C. HEIDELBERGER: Biochem. Pharmacol. **1**, 328 (1959).
COHEN, S. S., J. G. FLAKS, H. D. BARNER, M. R. LOEB, and J. LICHTENSTEIN: Proc. nat. Acad. Sci. (Wash.) **44**, 1004 (1958).
COOPER, P. D.: Virology **22**, 186 (1964).
CURRERI, A. R., F. J. ANSFIELD, F. A. McIVER, H. A. WAISMAN, and C. HEIDELBERGER: Cancer Res. **18**, 478 (1958).
DUSCHINSKY, R., E. PLEVEN, and C. HEIDELBERGER: J. Amer. chem. Soc. **79**, 4559 (1957).
FRIEDKIN, M., and D. ROBERTS: J. biol. Chem. **220**, 653 (1956).
GOLLIN, F. F., F. J. ANSFIELD, A. R. CURRERI, C. HEIDELBERGER, and H. VERMUND: Cancer **15**, 1209 (1962).
GORDON, M. P., and M. STAEHELIN: Biochim. biophys. Acta (Amst.) **36**, 351 (1959).
GOTTSCHLING, H., and C. HEIDELBERGER: J. molec. Biol. **7**, 541 (1963).
GRUNBERG-MANAGO, M., and A. M. MICHELSON: Biochim. biophys. Acta (Amst.) **87**, 593 (1964).
HARTMANN, K.-U., and C. HEIDELBERGER: J. biol. Chem. **236**, 3006 (1961).
HEIDELBERGER, C.: Progr. Nucleic Acid Res. and Mol. Biol. **4**, 1 (1965).
—, and S. W. ANDERSON: Cancer Res. **24**, 1979 (1964).
—, and F. J. ANSFIELD: Cancer Res. **23**, 1226 (1963).
—, N. K. CHAUDHURI, P. B. DANNEBERG, D. MOOREN, L. GRIESBACH, R. DUSCHINSKY, R. J. SCHNITZER, E. PLEVEN, and J. SCHEINER: Nature (Lond.) **179**, 663 (1957a).
—, K. C. LEIBMAN, E. HARBERS, and P. M. BHARGAVA: Cancer Res. **17**, 399 (1957b).
—, L. GRIESBACH, B. J. MONTAG, D. MOOREN, O. CRUZ, R. J. SCHNITZER, and E. GRUNBERG: Cancer Res. **18**, 305 (1958a).
— —, O. CRUZ, R. J. SCHNITZER, and E. GRUNBERG: Proc. Soc. exp. Biol. (N. Y.) **97**, 470 (1958b).
—, G. KALDOR, K. L. MUKHERJEE, and P. B. DANNEBERG: Cancer Res. **20**, 897 (1960).
—, J. BOOHAR, and G. D. BIRNIE: Biochim. biophys. Acta (Amst.) **91**, 636 (1964a).
—, D. G. PARSONS, and D. C. REMY: J. med. Chem. **7**, 1 (1964b).
—, J. BOOHAR, and B. KAMPSCHROER: Cancer Res. **25**, 377 (1965).
HOROWITZ, J., and E. CHARGAFF: Nature (Lond.) **184**, 1213 (1959).
—, J. J. SAUKKONEN, and E. CHARGAFF: J. biol. Chem. **235**, 3266 (1960).
HOULABEK, V.: J. molec. Biol. **6**, 164 (1962).
JOHNSON, R. O., W. A. KISKEN, and A. R. CURRERI: Surg. Gynec. Obstet. **120**, 530 (1965).
KAUFMAN, H. E., and C. HEIDELBERGER: Science **145**, 585 (1964).
KRAMER, G., H. G. WITTMANN, and H. SCHUSTER: Z. Naturforsch. **19b**, 46 (1964).

LANGEN, P., u. G. ETZOLD: Biochem. Z. **339**, 190 (1963).
LINDNER, A.: Cancer Res. **19**, 189 (1959).
MUKHERJEE, K. L., and C. HEIDELBERGER: J. biol. Chem. **235**, 433 (1960).
—, J. BOOHAR, D. WENTLAND, F. J. ANSFIELD, and C. HEIDELBERGER: Cancer Res. **23**, 49 (1963a).
—, A. R. CURRERI, M. JAVID, and C. HEIDELBERGER: Cancer Res. **23**, 67 (1963b).
MUNYON, W., and N. P. SALZMAN: Virology **18**, 95 (1962).
NAKADA, D., and B. MAGASANIK: J. molec. Biol. **8**, 105 (1964).
REYES, P., and C. HEIDELBERGER: Mol. Pharmacol. **1**, 14 (1965).
RUTMAN, R. J., A. CANTAROW, and K. E. PASCHKIS: Cancer Res. **14**, 119 (1954).
RYAN, K. J., E. M. ACTON, and L. GOODMAN: J. org. Chem. **31**, 1181 (1966).
SAMP, R. J., and F. J. ANSFIELD: 1966. J. Amer. med. Ass., In press.
SUTIC, D., and B. DJORDJEVIC: Nature (Lond.) **203**, 434 (1964).
SZYBALSKI, W., N. K. COHN, and C. HEIDELBERGER: Fed. Proc. **22**, 532 (1963).
WAHBA, A. J., and M. FRIEDKIN: J. biol. Chem. **236**, PC 11 (1961).
—, R. S. GARDNER, C. BASILIO, R. S. MILLER, J. F. SPEYER, and P. LENGYEL: Proc. nat. Acad. Sci. (Wash.) **49**, 116 (1963).
WEMPEN, I., and J. J. FOX: J. med. Chem. **7**, 207 (1964).

Thymidylat-Synthese in Leukämiezellen unter der Einwirkung von Folsäure-Antagonisten

Von W. Wilmanns

Medizinische Universitäts-Klinik, Tübingen

Uracil-Antagonisten, deren Wirkungsmechanismus und therapeutische Bedeutung von HEIDELBERGER gezeigt wurde, und Folsäure-Antagonisten hemmen die de novo-Synthese von Thyminmethylgruppen. Von den Folsäure-Antagonisten wissen wir, daß sie zwar in extrem niedrigen Konzentrationen (10^{-9} bis 10^{-8} M) die FH_2-Reduktase hemmen, daß aber maligne Zellen resistent werden können und normale Zellen vor cytotoxischen Nebenwirkungen geschützt werden können, dadurch, daß sekundär die Neubildung von FH_2-Reduktase induziert wird. Ob eine therapeutische Methotrexatdosis für die Zelle lethal ist oder ob eine Resistenz sich entwickelt, dürfte davon abhängen, ob innerhalb eines Zellteilungscyclus eine Neuinduktion der zunächst gehemmten FH_2-Reduktase stattfinden kann. Je niedriger die Aktivität dieses Enzyms in den Zellen ist, um so geringere Methotrexatdosen sind zur Erzielung eines vollständigen Hemmeffektes erforderlich. Die Bestimmung der FH_2-Reduktase, der Geschwindigkeit der Enzymneuinduktion und der intracellulären Methotrexatkonzentration begünstigen in Einzelfällen eine gezielte Leukämiebehandlung mit Folsäure-Antagonisten auch bei Erwachsenen. Dabei ist die gleichzeitige Bestimmung der Thymidinkinase von Bedeutung, da durch Aktivitätsanstieg dieses Enzyms wahrscheinlich eine Kompensation der toxischen Wirkungen von Hemmstoffen der Thyminmethylgruppensynthese — in diesem Fall also von Uracil- und Folsäureantimetaboliten — möglich ist. Diese Aktivitätssteigerung läßt sich durch verminderte Bildung von Thymidintriphosphat (TTP) erklären, denn TTP hemmt durch Rückkoppelung die Thymidinkinase. Untersuchungen der Enzymaktivitäten von FH_2-Reduktase und Thymidinkinase in den Knochenmarkzellen bei einer Patientin, die aus suicidaler Absicht 150 bis 230 mg Methotrexat einnahm, zeigten folgende enzymatische Regulationsmechanismen bei der

Kompensation des Hemmeffektes der Folsäure-Antagonisten auf die DNS-Synthese.

1. Anstieg der Thymidinkinase im Stadium der FH_2-Reduktaseblockierung, hervorgerufen durch Fortfall eines hemmenden Rückkoppelungsmechanismus von TTP auf die Thymidinkinase.

2. Neubildung von FH_2-Reduktaseenzymprotein, für die als Ursache ein Induktionsmechanismus anzunehmen ist.

Auf Leukämiezellen übertragen bedeuten diese Befunde, daß hohe Thymidinkinaseaktivitäten die Resistenz gegenüber einer Behandlung mit Folsäureantagonisten begünstigen. Da akute Leukämiezellen häufig eine deutlich erniedrigte Thymidinkinaseaktivität im Vergleich zu normalen Leukocytenvorstufen haben, erhebt sich die Frage, ob es sinnvoll ist, bei einer Leukämiebehandlung mit Folsäure-Antagonisten oder bei der Tumorbehandlung mit Uracil-Antagonisten die normale Zellreifung durch Thymidininfusionen vor cytotoxischen Nebenwirkungen zu schützen und ob dadurch in Einzelfällen eine größere selektive Wirksamkeit der genannten Antimetaboliten zu erreichen ist.

Hemmstoffe des Abbaus von Fluordesoxyuridin und Joddesoxyuridin

Von P. LANGEN

*Institut für Biochemie der Deutschen Akademie der Wissenschaften
zu Berlin, Berlin-Buch*

Die Anwendbarkeit des Antimetaboliten Fluordesoxyuridin wird ebenso wie diejenige des Joddesoxyuridin durch den raschen Abbau im Körper beeinträchtigt. Dieser Abbau wird durch zwei verschiedene Pyrimidinnucleosid-Phosphorylasen eingeleitet, welche die Nucleoside zu stärker toxischen (Fluoruracil) oder inaktiven (Joduracil) Produkten aufspalten: 1. die Thymidin-Phosphorylase (E. C. 2. 4. 2. 4.), die in Tumoren und normalen Geweben des Menschen, in Pferdeleber und in verschiedenen Normalgeweben von Ratte und Maus vorkommt und 2. die Uridin-Desoxyuridin-Phosphorylase (E. C. 2. 4. 2. 3.), die bei Rind, Schwein, Hund und Katze und in allen bsiher untersuchten transplantablen Tumoren von Ratte und Maus vorkommt.

Wir haben daher nach Verbindungen gesucht, die eine Hemmung dieser Enzyme und damit möglicherweise eine Verbesserung der therapeutischen Anwendbarkeit von Fluordesoxyuridin und Joddesoxyuridin bewirken. Dabei erwies sich die Uridin-Desoxyuridin-Phosphorylase durch unnatürliche Nucleoside vom Typ des 2'-Desoxyglucopyranosyl-thymins und 2'-Desoxyxylopyranosyl-thymins als stark hemmbar. Die Hemmwirkung ist an die Pyranosestruktur des Zuckers zusammen mit einer im Vergleich zu den natürlichen Nucleosiden umgekehrten Konfiguration der OH-Gruppe am C-3' der Zuckerkette gebunden, da das pyranoide Isomere des Thymidins und 2'-Desoxyxylofuranosyl-thymin keine oder nur eine geringe Hemmwirkung besitzen. Zum Nachweis einer in-vivo-Wirksamkeit haben wir den Einfluß von 2'-Desoxyglucopyranosyl-thymin auf den Einbau von ^{131}J-Joddesoxyuridin in die DNS bei Katzen untersucht. Bei einer Anwendung von 11 mg/kg Joddesoxyuridin ließ sich durch die 2fach molare Menge an Desoxyglucopyranosyl-thymin eine 3- bis 7fache Förderung des Einbaus von Joddesoxyuridin in die DNS von Darm, Knochenmark und Milz erreichen. Das zeigt, daß sich mit dieser Verbindung der gewünschte Effekt im Falle des Joddesoxyuridins auch in vivo durchaus erreichen läßt.

Die Thymidin-Phosphorylase, deren Hemmung besonders wichtig ist, da sie beim Menschen für die Spaltung der oben genannten Carcinostatica verantwortlich ist, wird durch diese unnatürlichen Nucleoside nicht beeinflußt. Bei den Versuchen zur Hemmung dieses Enzyms gingen wir von der Tatsache aus, daß es einer Produkthemmung durch die bei der Reaktion entstehenden Pyrimidinbasen unterliegt. Durch Einführung einer NH_2-Gruppe am C-6 des Thymins, 5-Jod- oder Bromuracils gelang es uns, die Hemmwirkung bis zum 9fachen zu verstärken. Die Einführung einer Hydracinogruppe am C-6 des Thymins an Stelle der NH_2-Gruppe ergibt eine Abschwächung der Wirkung im Vergleich zum 6-Aminothymin, jedoch ist das 6-Hydracino-thymin immer noch zweimal wirksamer als Thymin. In vivo ergab sich in Versuchen mit beiden Verbindungen an Mäusen eine Steigerung des Joddesoxyuridineinbaus in die DNS um 50%. Da diese Verbindungen ebenso starke in-vitro-Hemmstoffe der Thymidin-Phosphorylase sind, wie die oben genannten unnatürlichen Nucleoside in bezug auf die Uridindesoxyuridin-Phosphorylase, ist die Ursache für diese im Vergleich zu der entsprechenden Wirkung des 2'-Desoxyglucopyranosyl-thymins bei Katzen nur sehr geringe Steigerung noch unklar.

Diskussion

HEIDELBERGER: I think the work of Dr. LANGEN ist very important. Possibly it has some practical advantages in enhancing the therapeutic activity of some of these nucleosides. Just to give credit again to people in this room, the original work we did on the enzyme nucleoside phosphorylase, was carried out by Dr. HANS KRÖGER.

HOLLDORF (Freiburg i. Br.): Dr. HEIDELBERGER, I have several comments on your lecture. First, I think, it is an open problem how de novo synthesis and precursor utilization (salvage pathway) in pyrimidine nucleotide biosynthesis are correlated. Hitherto it is impossible to say how many nucleotides are formed via de novo synthesis and how many preformed nucleosides or bases are utilized for nucleotide synthesis. For the evalution of drug action on either of these two pathways one should know the regulation mechanisms which coordinate both processes. My second comment concerns the inhibition of formate incorporation into thymidine nucleotides of the DNA in ascites cells. You demonstrated the inhibition of this incorporation by fluorosubstituted analogues. We have studied the influence of the free base thymine and of the nucleoside thymidine on the incorporation of formate. The inhibition curves for these compounds are very similar to the curves which you obtained with the analogues. The effective concentrations are almost the same. The third point on which I would like to make a comment is the role of thymidylate synthetase in the cell. It is generally accepted that the thymidine nucleotides, which are formed by this enzyme, have a key function in DNA synthesis, but one should not neglect their importance in the control of other metabolic processes. E.g. the nucleoside diphosphate reductases, which convert ribonucleoside diphosphates to deoxyribonucleoside diphosphates, are repressed by thymidine nucleotides. In studies with bacteria we were able to show, that a decreased synthesis of thymidine nucleotides by a reduced activity of thymidylate synthetase in a mutant strain is still sufficient for normal DNA synthesis and growth, but the level of thymidine nucleotides in these cells is not sufficient for the repression of the reductase systems.

HEIDELBERGER: Well, I think these points are all very interesting. I certainly agree that it will be good to have quantitative data on all these matters. It is quite clear, however, that in mammalian cells the preformed pathway going through thymidylic acid is considerably more important than the salvage pathway starting from thymidine, because in these cells there is very little thymidine present. Dr. POTTER has done very careful studies on this. About the only source of thymidine would be in the diet, which is quite low, or possibly, when tumors invade other tissues and break them down they may get some thymidine for this pathway. Everyone knows that the pyrimidine nucleotide biosynthesis is a terribly complicated business. Dr. POTTER once made a three-dimensional graph of all the interrelationships of all the uridine and cytidine nucleotides and it is really an enormously complicated situation, and certainly these things interact. Dr. POTTER perhaps more than anyone else has studied these quantitative interactions. However, there is good evidence that the inhibition of thymidylate synthetase is killing cells, and these cells undergo thymineless death. If one does a look at HeLa-cells, which are inhibited by FUDR or F_3TDR, they die; the cells become larger in size, the DNA content per cell stays the same, but the RNA and protein content of the cells increases and this has also been shown spectrophotometrically in ascites cells treated in vivo with fluorouracil. These cells become enormously larger, the DNA is the same, unbalanced growth takes place, RNA and protein synthesis continue and the cells are "kaputt". So I certainly agree that all of these things require further studies.

SACHSENMAIER (Heidelberg): Dr. HEIDELBERGER, I have a question concerning the biochemistry of the riboside of 5-fluorouracil (FUR). As you know we have used this substance as a tool to study the role of RNA synthesis in the control of mitosis in a synchronous slime mold, Physarum polycephalum. We found that FUR was a strong inhibitor of mitosis when administered during particular stages of the division cycle. This inhibitory effect was reversible with uridine but not with thymidine. The effect on mitosis was paralleled by a pronounced reduction of the rate of RNA synthesis. What do you think is the mechanism of the effect of FUR on the rate of RNA synthesis?

HEIDELBERGER: I am afraid I really have no good ideas on that. We have not worked as much with FUR in our own laboratory as we should have, primarily because it is so disappointing from the chemotherapeutic point of view. One reason for the extreme toxicity of FUR in ascites cells is that Dr. HARBERS showed that it is taken up into cells against a concentration gradient. In ascites cells there is a concentrative uptake of ribonucleosides, which is not found with the free bases and with the deoxyribonucleosides. However, in all our experiments with FUR and FUDR, at least in ascites cells, in order to get any inhibition of RNA-synthesis either in vivo or in vitro, it was necessary to take the dose higher than the lethal dose. I just cannot say what the explanation is. There are many possibilities of feedback interactions and of similar things.

WEIL (Strasbourg): I just want to comment on a point of your lecture, Dr. HEIDELBERGER: You said that the translation process in protein biosynthesis should not be altered by 5-fluorouracil, because 5-fluorouracil has the same coding properties as uracil. We have obtained soluble-RNA (s-RNA's) from yeast, where up to 50% of the uracil is replaced by 5-fluorouracil (when growing the yeast on a culture medium containing this analogue). We have seen that these s-RNA have altered biological properties: their accepting capacity (the attachment of the amino acids to their corresponding s-RNA) is either increased or decreased according to the amino acid studied, and their transferring capacity (the transfer of the amino acids from the aminoacyl-s-RNA to the polypeptide chains) is also altered. Therefore I think that the action of 5-fluorouracil should not be considered only as the result of an interference of the analogue with DNA biosynthesis. In fact the action of 5-fluorouracil may also be due to its incorporation into messenger-RNA resulting in the production of inactive enzymes (as shown particularly by F. GROS), and perhaps also to its incorporation into s-RNA causing alterations of protein biosynthesis (as 5-fluorouracil-containing s-RNA have accepting and transferring properties which are different from those of normal s-RNA).

HEIDELBERGER (zu WEIL): Well, I put my slides together before you told me about your results earlier this morning. I certainly did not mean to give the impression that the only effect of fluorouracil in biology is the inhibition of thymidylate synthetase. The point that I wanted to make is that in tumor inhibition, as far as we know, the main effect is due to thymidylate synthetase. Now, actually the phenotypic reversions of BENZER can also be explained better perhaps on the basis of interference in the translation rather than the transcription level. From what I understood from you this morning, these changes particularly in activating enzymes and transfer activity with various amino acids in the fluorouracil containing s-RNA were quantitative rather than qualitative.

HOLLDORF: Herr WILMANNS nimmt an, daß die Thymidinkinase durch Thymidintriphosphat allosterisch gehemmt wird. Ist anzunehmen, daß diese Hemmung dauernd besteht oder ist eine Abhängigkeit vom Zellteilungscyclus anzunehmen? Zumindest bei Mikroorganismen sind allosterische Effekte sehr rasch und in der Regel nur kurz wirksame Vorgänge. Zum anderen ist fraglich, ob in der Zelle die für die Hemmung erforderlichen Konzentrationen an Thymidintriphosphat erreicht werden können. Dieses Nucleotid kommt in so geringer Konzentration vor, daß es noch nie jemand in der Zelle genau bestimmt hat. Die Konzentrationen liegen unter 10^{-8} bis 10^{-9} M.

WILMANNS (Tübingen): Leider ist es bisher nicht möglich, die Thymidintriphosphatkonzentration in den Zellen zu bestimmen. Thymidintriphosphat wird, wie ich erwähnte, über die Thyminmethylgruppensynthese (de novo-Synthese) gebildet und häuft sich bis kurz vor der präsynthetischen Phase, in der sich die Polymerase vermehrt, an. Ich halte es wohl für möglich, daß eine Anhäufung von Thymidintriphosphat bei vermehrter de novo-Synthese, wie sie in Leukämiezellen vorhanden ist, eine Feedback-Hemmung auf die Thymidinkinase ausüben kann. Diese Annahme erscheint um so mehr berechtigt, als bei Hemmung der FH_2-Reduktase durch Methotrexat die Aktivität der Thymidinkinase ansteigt und wieder auf niedrige Werte abfällt nach Abklingen des Methotrexathemmeffektes auf die FH_2-Reduktase.

Onkogene Viren

Von A. GRAFFI

Institut für Krebsforschung der Deutschen Akademie der Wissenschaften zu Berlin, Experimenteller Bereich, Berlin-Buch

Mit 33 Abbildungen

Ziel meiner Ausführungen soll in erster Linie sein, durch einen mit groben Strichen gezeichneten Überblick über das Gesamtgebiet der onkogenen Viren und der virusbedingten Geschwülste das Terrain für die nachfolgenden, mehr in das Detail gehenden Referate vorzubereiten. Dabei erscheint es mir notwendig, vor allem die gesicherten Fakten dieser Thematik möglichst in ihrer ganzen Vielfalt kurz darzulegen, was mit sich bringt, daß auf viele, schon seit längerer Zeit bekannte Tatsachen zurückgegriffen werden muß.

I. Historische Entwicklung des Problems Virus und Krebs[1—16]

Nachdem bereits im 17. Jahrhundert von SENNERT in Prag und LUSITANUS in Lissabon die Kontagiosität des Krebses erwogen wurde und in der ersten Blütezeit der Bakteriologie verschiedene Ärzte bestimmte Bakterien, Hefen und Pilze als Krebsursache ansahen, wurde 1903 von BORREL[2] erstmalig die Hypothese einer möglichen Virusätiologie maligner Tumoren klar umrissen. Es existierte um diese Zeit auch bereits ein Beweis für die subcelluläre Übertragbarkeit einer zumindest geschwulstähnlichen Erkrankung, nämlich der Myxomatose des Kaninchens durch SANARELLI[12] in Montevideo (1898). Als Meilensteine der weiteren Entwicklung sind vor allem die Entdeckungen von ELLERMANN und BANG[4] (1908), ROUS[11] (1911), SHOPE[13,14] (1932), BITTNER[1] (1936) und GROSS[5,6,7] (1951) anzusehen. Der Beweis, daß Viren echte Geschwülste hervorrufen können, hat zur allgemeinen Virustheorie der malignen Tumoren einschließlich der des Menschen geführt (GYE[8], OBERLING[9,10], STANLEY[15], ZILBER[16]), die auch gegenwärtig hohe Aktualität besitzt.

II. Die onkogenen Viren und die durch sie erzeugten Tumoren[1,4,5,11,13,14,17—32]

In der Tab. 1 sind die vom Standpunkt der experimentellen Forschung, speziell im Hinblick auf die Frage des Mechanismus der Cancerogenese wichtigsten Formen der onkogenen Viren und der durch sie erzeugten Tumorarten wiedergegeben, wobei auf die Unvollständigkeit dieser Aufzählung besonders hingewiesen werden

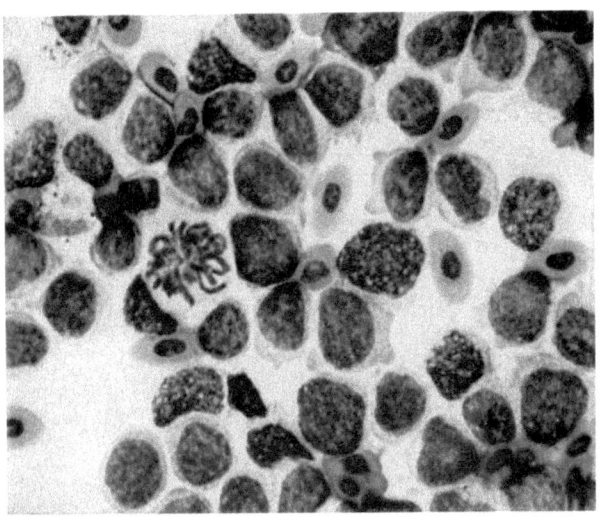

Abb. 1. Blutausstrich eines an Myeloblastose erkrankten 20 Tage alten Kükens, das im Alter von 3 Tagen mit dem Virus inoculiert wurde (nach THORELL, 1958)

muß. Es fehlen vor allem eine große Zahl von Viren, die bei verschiedenen Säugetieren und dem Menschen Warzen, Papillome und Fibrome und bei Kaltblütern (Amphibien und Fischen) verschiedenartige Tumoren auslösen, die jedoch zur Zeit weniger im Blickpunkt der Forschung liegen. Erst kürzlich wurde von HARVEY über ein Agens berichtet, das bei Mäusen Sarkome hervorruft und mit dem Polyomavirus sicher *nicht* identisch ist. Von MOLONEY wurde ein Virus nachgewiesen, das ebenfalls bei Mäusen Rhabdomyosarkome hervorruft (Leukämie-Sympos. Philadelphia, Okt. 1965). Insgesamt umfaßt die Palette der Viren, die gut- oder bösartige Geschwülste auszulösen vermögen, über 35 verschiedene Typen.

Die in der Tab. 1 angeführten onkogenen Viren wurden in drei Gruppen, nämlich in RNS- sowie kleine und große DNS-Viren auf-

Tabelle 1. *Onkogene Viren und Virustumoren*

Virus	Natürliches Vorkommen	Onkogene Wirkung auf	Tumortyp	Autor	Literaturzitate
			I. RNS-Viren		
Rous-Virus (RSV)	Huhn	Huhn Ente Fasan Taube Ratte Goldhamster Kaninchen Maus Affe Mensch (in vitro)	verschiedene Sarkome	Rous 1911	11, 16, 17, 19, 20, 21, 29, 30, 35, 35b
Hühnerleukosevirus	Huhn	Huhn	Lymphomatose Myeloblastose Erythroblastose Nierentumoren	Ellermann und Bang 1908	4, 45, 47, 22, 19, 24a
Milchfaktor Mäuseleukämieviren	Maus Maus	Maus Maus Ratte (Hamster)	Mammacarcinome lymphatische myeloische reticulumcelluläre und erythroblastische Leukämien	Bittner 1936 Gross 1951	1, 18, 83 5, 6, 33 25, 26, 32 24, 34, 36

II. Kleine DNS-Viren

Polyomavirus	Maus	Maus Ratte Goldhamster Kaninchen Meerschweinchen Frettchen	Parotistumore diverse Sarkome Fibrome Lipome Carcinome in verschiedenen Organen	GROSS 1953 STEWART und EDDY 1957	5, 7, 37, 19, 26
SV 40-Virus (Simianvirus)	Rhesusaffe	Goldhamster Mastomys	Sarkome Ependymome	SWEET und HILLEMAN 1960 EDDY 1961	23a
Shope-Papillom	Kaninchen Cottontail	Kaninchen Cottontail	Papillome (Carcinome der Haut)	SHOPE 1933	14
Warzenvirus	Mensch	Mensch	Warzen		
Adenovirus (Typ 7, 12, 18, 31)	Mensch	Goldhamster Maus Ratte Mastomys	Sarkome Lungentumoren Lymphosarkome	TRENTIN 1962 HUEBNER et al. 1962	38, 31

III. Große DNS-Viren

Shope-Fibromvirus	Kaninchen Cottontail	Kaninchen Cottontail Hauskaninchen	Fibrome (Sarkome)	SHOPE 1932	13
Fibrom- und Schleimhautpapillomviren	Hund Pferd Rind Mensch	Hund Pferd Rind Mensch	Fibrome und Papillome der Haut und Schleimhaut		siehe [5]
Molluscum Contagiosum			benigne Hauttumoren		siehe [5]
Nierencarcinomvirus	Frosch	Frosch Salamander	Nierencarcinome Knochentumoren	LUCKÉ 1934	31a

geteilt, eine Klassifizierung, die sich in der letzten Zeit vielfach eingebürgert hat, da mit dieser Einteilung auch viele gemeinsame biologische Eigenschaften speziell hinsichtlich der Wirkungsweise umfaßt werden.

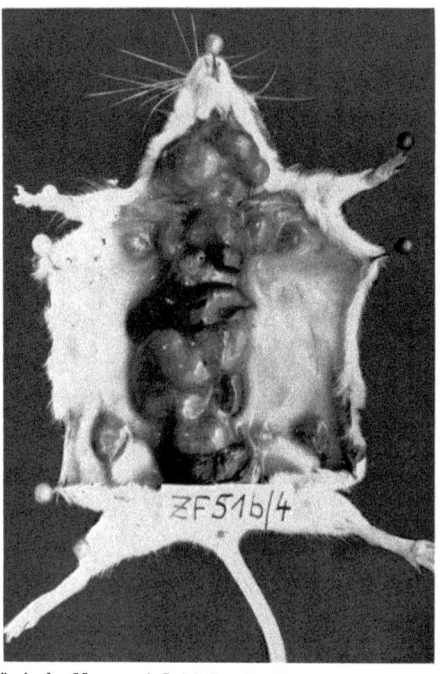

Abb. 2. Chloroleukämie der Maus nach Injektion des Virus der myeloischen Mäuseleukämie. Starke Vergrößerung und dunkelgrüne Verfärbung aller Lymphknoten, vor allem am Hals und in der Achselgegend (nach GRAFFI, FEY, BIELKA und SCHARSACH, 1954)

Zunächst ist aus dieser Tabelle die große Vielfalt virusbedingter Tumoren in bezug auf organmäßige Herkunft sowie in histologischer Beziehung erkennbar (Abb. 1 bis 12). Das Spektrum dieser Tumoren, das gut- und bösartige Geschwülste, Leukosen, Sarkome und Carcinome der verschiedensten Organsysteme und Reifegrade umfaßt, ist fast so breit wie bei den durch chemische Substanzen (Kohlenwasserstoffe, Nitrosamine) ausgelösten Geschwülsten. Diese Vielfalt der Tumorformen wird einmal durch die Vielzahl der Virusarten bedingt, andererseits aber auch dadurch, daß dasselbe Virus (z. B. das Polyomavirus) ein breites Spektrum sehr verschiedener Tumoren sowohl in histologischer Hinsicht als auch bezüglich der

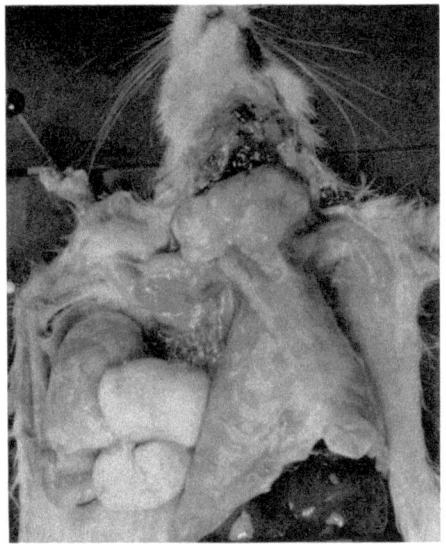

Abb. 3. Leukämie und Rethotelsarkom (rechte Achselhöhle) bei einer Ratte nach Applikation des Virus der myeloischen Leukämie der Maus an neugeborene Tiere (nach GRAFFI und GIMMY)

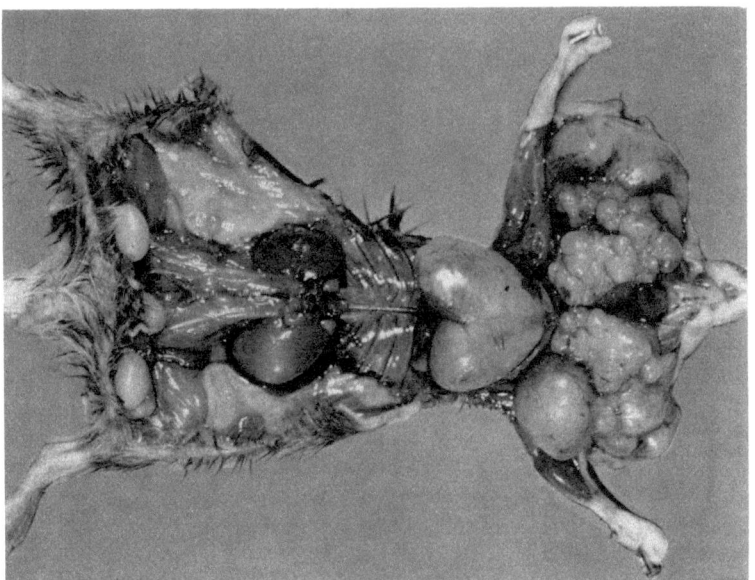

Abb. 4. Makroskopisches Bild einer Maus mit Speicheldrüsentumoren, Thymomen und Nierensarkomen (Original von STEWART)

organmäßigen Herkunft auszulösen vermag. Eine ähnliche Vielfalt in der Wirkung der onkogenen Viren finden wir bei vielen Typen auch hinsichtlich der Species. So wirkt beispielsweise das Rous-Virus[16,17,35,36b] nicht nur bei verschiedenen Vogelarten, sondern

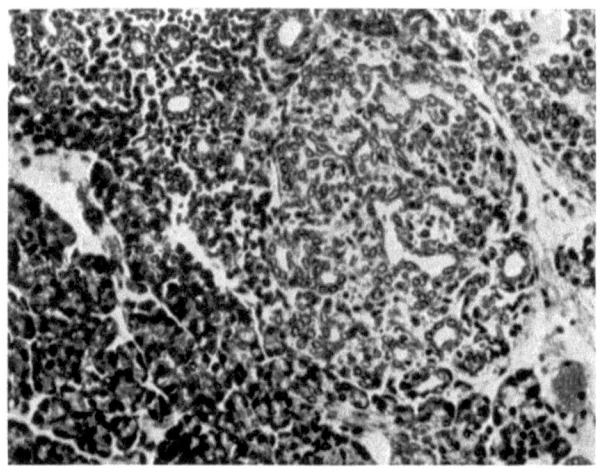

Abb. 5. Histologisches Bild eines Parotisdrüsencarcinoms der Maus durch Applikation von Parotistumor-Polyomavirus (nach Untersuchungen von GROSS 1953)

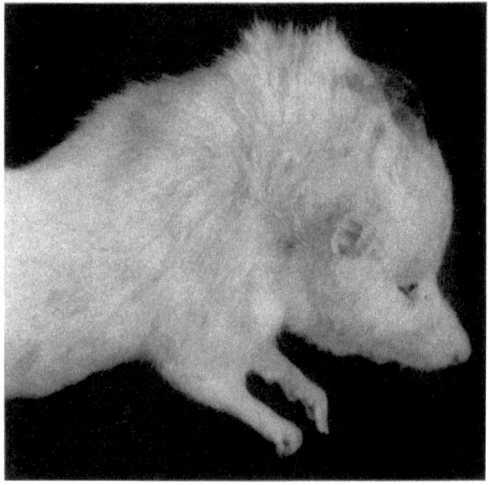

Abb. 6. Riesiger Konglomerattumor aus lipomatösen, sarkomatösen und osteosarkomatösen Anteilen auf dem Schädeldach und in der Nackenpartie einer jungen Ratte nach subcutaner Applikation von Polyomavirus (Unterstamm BB-T2) (nach GRAFFI und GIMMY)

nach neueren Ergebnissen auch auf ein breites Spektrum von Säugetieren (Maus, Ratte, Kaninchen usw.) und vermag in vitro sogar menschliche Zellen in bösartige umzuwandeln. Die onkogene Wir-

Abb. 7. Tumoren am Schädeldach einer Ratte (links zwei Osteosarkome, rechts ein Lipom) nach Applikation von Polyomavirus aus der Gewebekultur nach Beimpfung derselben mit infektiöser DNS aus Polyoma-infizierten Mäuseembryonalkulturen (nach GRAFFI und FRITZ, 1959)

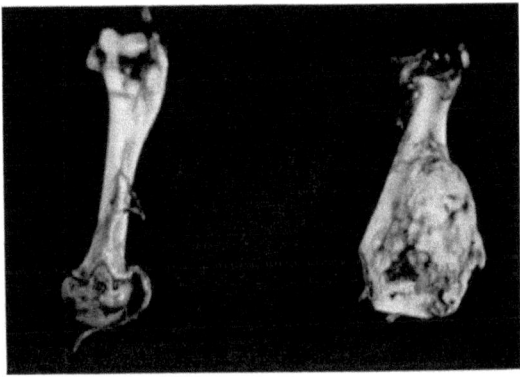

Abb. 8. Durch Polyomavirus (Unterstamm BB-T2) bei der Ratte im Bereich des Humerus (rechte Bildhälfte) induziertes Knochensarkom, links zum Vergleich normaler Humerus (nach GRAFFI und GIMMY, 1959)

kung des Polyomavirus erstreckt sich mit etwas unterschiedlichem Tumorspektrum auf sämtliche in der experimentellen Forschung benutzten Nagetiere, von der Maus bis zum Meerschweinchen[37].

In der letzten Zeit wurden Viren entdeckt, für die eine onkogene Wirkung bei den Herkunftsspecies noch nicht sicher nachgewiesen werden konnte, während sie bei der künstlichen Übertragung auf andere Tierarten eine starke geschwulstauslösende Wirkung zeigten.

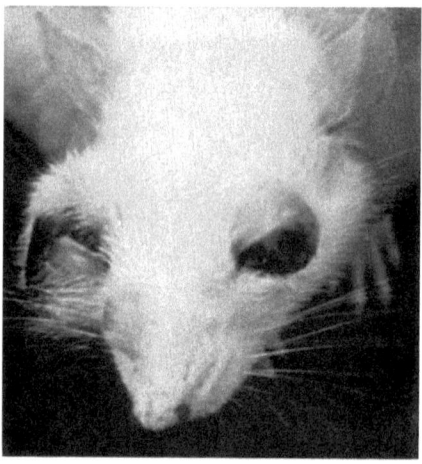

Abb. 9. Doppelseitige intra- und retrobulbäre Sarkome und Knochentumoren bei einer Ratte nach subcutaner Applikation des Polyomavirus (Unterstamm BB-T2) im neugeborenen Zustand unter die Rückhaut (nach GRAFFI und GIMMY)

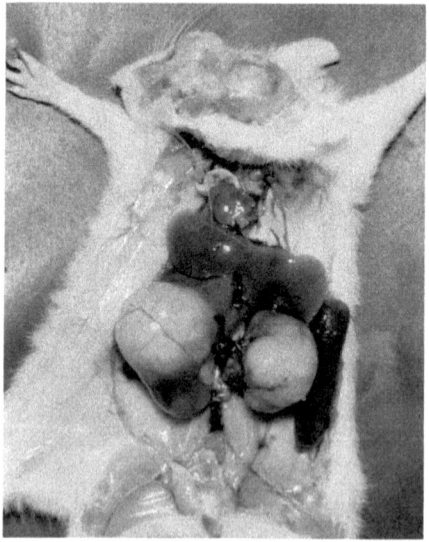

Abb. 10. Doppelseitige Nierensarkome und multiple Knochentumoren am Schädeldach bei einer Ratte nach Injektion von Polyomavirus (Unterstamm BB-T2) (nach GRAFFI, GIMMY, BAUMBACH und SCHRAMM)

Onkogene Viren 193

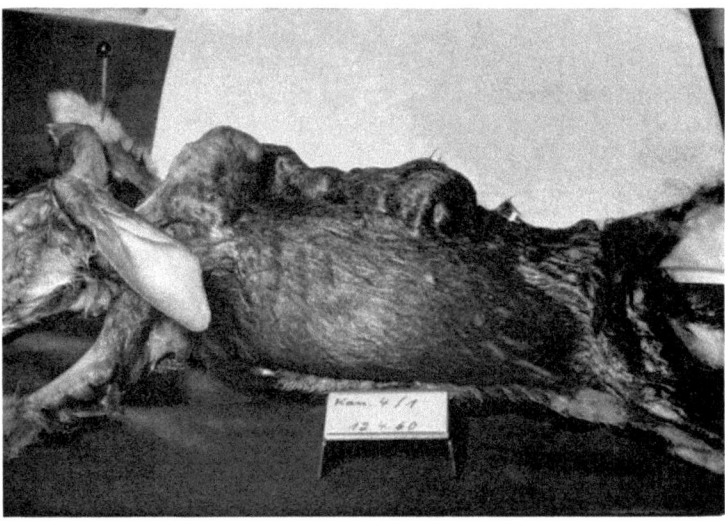

Abb. 11. Subcutane Sarkome, Fibrome und Lipome der Rückengegend eines Kaninchens nach Infektion mit Polyomavirus (Unterstamm BB-T2). Latenzzeit 54 Tage (nach GRAFFI, GIMMY und BAUMBACH, 1960)

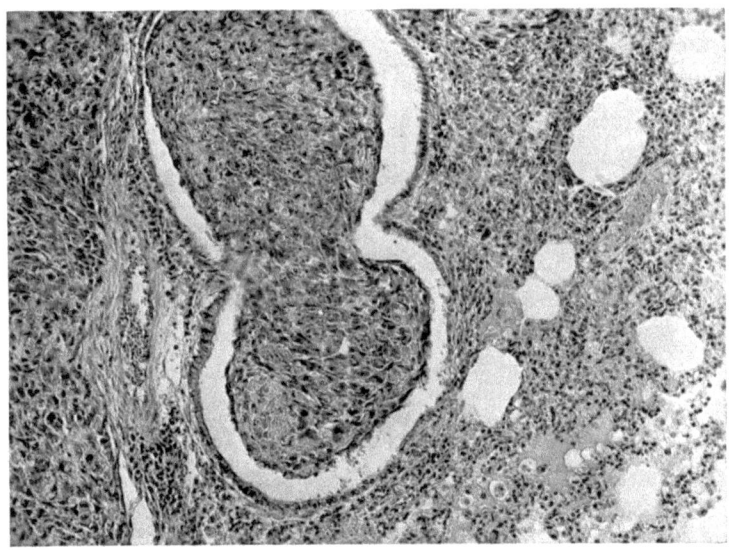

Abb. 12. Sarkommetastase in der Lunge (kleiner Bronchus) eines Kaninchens von einem subcutanen Sarkom nach Applikation von Polyomavirus (Unterstamm BB-T2) ausgehend (nach GRAFFI, GIMMY und BAUMBACH)

17. Mosbacher Colloquium

Dies gilt z. B. für verschiedene vom Menschen sich ableitende Adenoviren (z. B. Typ 12 und 18), die bei dem Hamster und anderen Nagern in vivo Tumoren hervorrufen[38,31] sowie für SV40, das vom Rhesusaffen abstammt und speziell bei Goldhamstern onkogen wirksam ist[23a]. Bei diesen Viren ist also die onkogene Wirkung erst durch den Kunstgriff des Experiments sichtbar geworden, und es ergibt sich die Frage, in welchem Ausmaß Gleiches nicht auch für weitere Viren gilt, die bei ihrem natürlichen Wirt ganz andere Erkrankungen auslösen oder vielleicht als latente Infektion vorkommen, ohne pathologische Erscheinungen zu verursachen.

III. Ultrastruktur und Bildungsweise der onkogenen Viren auf Grund elektronenmikroskopischer Untersuchungen[39—74]

Mit Hilfe der verschiedenen Methoden der Elektronenmikroskopie (Dünnschnittechnik, Negativkontrastierung, Histochemie im elektronenmikro-

Tabelle 2. *Ultrastruktur der onkogenen Viren*

Virusart	NS-Typ	Capsel-symmetrie	Envelope	Größe in mµ	Intracell. Bildungsort	*Typen-signatur
Polyoma		kubisch	nackt	--45	Zellkern	
SV 40		,,	,,	45	,,	
Kan.-Papillom	Papova	,,	,,	33	,,	
Shope Warzenvirus	DNS	,,	,,	50—60	,,	DCN
Adeno Typ 7						
,, 12		,,	,,	70	,,	
,, 18						
,, 31						
Pox-Viren: Kan.-Fibrom Shope Molluscum Contagiosum	DNS (RNS)	helikal	2—4 Hüllen	250	Cytoplasma (Viroplasmen)	DHE
Rous-Virus und Hühnerleukoseviren		helikal ?	1—2 Hüllen	70—110	Cytoplasma-Zellmembran	
Milchfaktor	RNS	helikal ?	1—2 Hüllen	A-Partik. 50—70 B-Partik. 100	,,	RHE
Mäuseleukämieviren		helikal ?	2 Hüllen	etwa 100	,,	

* s. Cold Spring Harbor Sympos. Quant. Biol. XXVII (1962)

skopischen Bereich usw.) ist es möglich, über die Ultrastruktur der Viren und die verschiedenen Anteile des Virions (Nucleoid, Capsomerenkapsel, äußere Hüllmembranen usw.) detaillierte Informationen zu erzielen. An Hand

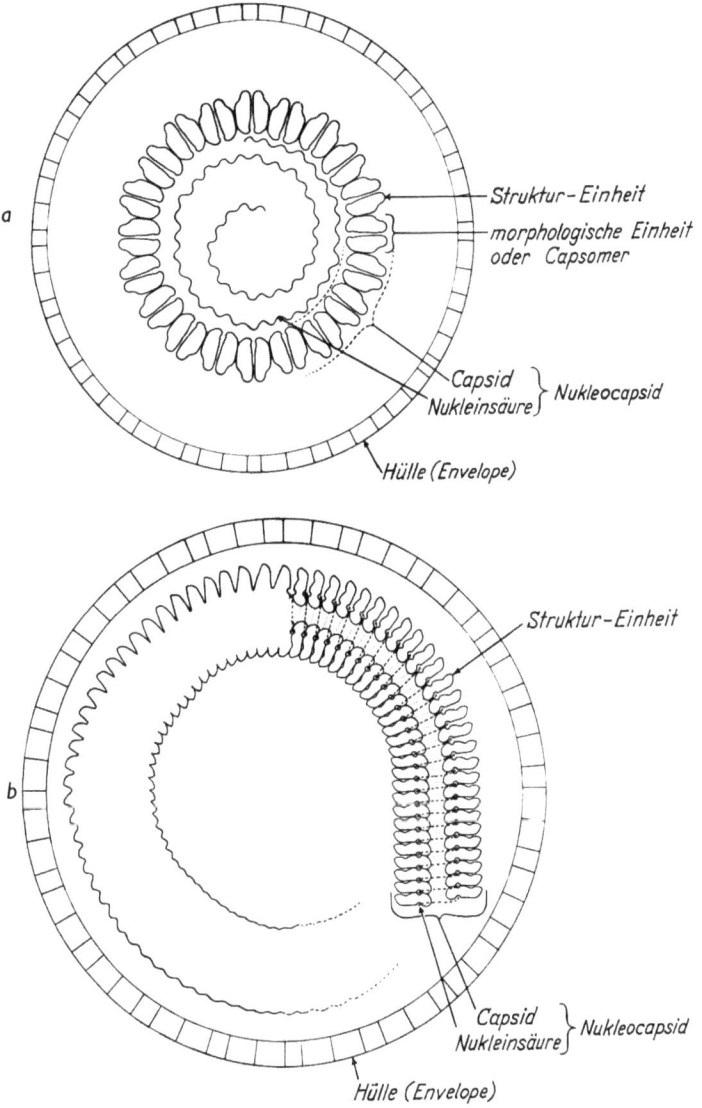

Abb. 13. Schematische Darstellung von kubischen (a) und helikalen (b) Viren. CASPAR et al. Cold Spring Harbor Symp. (1962)

von Unterschieden des strukturellen Aufbaus des Virions unter Einbeziehung der unterschiedlichen Natur der Nucleinsäure des Virusgenoms (Nucleoid) als Differenzierungsmerkmal wurde 1962 von LWOFF, HORNE und TOURNIER ein Einteilungssystem für Viren im allgemeinen entworfen, das auch der nachfolgenden Klassifizierung der onkogenen Viren zugrunde gelegt werden soll (Tab. 2). Nach dieser Einteilung werden sowohl bei der Gruppe der RNS- (R) als auch DNS-Viren (D) entsprechend der Capsomerenanordnung und Kapselsymmetrie (s. Abb. 13) eine helikale (H) und eine kubische (C) Unterordnung unterschieden, die weiter je nach Vorhandensein oder Fehlen äußerer, die Capsomerenkapsel sekundär umschließender Hüllen (Envelope) in eine nackte (N) und eine mit Hüllen versehene Gruppe (E) aufgegliedert werden. Aus den Anfangsbuchstaben dieser Merkmale ergeben sich die Typensignaturen für die verschiedenen Virusarten.

Nicht für alle onkogenen Viren sind bereits genügend Daten über die erwähnten Strukturmerkmale vorhanden, um sie mit völliger

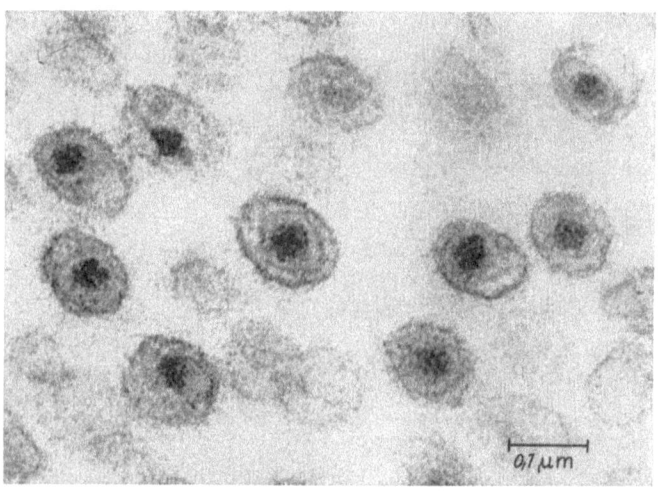

Abb. 14. Ultradünnschnitt des Myeloblastosevirus in einer gereinigten Präparation (Zentrifugation in Ringer-Lösung und Wasser). OsO$_4$-Fixierung, Bleihydroxydfärbung; 110000 ×
(Original von BONAR et al.)

Sicherheit einordnen zu können, so daß einige Klassifizierungen noch mit einem Fragezeichen versehen werden müssen. Dies gilt z. B. für die Gruppe der RNS-Viren (Rous-Sarkom), die Viren der Mäuse- und Hühnerleukämien (Abb. 14) und den Milchfaktor, für welche die Kapselsymmetrie noch nicht völlig sichergestellt ist, die jedoch höchstwahrscheinlich der RHE-Gruppe angehören, d. h. also dem RNS-Typ mit helikaler Kapselsymmetrie und zusätzlicher

Sekundärumhüllung. Sie stellen vom morphologischen Standpunkt eine der Hauptgruppen der onkogenen Viren dar. Die zweite

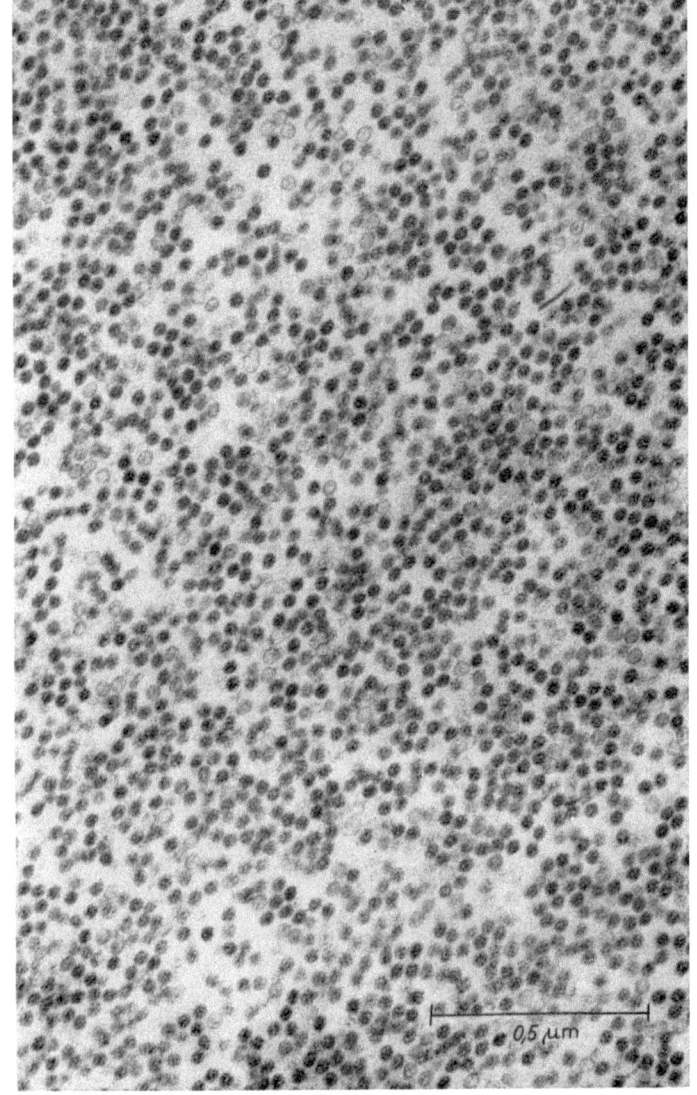

Abb. 15. Shope-Papillomvirus. Viruspartikel aus dem Papillomsaft, isoliert nach BEARD und WYCKOFF, 77000 × (aus OBERLING [73])

große Gruppe onkogener Viren, die sog. kleinen DNS-Viren (Papova-Gruppe[32a] und Adenoviren) gehören zum DCN-Typ, d. h. zu den nackten, kubischen DNS-Viren (Abb. 15). Die dritte Gruppe onkogener Viren, nämlich die großen DNS-Viren, die teilweise der Poxgruppe nahestehen und zu denen das Virus des Shopeschen Kanin-

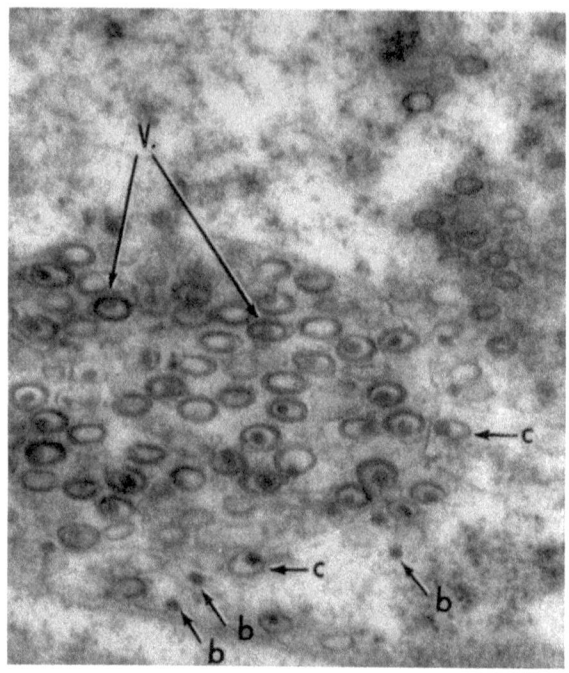

Abb. 16. Bildung des Luckéschen Nierencarcinomvirus des Leopardfrosches im Zellkern einer Tumorzelle. Neben leeren Hüllen und freien Nucleoiden (b) auch komplette Virions. 100000 ×
[nach D. W. FAWCETT: J. biophys. biochem. Cytol. **2**, 725 (1956)[

chenfibroms, des Molluscum contagiosum, des Nieren-Ca des Leopardfrosches (Abb. 16) sowie weitere Erreger von Fibromen und Schleimhautpapillomen gehören, sind mit großer Wahrscheinlichkeit dem DHE-Typ (DNS-Typ, helikal mit Envelope) zuzuordnen.

Es ergibt sich also, daß auf Grund ihrer Ultrastruktur die onkogenen Viren eine sehr heterogene Gruppe darstellen, womit der Ansicht von einem einheitlichen Krebserreger jeder Boden entzogen ist. Auch die sehr unterschiedliche Größe der onkogenen Viren mit Durchmessern von 35 bis 250 mμ spricht im gleichen Sinne.

Gleichzeitig ergibt sich aus den ultrastrukturellen Daten der onkogenen Viren sehr eindeutig, daß onkogene Wirksamkeit in keiner Weise mit einer strukturellen Besonderheit verbunden ist, zumindest was den elektronenmikroskopisch erforschbaren Größenbereich anbelangt, da in den gleichen Gruppen, in denen onkogene Viren zu finden sind, gleichzeitig auch nicht onkogene, rein cytocide und entzündungsauslösende Typen angetroffen werden.

Eine beträchtliche Vielfalt der morphologischen, mit Hilfe des Elektronenmikroskopes erfaßbaren Erscheinungen finden wir auch bei dem Vermehrungscyclus der verschiedenen onkogenen Viren in der Zelle, d. h. bei der sog. lytischen oder auch cytociden Virus-Zellbeziehung (Abb. 16 bis 23).

Diese vielfältigen Erscheinungen können wie bei der Virusvermehrung im allgemeinen in folgende drei Phasen eingeteilt werden:

1. in die auch hier passiv durch Phagocytose erfolgende, oft durch bestimmte Oberflächenaffinitäten zwischen Virushülle und Zellmembran (Receptoren) gelenkte Aufnahme des Virions in die Zelle;

2. im nächsten Stadium, der sog. Eklipse, wird dann durch Ablösung der Hüllmembranen und Kapsel die prägende Virusnucleinsäure freigesetzt, die anschließend ihre genetische Wirkung entfaltet, und zwar einerseits bei der Autoreduplikation der Virusnucleinsäure und andererseits über eine Messenger-RNS bei der Synthese von spezifischen Fermenten für den Aufbau der Virus-NS sowie bei der Synthese der Virushüllproteine. Bei dieser Proteinsynthese für den Virusaufbau wird der proteinsynthetisierende Apparat der Zelle (Ribosomen, s-RNS, pH 5-Enzyme) in Anspruch genommen. In dieser Phase ist in der Zelle weder elektronenmikroskopisch noch mit Hilfe des Infektionstestes aktives Virus, d. h. komplette Virions, nachweisbar;

3. der abschließende Akt der Virusvermehrung ist dann die Freisetzung der fertigen Virions in den extracellulären Raum, die entweder durch eine Ausschleusung bzw. Abschnürung der Partikel an der Zelloberfläche oder durch totalen Zellzerfall erfolgt.

Alle drei Phasen zeigen nun bei den verschiedenen Typen sowohl im zeitlichen Ablauf als auch cytotopographisch (Zellkern oder Cytoplasma) sowie in bezug auf die morphologisch erfaßbaren Details beträchtliche Unterschiede, woraus ein außerordentlich mannigfaltiges, elektronenmikroskopisch erfaßbares Bild des Gesamtablaufes der Virusvermehrung resultiert. Nichtsdestoweniger sind auch hinsichtlich der Virusvermehrung gleichsam Gruppenspezifitäten bei den verschiedenen Arten der onkogenen Viren nachweisbar, für die im folgenden einige Beispiele angeführt werden.

Bei dem zu den großen DNS-Viren der Poxgruppe gehörenden Virus des Shopeschen Fibroms, das wir deshalb an den Beginn unserer Beschreibung stellen wollen, weil hier die einzelnen Phasen der Virusvermehrung sich elektronenmikroskopisch besonders klar abzeichnen (OBERLING, BERNHARD und FEBVRE[54,55,41]), findet man ungefähr 4 Std nach der Infektion von Kaninchenfibroblasten in der Gewebekultur als erstes Zeichen der Virusvermehrung feulgen-

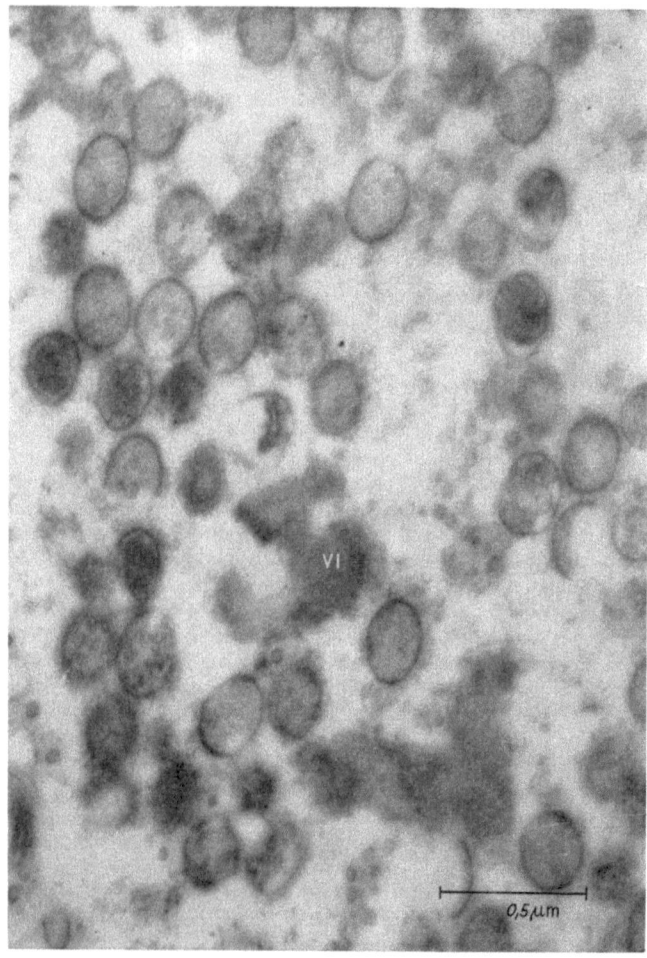

Abb. 17. Shopesches Fibromvirus. Unreife Viruspartikel im Cytoplasma neben Viroplasma (*vi*). 40000× (Original von BERNHARD)

positive, also DNS-haltige Bezirke im kernnahen Cytoplasma, in denen sich etwas später noch RNS anreichert. Der Vermehrung der Virus-DNS in diesen elektronenmikroskopisch als Verdichtungen des Cytoplasmas darstellbaren Einschlüssen folgt also eine durch die DNS induzierte Synthese von viraler Messenger-RNS für die Bildung der diversen Virusproteine, vor allem für die Virionhüllen. In diesen cytoplasmatischen Einschlüssen, den sog. Viroplasmen (Virusbildungsstätten), werden nach 8 bis 10 Std erstmalig Partikel von Virusgröße mit einfacher Membran ausgebildet (Abb. 17), in deren Innerem sich durch Substanzverdichtung ein hauptsächlich aus DNS bestehendes Nucleoid bildet. Der Abschluß der Virionsynthese wird durch die Ausbildung von 1 bis 3 weiteren Hüllmembranen abgeschlossen. Die Virusneubildung in der Zelle erfolgt mehrere Tage lang bis etwa eine Woche nach der Infektion. Die Zelle zerfällt dann, und dadurch werden die Virions freigesetzt.

Bei dem ebenfalls zu den großen DNS-Viren zugehörigen Virus des Nieren-Ca des Leopardfrosches vollzieht sich der Hauptprozeß der Virussynthese nicht im Cytoplasma wie bei dem Shopeschen Fibromvirus, sondern im Zellkern (Abb. 16), wobei die getrennte Synthese und Ausbildung von Capsomerenhülle und Nucleoid besonders schön elektronenmikroskopisch zu beobachten ist, indem beide Formbestandteile zunächst unabhängig nebeneinander wahrgenommen werden und in einer nachfolgenden Phase sogar das „Einwandern" der Nucleoide in die Proteinhüllen sichtbar wird.

Die kleinen DNS-Viren der Papovagruppe [69-74,41,40] werden alle im Zellkern gebildet. Eine weitere Gemeinsamkeit der Viren dieser Gruppe ist es, daß sie nicht durch einen spezifischen Ausschleusungsmechanismus, der immer an der Zelloberfläche stattfindet, in den extracellulären Raum gelangen, sondern speziell bei dem Polyomavirus durch Zellzerfall infolge der cytociden Viruswirkung, die in erster Linie auf der zahlenmäßig sehr starken Virusneubildung in der Zelle beruhen dürfte (Abb. 18). Infolge dieser Virusfreisetzung durch Zellzerfall kommt es nicht zu einer sekundären Umhüllung des Virions, wie dies bei dem Milchfaktor und den Leukämieviren beim Ausschleusen an der Zelloberfläche der Fall ist, sondern es bleibt bei der Struktur einer nackten Capsomerenkapsel mit eingeschlossenem Nucleoid. Aus diesem Grunde ist bei Polyoma- und den übrigen Viren der Papovagruppe durch Negativkontrastierung, z. B. Phosphorwolframsäure, die Capsidarchitektur besonders gut darstellbar, so daß Zahl, Anordnung (Symmetrieebenen) der Capsomeren sowie ihre Strukturuntereinheiten, die dem letzten Bauelement des Capsids, nämlich einzelnen Eiweißmolekülen entsprechen, besonders deutlich dargestellt werden (Abb. 19), indem das stark kontrastie-

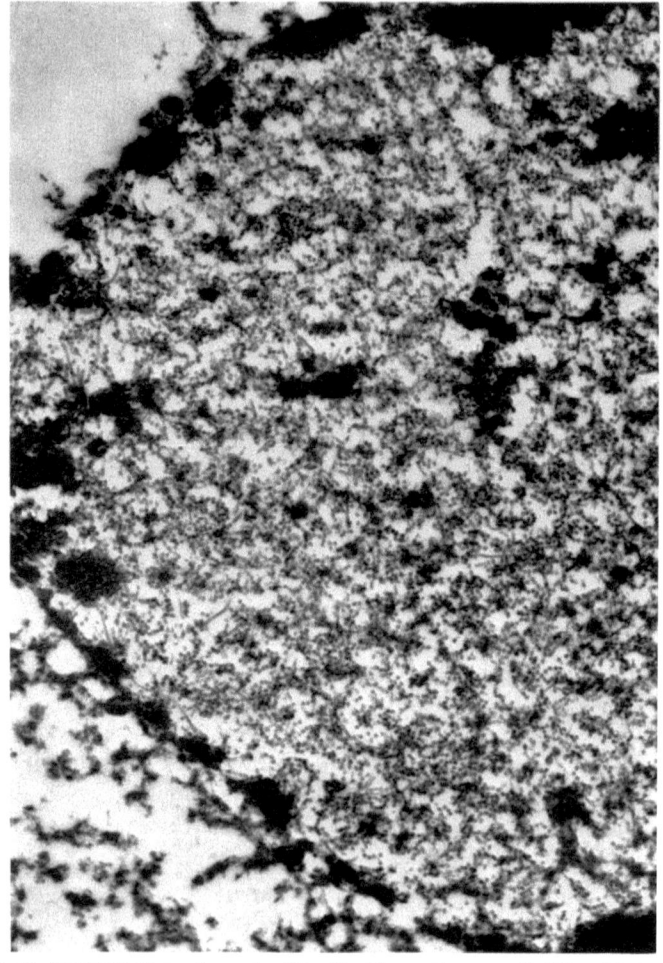

Abb. 18. Ausschnitt eines embryonalen Mäusefibroblasenzellkernes, vollgefüllt mit Polyomaviruspartikeln (Unterstamm BB-T2). Außer runden Teilchen einzelne fadenförmige Elemente. Elektronenmikroskopisches Bild im ultradünnen Schnitt; 35000fache Vergrößerung (nach BIERWOLF und GRAFFI, 1960)

rende Schwermetall in die Fugen zwischen die Capsomeren und Untereinheiten eindringt und nach dem Austrocknen ein negatives

Bild liefert. Im Gegensatz hierzu wird die Darstellung der Capsidstruktur bei Viren mit sekundären von der Zelloberfläche gelieferten Hüllen, wie sie bei allen onkogenen RNS-Viren vorhanden sind (Leukämie-, Rous-Virus usw.) unmöglich oder zumindest stark behindert und gelingt nur ausnahmsweise nach künstlicher Absprengung dieser Sekundärumhüllungen. Bei den kleinen DNS-Viren bestehen bei den verschiedenen Formen dieser Gruppe bei

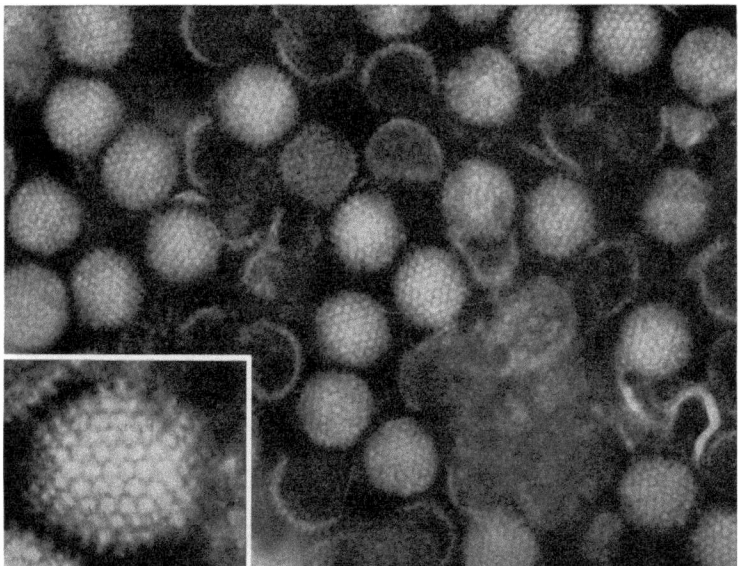

Abb. 19. Adenovirus Typ 12, Negativkontrast. ALMEIDA und HOWATSON. J. Cell Biol. **1953**, 16

grundsätzlich analogem Aufbau der Capsidhulle gewisse Unterschiede in der Zahl der Capsomeren hinsichtlich der Symmetrieebenen ihrer Anordnung und wahrscheinlich auch in der Konfiguration der molekularen Untereinheiten des Capsomers, nämlich der sie darstellenden Proteinmoleküle. Zuweilen können beispielsweise bei Polyomavirus auch leere Capsidhülsen sowie fadenförmige Gebilde mit dem Durchmesser eines Polyomavirions beobachtet werden, die als Fehlbildungen der Virussynthese anzusehen sind (s. Abb. 18).

Beim Shopeschen Papillomvirus[56,57,58,41] ist als Besonderheit zu vermerken, daß die Bildung von infektiösen Virions in vivo an ganz spezielle biochemische Bedingungen in der Zelle, nämlich einen hohen Differenzierungs- und Reifungszustand der Zellen mit völlig fehlender Wachstumstätigkeit gebunden ist. Virions findet man nämlich sowohl elektronenmikroskopisch als auch mit der Immunofluorescenzmethode unter Verwendung virusspezifischer Antikörper, in den Papillomen der Cottontailkaninchen ausschließlich im Stratum granulosum und in der bereits abgestorbenen Keratin-

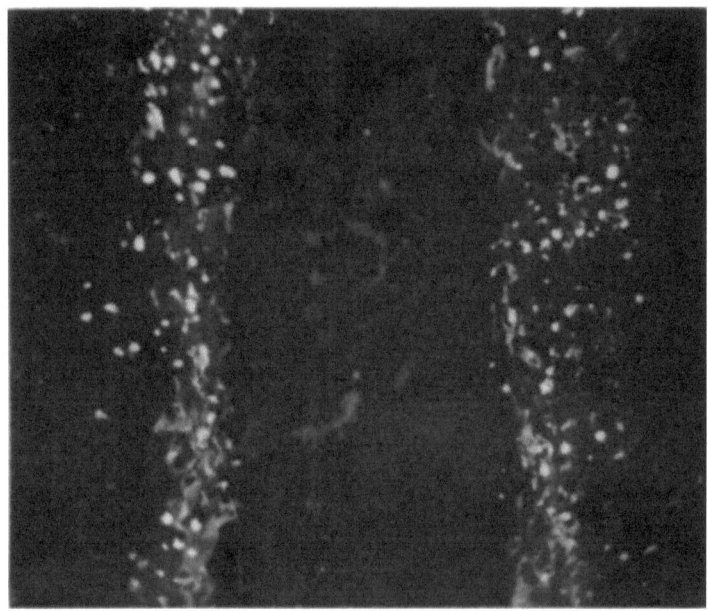

Abb. 20. Shope-Papillom. Gefrierschnitt eines bei einem Wildkaninchen spontan aufgetretenen Papilloms, gefärbt mit fluoresceingekoppelten Kaninchenantikörpern gegen Shope-Virusantigen (leuchtend weiße, runde und ovale Bezirke im Bild entsprechen apfelgrünen im Fluorescenzmikroskop). Virusantigene sind in den Kernen und Kernresten der keratinisierten Epithelzellen lokalisiert und fehlen in den proliferierenden Epithelzellen. Das keratinisierte Epithel ist in zwei vertikalen Bändern mit leuchtend weißen Bezirken (Virusantigen) angeordnet. Das Keratin im Cytoplasma einiger Epithelzellen ist grau (blau im Mikroskop) infolge von Autofluorescenz. Das benachbarte, nicht keratinisierte Epithel zeigt kein Virusantigen und ist nicht sichtbar. 250 × (Original von MELLORS, einschl. Legende)

schicht der verdickten Epidermis (Abb. 20). Da jedoch die infolge der Virusinfektion gesteigerte Proliferationstätigkeit der Basalschicht für die Geschwulstbildung verantwortlich ist, da ausschließlich hier Mitosen stattfinden, kann mit Sicherheit angenommen werden, daß zumindest das Virusgenom auch in den Zellen der

Basalschicht sowie des Stratum granulosum in zwar onkogen wirksamer, aber morphologisch und immunologisch nicht nachweisbarer, also maskierter Form enthalten ist, wahrscheinlich als freie Virusnucleinsäure. Die Wachstumsstimulierung ist also hier sicher der Wirkung dieser Virusnucleinsäure zuzuschreiben. Andererseits kann geschlossen werden, daß die erhöhte Wachstumstätigkeit der Epidermiszellen der Ausbildung infektiöser Virions vielleicht durch eine Art Interferenzphänomen im Bereich der Eiweißsynthese so stark entgegenwirkt, daß infektiöses Virusmaterial erst in den nicht mehr wachsenden, im Absterben begriffenen Zellen der Ceratohyalinschicht entstehen kann. Eine weitere Besonderheit des Shopeschen Papillomvirus ist seine Bildung in enger morphologischer Korrelation zum Nucleolus der Zellen. Es sei weiterhin vermerkt, daß die Virusbildung nur im Papillomstadium der Geschwulstbildung und auch hier mit Regelmäßigkeit nur bei der Cottontailrasse zu finden ist, während in den durch maligne Entartung aus den Papillomen entstehenden Carcinomen die Bildung infektiöser Virions völlig fehlt, ähnlich wie dies bei den übrigen Papovaviren in den geschwulstmäßig transformierten Zellen sowohl in vivo als auch in vitro der Fall ist.

Eine weitere Form der morphologisch darstellbaren Virusbildung in der Zelle findet man bei den onkogenen RNS-Viren, zu denen die Viren der Hühnerleukosen[40,41,45-47,49,51], das Rous-Virus[40,41,50,52,53], der Milchfaktor[40,51,49-61] und die verschiedenen Mäuseleukämieviren[40,41,62-68] gehören und die eventuell mit den Myxoviren (Influenza, Masern) gewisse Ähnlichkeiten haben. Mit relativ geringen Abweichungen vollzieht sich bei diesen Viren die Neubildung von Virions in folgender Weise: Nach der Aufnahme infektiöser Partikel in für Virusvermehrung kompetente Zellen, gleichgültig ob in vivo oder in vitro, sind zunächst, zumal in der Zeit der Eklipse, irgendwelche auffallenden Strukturveränderungen in den Zellen nicht nachweisbar. Am Zellkern ist bei diesen Viren zu keinem Zeitpunkt irgendeine morphologische, elektronenmikroskopisch bisher erfaßbare Veränderung zu erkennen, die auf seine Mitwirkung bei der Virusvermehrung schließen lassen könnte. Dagegen besitzt hier meist die äußere Zellmembran, viel seltener andere cytoplasmatische Membransysteme, die dominierende Bedeutung bei der morphologischen Gestaltung neuer Virions. In einem als „budding" oder Knospung bezeichneten Prozeß,

beginnend mit einer lokal begrenzten Substanzvermehrung und
Verdickung der äußeren Zellmembran schnüren sich über eine

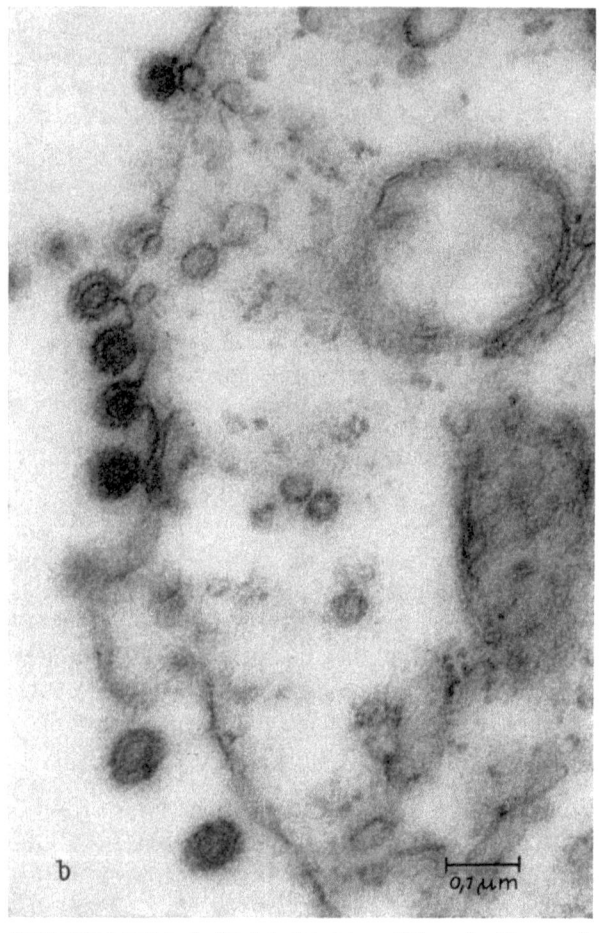

Abb. 21. Erythroblastosevirus. 7. Gewebekulturpassage. Bildung des Virus an der Zellmembran. 100000× (Original von HEINE und BEARD)

doppelkonturierte, halbkugelförmige kleine Vorwölbung vom Durchmesser eines Virions als Zwischenstadium Viruspartikel mit doppelter Hüllmembran und einem Nucleoid ab (Abb. 21, 22), die dann in den extracellulären Raum abgegeben werden. Die Verdichtung

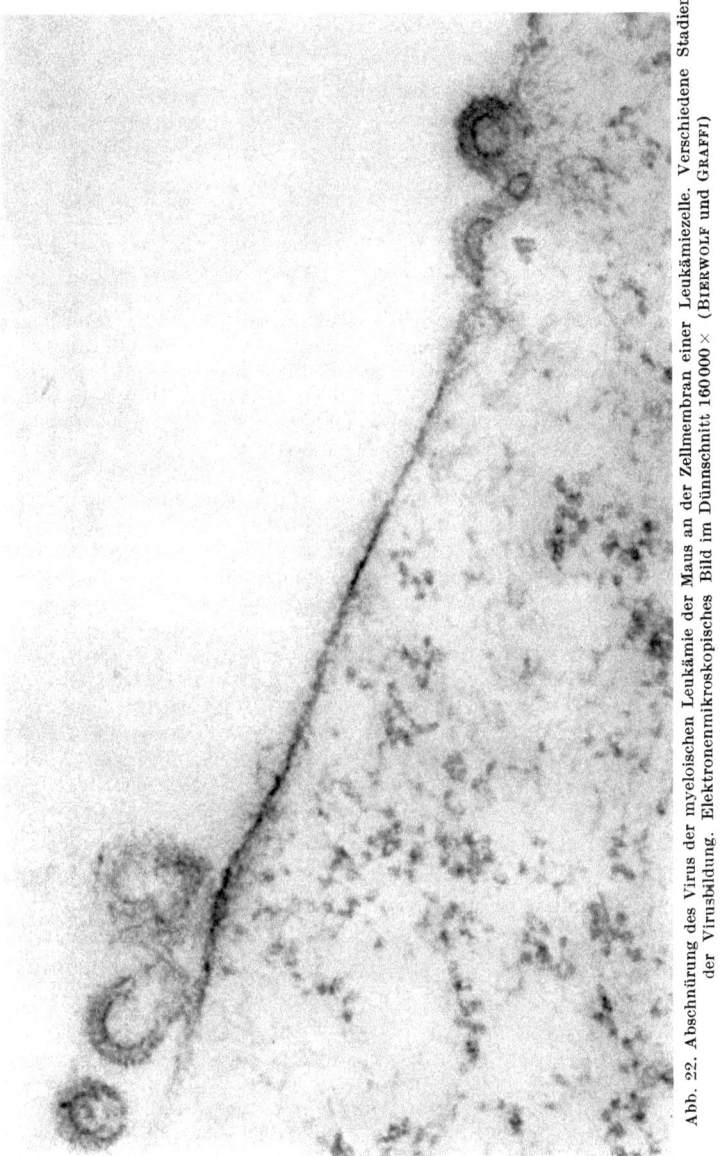

Abb. 22. Abschnürung des Virus der myeloischen Leukämie der Maus an der Zellmembran einer Leukämiezelle. Verschiedene Stadien der Virusbildung. Elektronenmikroskopisches Bild im Dünnschnitt 160 000 × (BIERWOLF und GRAFFI)

und Ausreifung des zentralen Nucleoids findet teils bereits während der Knospung, teilweise, z. B. beim Milchfaktor, häufig erst nach der Abschnürung statt. In bestimmten Fällen, beispielsweise bei der von uns bearbeiteten myeloischen Leukämie der Maus oder auch beim Milchfaktor scheinen einzelne Partikel auch im Cytoplasma, möglicherweise im Bereich der Membranen des endoplasmatischen Reticulums gebildet zu werden.

In vielen Fällen sind jedoch cytoplasmatische Viruspartikel auf eine sekundäre Phagocytose bereits über die Zellmembran ausgeschleuster Elemente zurückzuführen. Sie werden in diesem Fall häufig, wie z. B. bei der Myeloblastose der Hühner, in Form größerer Einschlüsse, der sog. ,,gray bodies''[47], angereichert, die auf Grund der starken ATPase-Reaktion des Myeloblastosevirus (s. [105]) ebenfalls eine positive ATPase-Reaktion zeigen. Die Ausschleusung an der Zellmembran bringt es mit sich, daß Anteile der normalen Zelloberfläche mit ihren zum Teil sehr spezifischen Ausstattungen an Receptoren, Antigenen und Fermenten in die äußeren Hüllen des Virions mit eingehen, wodurch die Viruspartikel immunologische und biochemische Eigenschaften der Wirtszelle mitbekommen. Auf diese Weise kommt die hohe ATPase-Aktivität des Myeloblastosevirus zustande, da der sie reproduzierende Myeloblast eine hohe ATPase-Aktivität in seiner äußeren Membran besitzt, im Gegensatz zu dem Erythroblasten, demzufolge das Erythroblastosevirus eine ATPase-Aktivität praktisch vermissen läßt (BEARD). Wir vermuten, daß die in bezug auf die Bildungszelle herkunftsgemäße, biochemische Prägung der sekundären Virionhüllen speziell bei den Mäuseleukämien eventuell auch zu dem Wirkungsmechanismus des Virus enge Beziehungen besitzt, in der Weise, daß daraus gelenkte Histotropismen resultieren, die die Tatsache erklären, daß durch ein aus reifen myeloischen Leukämien isoliertes Virusmaterial stark bevorzugt wiederum reifzellige myeloische Leukämien ausgelöst werden, während Virusmaterial, das von lymphatischen Leukämiezellen produziert wurde, weit überwiegend lymphatische Leukämien induziert. Entsprechendes konnten wir weiterhin auch für Erythroblastenleukämien nachweisen (FEY und GRAFFI, 1965).

Nach DMOCHOWSKI soll das Lymphomatosevirus auch innerhalb der Mitochondrien der malignen Lymphoblasten gebildet werden. Beim Milchfaktor wurden von BERNHARD[39,40] intracelluläre, im Cytoplasma oft haufenartig zusammengeballte, kleinere Partikel von etwa 60 mµ Durchmesser, die in Form eines Doppelringes einen kontrastarmen Innenraum umschließen, als sog. A-Partikel von den extracellulären, etwas größeren, mit einer weiteren als Envelope anzusehenden Membran ausgestatteten Partikel, die gleichzeitig ein eindeutiges zentrales oder randständiges, elektronendichtes Nucleoid besitzen, als B-Partikel unterschieden. Es spricht vieles dafür, daß die B-Partikel aus den A-Partikeln durch deren Ausschleusung über Mikrovilli der Zelloberfläche, wobei sie eine zusätzliche Hülle an der Zellmembran empfangen, entstehen (Abb. 23).

Eine generelle Besonderheit der Reproduktionsweise der onkogenen RNS-Viren der RHE-Gruppe (Leukämieviren, Rous-Virus, usw.) ist ihre Bildung *ohne* tödliche Schädigung der Zellen, also praktisch ohne sichtbaren cytociden Effekt sowie der Umstand, daß sie allem Anschein nach nicht nur von Normalzellen und im präneoplastischen Zustand gebildet werden, sondern auch *nach* der

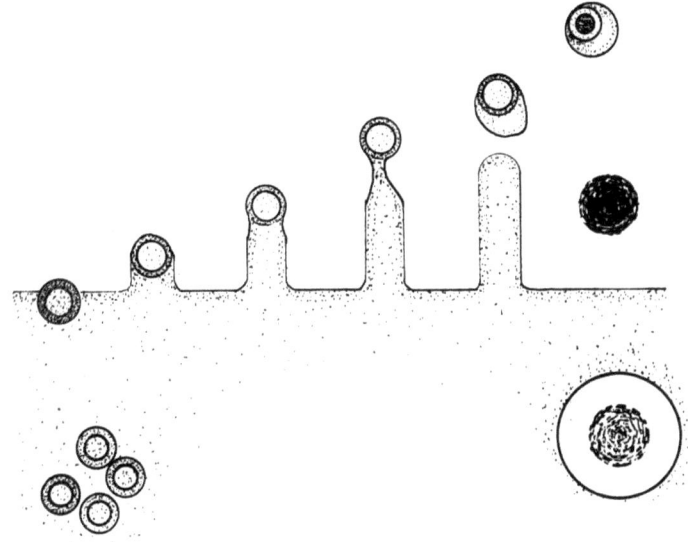

Abb. 23. Schematischer Rekonstruktionsversuch des Durchtrittes von Bittner-Viren durch die Zellwand und deren Umwandlung in die extracelluläre Form. Rechts unten und Mitte normale Milchpartikeln. Nach BERNHARD aus: Klin. Wschr. **35**, 251 (1957)

malignen Transformation (s. Kap. VII), die hier häufig (Rous-Sarkom, Hühnerleukosen) in sehr kurzer Zeit (wenige Tage) erfolgen kann. Es ist eventuell sogar möglich, daß die laufende Reproduktion des Virus, zumindest seines Genoms, für die Aufrechterhaltung der Malignität der Zellen unbedingt notwendig ist. In beiden Punkten, nämlich fehlende cytocide Wirkung und Fortdauer der Virusvermehrung auch nach der malignen Umwandlung, ist also ein charakteristischer Unterschied zwischen DNS-Viren und RNS-Viren zu verzeichnen. Als weitere gewisse Besonderheit der onkogenen Viren insgesamt im Vergleich zu den meisten nicht onkogenen Typen kann außerdem ein relativ langsam ablaufender Vermehrungscyclus der ersteren angesehen werden.

IV. Epidemiologie[75—84]

Die Epidemiologie, d.h. die Verbreitung und Übertragung der onkogenen Viren innerhalb der empfänglichen Species, gleichgültig, ob die Einzelindividuen darauf mit maligner Entartung reagieren oder nur als Virusträger fungieren, ist vor allem an die Vermehrungsweise in quantitativer und qualitativer Hinsicht sowie an die Ausbreitung des Virus im Organismus und seine Ausscheidung aus dem Organismus gekoppelt. Letztere Phänomene wiederum werden nicht nur durch die jeweiligen Zell-Virusbeziehungen bedingt, sondern außerdem vor allem noch durch die immunologische Situation des Virusträgers sowie des Virusempfängers und der gesamten im Streubereich liegenden Population, wobei Phänomene wie latente Durchseuchung bei partieller oder hochgradiger Immunität, hochgradige Empfindlichkeit infolge Immunotoleranz sowie völliger Mangel an Antikörpern infolge noch fehlenden Kontaktes mit dem betreffenden Virus eine entscheidende Rolle spielen. Die jeweiligen Kombinationen dieser verschiedenen Möglichkeiten ergeben das bunte epidemiologische Bild der onkogenen Viren, das nachfolgend an Hand einiger typischer Beispiele dargelegt werden soll.

Während das Rous-Sarkom sowie die Erythro- und Myeloblastose der Hühner spontan nur ganz sporadisch und isoliert vorkommen und Kontaktinfektionen oder auch andersartige horizontale Übertragungen z. B. durch Insekten von Tier zu Tier nur sehr selten und unter ganz besonderen Bedingungeu beobachtet werden (s. [19]), zeigt die Lymphomatose der Hühner, die wie bereits dargestellt wurde, durch ein sehr nahe verwandtes Virus induziert wird, häufig den Charakter einer echten Infektionskrankheit[79]. Ihr Auftreten ist in vielen Fällen epidemisch und ganze Hühnerzuchten können durch sie in kurzer Zeit dezimiert werden. Dabei erfolgt die Infektion von Tier zu Tier überwiegend durch direkten Kontakt über Kot, Urin und Speichel, in denen bei erkrankten Tieren infektiöses Virus in großen Mengen ausgeschieden wird. Die hohe Virusausscheidung in den Exkreten hängt zum großen Teil mit der massiven Miterkrankung aller inneren Organe, speziell der Leber, Milz, Niere und des Darmtraktes zusammen, während die Virusausscheidung bei lokalisierten Rous-Sarkomen oder bei der sich vorzugsweise im Blut und Knochenmark abspielenden Myelo- und Erythroblastose weitgehend fehlt. Von entscheidender Bedeutung für das Zustandekommen einer epidemischen Ausbreitung der Lymphomatose ist jedoch der Immunitätszustand der betreffenden Hühnerzucht (Abb. 24). Sehr viele Hühnerfarmen sind latent mit dem Virus infiziert, erkranken jedoch deshalb nicht, weil sie durch neutralisierende Antikörper im Blut genügend immun sind. Das

Virus wird zum Teil einerseits postnatal durch Kontakt und andererseits vertikal von Generation zu Generation über das Ei an die Nachkommenschaft weitergegeben. Dadurch jedoch, daß gleichzeitig ebenfalls mit dem Ei auch Antikörper vom Muttertier auf die Küken übertragen werden, sind diese in der frühesten besonders anfälligen Jugendperiode vorerst ausreichend passiv immunisiert, so daß sie nicht erkranken. Durch eigene Antikörperproduktion in den ersten Lebensmonaten wird die Immunität noch weiter gesteigert, so daß trotz latenter Verseuchung die Erkrankung ausbleibt. Die Katastrophe tritt jedoch ein, wenn in eine Hühnerzucht *ohne* latente Virusdurchseuchung, d. h. völlig ohne Immunität, ein virus-

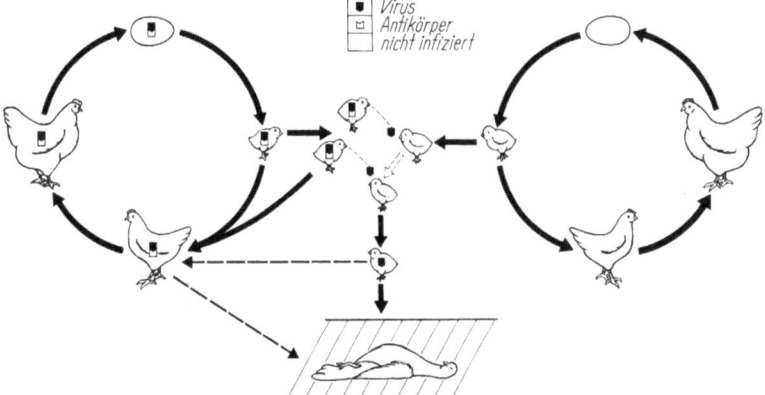

Abb. 24. Übertragung des Virus der visceralen Lymphomatose unter natürlichen Bedingungen
(aus BURMESTER [78])

ausscheidendes Huhn gelangt, an welchem sich dann die virusfreien nicht immunisierten Küken der Zucht anstecken, in wenigen Tagen erkranken und gleichzeitig eine Kontaktinfektionsquelle für den gesamten nicht immunen Hühnerbestand werden. Die der Krankheit nicht erliegenden Hühner bleiben als latent verseuchte und gleichzeitig als immune Tiere übrig und geben nun ihrerseits das Virus mit den entsprechenden Antikörpern über das Ei an ihre Nachkommen weiter, so daß der in der linken Hälfte des Schemas von BURMEISTER (Abb. 24) angegebene Kreislauf wieder von neuem beginnt. Bei den Mäuseleukämieviren[80,81,82] und dem Milchfaktor[78,75,76,83,19] dominiert ein anderer für mehrere onkogene Viren typischer Übertragungsmodus, nämlich die erstmals von BITTNER[1] für das Mammatumoragens der Maus und mehrere Jahre später von GROSS[5] für die Mäuseleukämie nachgewiesene vertikale Übertragung (Abb. 25). In beiden Fällen geht der Hauptweg dieser vertikalen Übertragung über die Muttermilch (BITTNER; LAW und MOLONEY, GRAFFI und KRISCHKE) und in geringerem Ausmaße außerdem noch diaplacentar (beim Leukämievirus) oder über das Ejaculat der Männchen (beim Bittner-Faktor). Für die natürlicherweise vorkommende vertikale Übertragung dieser Viren in speziellen Mäuselinien, die dann meist mit hoher spontaner Quote an Leukämien bzw. Mamma-Ca behaftet sind, ist eine besondere immunologische Situation, nämlich eine oft das ganze Leben dauernde

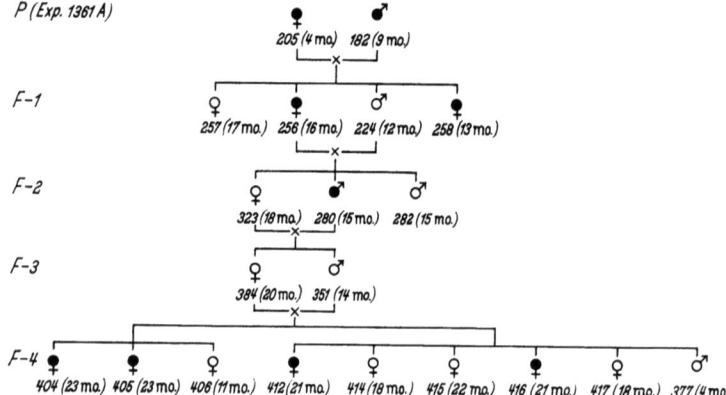

Abb. 25. "Vertical" (from one generation to another) transmission of leukemia in the C3H strain of mice. Only female 205 and her brother 182 had been injected (within a few hours after birth) with Ak leukemic extracts. Their offspring were untreated. Black color indicates the development of leukemia. It is evident that leukemia developed in several descendants of successive generations. (This strain of mice is essentially free from spontaneous leukemia. The inoculation of the parent-couple 205 and 182 with the leukemic agent apparently initiated a "vertical" transmission of this disease in this particular family of mice.)
GROSS, L.: Ciba Foundation Symposium on Leukemia Research, p. 76; J. and A. Churchill, London, 1954

Toleranz dem entsprechenden Virus gegenüber charakteristisch (HALLAUER[84a], PASTERNAK[83a], B. WAHREN[97a] u. a.). Sie ist darauf zurückzuführen, daß die Tiere bereits im neugeborenen Zustand, also im Stadium der immunologischen Nichtansprechbarkeit ziemlich massiv von der Mutter aus mit dem Virus infiziert werden. Dieses kann dadurch nachgewiesen werden, daß die Tiere auch im späteren Leben gegen das betreffende Virus nicht aktiv immunisierbar sind, während gegen andere ähnliche Viren, also z. B. ein bezüglich der Antigenität differentes Leukämievirus, die künstliche Erzeugung von Virus-neutralisierenden Antikörpern im Blut ohne weiteres gelingt. Eine Kontaktinfektion, also horizontale Übertragung ist sowohl beim Milchfaktor als auch bei der Mäuseleukämie ein bestenfalls äußerst seltenes Ereignis. Auch wird von erkrankten Tieren, trotzdem schon wenige Tage nach der Infektion neugeborener Tiere massenhaft Virus im Blut und in den Organen nachzuweisen ist, kaum infektiöses Material im Urin, Kot und Speichel ausgeschieden[80], und bei dem Zusammenleben künstlich infizierter und nicht infizierter Tiere von der Geburt an ist keine eindeutige Steigerung der Leukämiequote bei den nicht infizierten Tieren vorhanden[80].

Kaninchenpapillom und Fibrom sind endemisch bei Cottontailkaninchen oft etwas gehäuft vorkommende Erkrankungen, die überwiegend durch Insekten (Moskitos, Anopheles), von kranken Tieren auf die gesunden übertragen werden (s. [77]).

Das Polyomavirus ist bei außerordentlich vielen Mäusestämmen einschließlich der Wildmaus in Form einer latenten Infektion vorhanden. Daß Mäuse außerordentlich selten spontan an Polyomatumoren erkranken, trotz seiner starken onkogenen Wirkung auf diese Tierart bei der *künstlichen* Infektion, hängt damit zusammen, daß die Mäuse, die sich erst im Laufe der ersten Lebenswochen durch Kontakt, zum Teil über die Atmungsluft (SACHS[84]) also horizontal infizieren, reichlich Antikörper gegen das Polyomavirus zunächst von der Mutter über die Milch mitbekommen und anschließend selber entwickeln und gleichzeitig über das ebenfalls im Zuge der temporären Virusvermehrung gebildete neue tumorspezifische Zellantigen eine relative Immunität gegenüber der Polyomatumorzelle erlangen. Wird das Virus jedoch neugeborenen Mäusen oder auch anderen Tierarten in größeren Mengen appliziert, also im Zustand noch fehlender immunologischer Reaktionsfähigkeit, so kommt es nicht zur Immunität, sondern zur Immunotoleranz, sowohl gegenüber dem Polyomavirus als auch den durch das Polyomavirus transformierten malignen Zellen, womit infolge fehlender Abwehr die wichtigste Voraussetzung für eine Geschwulstbildung gegeben ist.

Diese wenigen Beispiele für die epidemiologischen Verhältnisse bei onkogenen Viren zeigen gleichfalls die große Vielfalt der Reaktionsweise von Einzelindividuen und ganzen Populationen auf diese Viren, wobei zu vermerken ist, daß ähnliches auch für nicht onkogene Viren zutrifft. Eine gewisse Besonderheit der onkogenen Viren liegt aber vielleicht dennoch in der bei sehr vielen Typen vorkommenden vertikalen Weitergabe des Virus sowie der Bedeutung einer Immunotoleranz primär gegen das Virus und sekundär über das neugebildete tumorspezifische Zellantigen eventuell auch gegen die durch das gleiche Virus induzierte Tumorzelle. Die Bedeutung künstlich provozierter Änderungen der immunologischen Reaktionslage für die Auslösung viraler Tumoren sei an zwei Beispielen dargelegt. Durch zusätzliche Röntgenbestrahlung lassen sich Shope-Fibrome leicht in bösartige Tumoren von Sarkomcharakter überführen. Mit dem Virus der myeloischen Leukämie der Maus können in der Kombination mit Ganzkörperbestrahlung erwachsene Tiere mit relativ kleinen Virusdosen, die für sich allein praktisch unwirksam sind, erfolgreich infiziert werden, so daß sie an Leukämie erkranken[80a]. Auch die hochprozentige Induktion von Leukämien bei bestimmten Mäusestämmen durch Ganzkörperbestrahlung, wobei in den entstehenden Lymphomen durch zellfreie Übertragung Virus nachweisbar ist[5], dürfte nach neuesten Ansichten weniger auf die unmittelbare Aktivierung eines Virus durch die Röntgenbestrahlung als vielmehr auf die Verminderung der immunologischen Abwehr und die dadurch ermöglichte Vermehrung eines bereits latent vorhanden gewesenen Virus zurückzuführen sein.

V. Immunologische Eigenschaften der onkogenen Viren[85—97a]

Wie praktisch alle Viren besitzen auch die onkogenen eine spezifische Antigenität, so daß gegen sie heterologe und homologe Antikörper erzeugt werden können, die sie präcipitieren oder agglutinieren und sie daher in bezug auf Infektiosität inaktivieren. Derartige inaktivierende Antikörper sind schon vor vielen Jahrzehnten gegen das Rous-Virus (ROUS, 1919) und die verschiedenen Hühnerleukoseviren[85] erzeugt worden. Wir selbst konnten gegen das Virus der myeloischen Leukämie neutralisierende Antikörper gewinnen[86,26]. Entsprechendes wie für diese RNS-Viren gilt auch für die DNS-Viren, z. B. Polyoma. Gegen dieses Virus erzeugte homologe und heterologe Antikörper, z. B. bei latent infizierten Mäusen, können dadurch nachgewiesen werden, daß sie die hämagglutinierende Wirkung des Polyomavirus durch vorausgehende Präcipitation der Viruspartikel hemmen[32]. Die Viruspräcipitation bzw. Agglutination durch spezifische Antiseren wurde bei Hühnerleukoseviren, speziell dem Myelo- und Erythroblastosevirus von

BEARD u. Mitarb.[85] an hoch gereinigten, aus dem Blutplasma gewonnenen Virussuspensionen auch elektronenmikroskopisch ver-

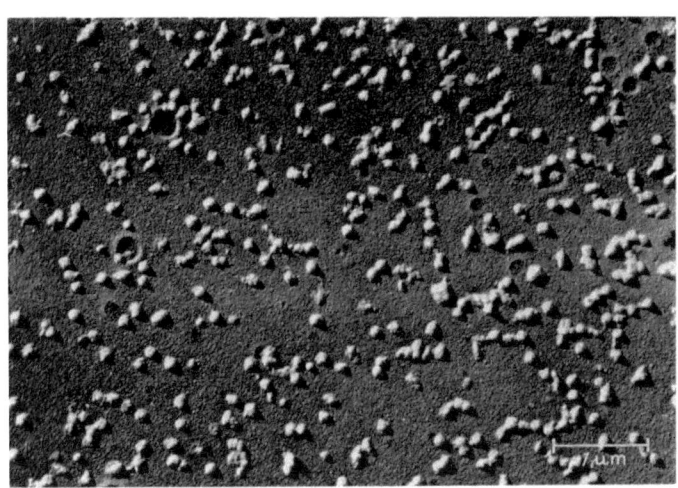

Abb. 26a. Myeloblastosevirus aus dem Plasma. Schrägbedampfung. 17000 × (Original von BEARD)

folgt (Abb. 26a und b). Die Antigenität einiger onkogener Viren, speziell der Virustypen mit sekundären Hüllen (RNS-Gruppe) ist

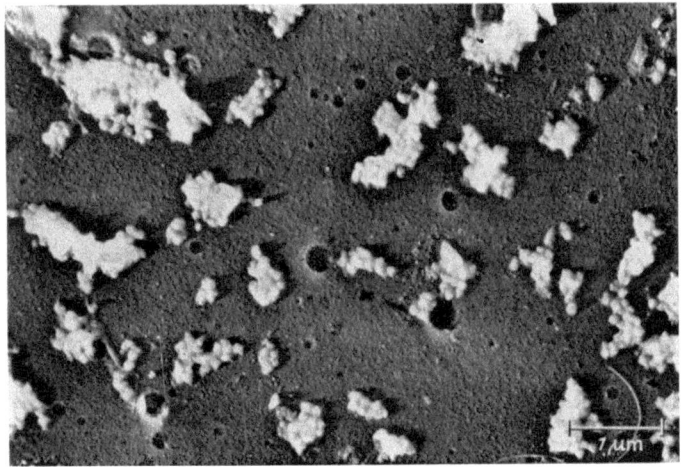

Abb. 26b. Myeloblastosevirus präcipitiert mit Anti-Myeloblastose-Virus-Kükenimmunserum. 17000 × (Original von BEARD et al.)

dadurch etwas vielschichtig, daß äußere Hüllproteine und Capsomerenproteine sehr wahrscheinlich unterschiedliche chemische Struktur und damit auch Antigenität besitzen. Es muß außerdem in Betracht gezogen werden, daß in die äußere Hülle bei denjenigen Viren, die an der Zelloberfläche abgeschnürt werden, sehr wahrscheinlich zusätzlich zu den von dem Virusgenom induzierten Hüllproteinen noch Membrananteile der betreffenden Zellart einbezogen werden, so daß das Virion dadurch auch gewisse Antigenqualitäten der Wirtszelle gewinnt. Damit wird die Frage der Antigenität verschiedener onkogener Viren sehr komplex und oft auch etwas verworren, und es ist beispielsweise für die Viren des Rous-Sarkom-Leukosekomplexes der Hühner noch nicht völlig eindeutig geklärt, ob Antigenverwandtschaften, die sich in immunologischen Kreuzreaktionen äußern, nur auf eine echte genetische Verwandtschaft dieser Virusgruppe, also die im Virusgenom selber verankerte genetische Information, zurückzuführen sind oder aber zum Teil auch durch die bei der Virusbildung inkorporierten Wirtszellantigene bedingt werden. Im ganzen sind jedoch in bezug auf immunologische Reaktionsweise keine grundsätzlichen Differenzen bei onkogenen und nicht onkogenen Viren vorhanden. Andererseits zeigen bis auf die erwähnten Gruppenreaktionen bei nahe verwandten Virusarten (Hühnerleukosen) die verschiedenen onkogenen Viren scharfe Spezifitätsunterschiede, so daß mit ziemlicher Sicherheit ausgesagt werden kann, daß zumindest was die Hüllproteine anbelangt keinerlei verwandtschaftliche Beziehung und damit auch kein gemeinsamer genetischer Informationsgehalt bei den verschiedenen Gruppen onkogener Viren vorhanden sind, also auch von dieser Seite aus die erwähnte Heterogenität der onkogenen Viren aufgezeigt wird.

Die für das Verständnis der Cancerogenese interessantere Seite der Tumorimmunologie resultiert aus der *Wechselbeziehung* zwischen Virus und Zelle nach dem Eintritt virulenter Partikel in die verschiedenen Zelltypen, vor allem bei solchen, die dabei eine maligne Transformation erfahren.

Es ist ohne weiteres ersichtlich, daß sich dabei Änderungen der Antigenität der virusbefallenen *Zelle* einstellen müssen, werden doch durch das Virus verschiedene neue Antigene im Zuge seiner Reproduktion sowie seiner Interaktion mit dem Zellgenom gebildet, die die Antigenqualität der Zelle im ganzen oder ihrer einzelnen Teile zwangsläufig gegenüber einer völlig normalen, nicht infizierten Zelle abändern müssen. Bei der Virusvermehrung in der Zelle werden zunächst einmal vom Virusgenom codierte Hüllproteine synthetisiert, entweder im Zellkern oder im Cytoplasma oder an der Zelloberfläche, die allein durch ihre Gegenwart der Zelle die Antigenqualitäten des intakten Virions als zusätzliche immunologische Qualität vermitteln. Diese Hüllproteine sind also an die Virusvermehrung in der Zelle gebunden, sind virusspezifisch und reagieren mit einem spezifischen, gegen das Virus gerichteten Antiserum. Sie sind mit den verschiedensten Methoden eines serologischen Antigennachweises wie Komplementbindungsreaktion, Cyto-

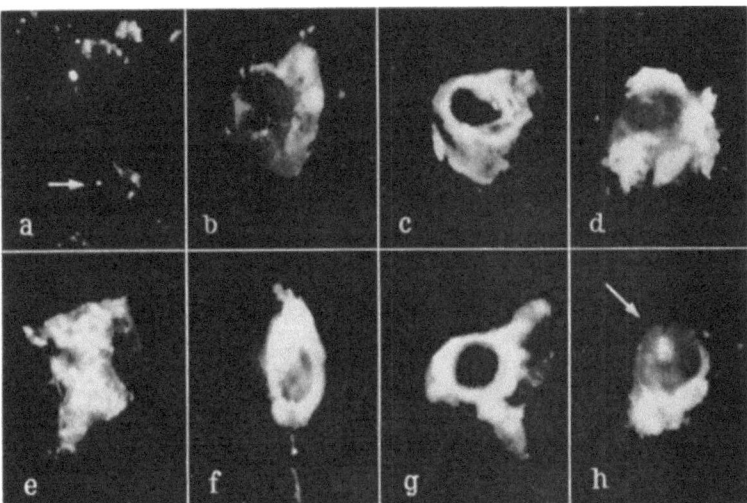

Abb. 27. Intracelluläre Lokalisation des Rous-Virus-Antigens, nachgewiesen mit fluorescierenden Antikörpern [aus MELLORS: Cancer Res. **20**, 744 (1960)]. a) Flügeltumor; feinpartikuläre Verteilung des Antigens im Cytoplasma der Sarkomzelle; b) Fein- und grob-partikuläre sowie homogene Verteilung des Antigens im Cytoplasma der Sarkomzelle; c) Virus-Antigen im aggregierten Massen im Cytoplasma der Zelle; d) wie c; e) Implantat in der Lunge. Partikuläre und homogene Anordnung des Antigens im Cytoplasma; f) Plumpe spindelförmige Zelle mit antigenreichem Cytoplasma; g) wie f, aber bizarr geformte Sarkomzelle; h) Virusantigen in exzentrischem, rundem Focus im Kern, außerdem im Cytoplasma

toxtest und in bezug auf ihre topographische Lagerung besonders günstig mit der Immunofluorescenzmethode darstellbar (Abb. 27). Dazu gehören beispielsweise die in der Ceratohyalinschicht der Cottontailkaninchenpapillome nachweisbaren im Zellkern gelagerten Antigene[90] (s. Abb. 20), die entweder fertige, infektiöse Virions oder die noch nicht mit dem Virusgenom assoziierten Capsomerenproteine darstellen. Bekanntlich werden jedoch bei Virusinfektionen durch das Virusgenom häufig auch Proteine induziert, die nicht in die Hüllproteine und in das sich bildende Virion eingehen, sondern meist als Fermente hauptsächlich der Synthese der viralen NS dienen (Polymerasen, RNS-Synthetasen, also Replikasen, Thymidinkinasen, usw.). Da auch diese Proteine durch zellfremdes genetisches Material induziert werden, sind sie auch in bezug auf ihre Antigenität zellfremd und verleihen durch ihre Gegenwart der virusführenden Zelle ebenfalls neue Antigeneigenschaften. Schließlich scheint die Bildung neuer, normalerweise nicht vorhandener Zellantigene, in virusinfizierten Zellen eventuell auch dadurch zustande kommen zu können, daß das Virusgenom mit dem Zellgenom, sei es durch Inkorporation oder Integration oder auch durch andersartige Wechselwirkungen (Depression und Redepression genetischer Informationen im zelleigenen Genom, Interferenz usw.) in bestimmter Weise reagiert. Durch derartige Mechanismen und Wechselwirkungen sich ausbildende neue Antigenqualitäten der Zellen nach Infektion mit onkogenen Viren scheinen nun für das Zustandekommen der sog. malignen Transformation der Zellen von großer Bedeutung zu sein.

Wie wir später (Kap. VIII) noch ausführlicher darlegen werden, gibt es bei den onkogenen Viren vor allem zwei Reaktionsweisen mit der Zelle, die vor allem bei den kleinen DNS-Viren sehr klar voneinander trennbar sind.

1. Die lytische oder cytocide Reaktion mit Zellschädigung und gleichzeitig starker Virusvermehrung und

2. die sog. Virusgenomintegration in das Zellgenom, die zur malignen Transformation der Zelle führt, wobei sowohl Zellschädigung als auch die Bildung von neuen Virions (bei DNS-Viren) unterbleibt. Beide Reaktionsweisen zwischen Virus und Zelle müssen bei der Betrachtung der Änderung der cellulären Antigenität im Anschluß an die Virusinfektion wegen mehrfacher Differenzen zwischen beiden gesondert betrachtet werden. Außerdem bestehen gewisse qualitative Unterschiede auch zwischen DNS- und RNS-Viren, aus welchem Grunde die Verhältnisse bei diesen beiden Virusgruppen zweckmäßigerweise ebenfalls getrennt beschrieben werden.

Betrachten wir zunächst an Hand eines Schemas* (Abb. 28) die Antigenverhältnisse nach Infektion kompetenter Zellen mit Polyomavirus (gleiches gilt prinzipiell auch für Adeno- und SV40-Virus[89,88,96,94]), vergleichend einerseits bei der lytischen Infektion der Zelle, also mit Virusneubildung, und andererseits bei der Transformation mit fehlender Bildung von Virions und wahrscheinlich synchroner Vermehrung des inkorporierten Virusgenoms mit der Zellteilung. Bei der *lytischen* Infektion, die mit dem Zelluntergang mit der Freisetzung neuer Virions endet, treten naturgemäß die virusspezifischen, durch die Hüllproteine (Capsomeren) bedingten *Virus*antigene überwiegend im Zellkern auf (V). Außerdem entsteht noch ein Neoantigen noch unbekannter Bedeutung, und nicht sehr zweckmäßig auch als Tumorantigen (T) bezeichnet, das gleichfalls im

---

* Die beiden Schemata, die die Antigenverhältnisse bei den verschiedenen Wirkungsweisen onkogener Viren wiedergeben, verdanke ich meinem Mitarbeiter Dr. med. habil. G. PASTERNAK.

Zellkern lokalisiert ist und mit der Immunofluorescenzmethode hier nachgewiesen werden kann und sicher auf die Wirkung des genetischen Substrats des Virus entweder direkt oder auf dem Umwege über eine Zwischenreaktion mit dem Zellgenom zurückzuführen ist. Bei der *zweiten* Reaktionsweise des Virus, nämlich der *malignen Transformation*, fehlt, wie erwähnt, die Neubildung von Virions und damit auch das Virusantigen in der Zelle. Hingegen ist das Neo-

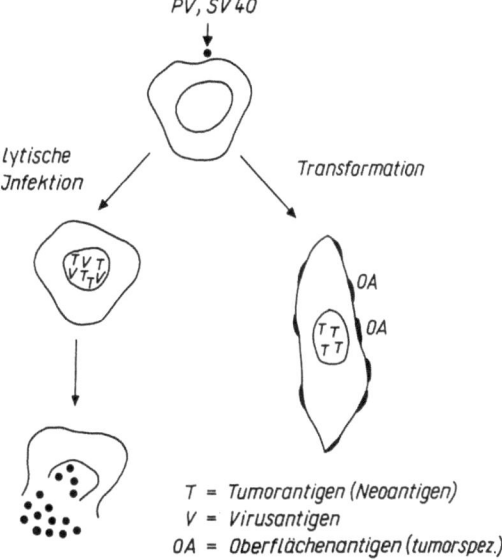

Abb. 28. Schematische Darstellung der Antigenverhältnisse der Zelle in der lytischen Phase und nach Transformation durch kleine DNS-Viren (Polyoma, SV 40, Adeno usw.)

oder Tumorantigen (T) in gleicher Weise wie bei der lytischen Infektion vorhanden, trotz der wahrscheinlichen Inkorporation des Virusgenoms. Schließlich wird in der transformierten Zelle noch ein drittes Antigen an der Oberfläche im Bereich der Zellmembran angetroffen (OA), das während der lytischen Phase der Infektion überhaupt nicht zu finden ist, das also nur aus der der Transformation zugrunde liegenden Interaktion zwischen Virusgenom und Zellgenom hergeleitet werden kann. Es ist tumorspezifisch in dem Sinne, daß es bei allen durch das gleiche Virus transformierten Zellen, also induzierten Tumoren, nachweisbar ist. Man kann mit ihm durch immunisierende Vorbehandlungen eine Transplantationsresistenz gegen die durch das gleiche Virus induzierten Tumoren auslösen. Es sei nochmals betont, daß es keinerlei immunologische Beziehung zu dem Virusantigen besitzt, d. h. mit einem Antiserum gegen die infektiösen Virions nicht reagiert. Es ist sehr wahrscheinlich, daß diese Besetzung der Zelloberfläche mit neuen zellfremden Antigenen (OA) infolge Änderung der normalen Receptoreneigen-

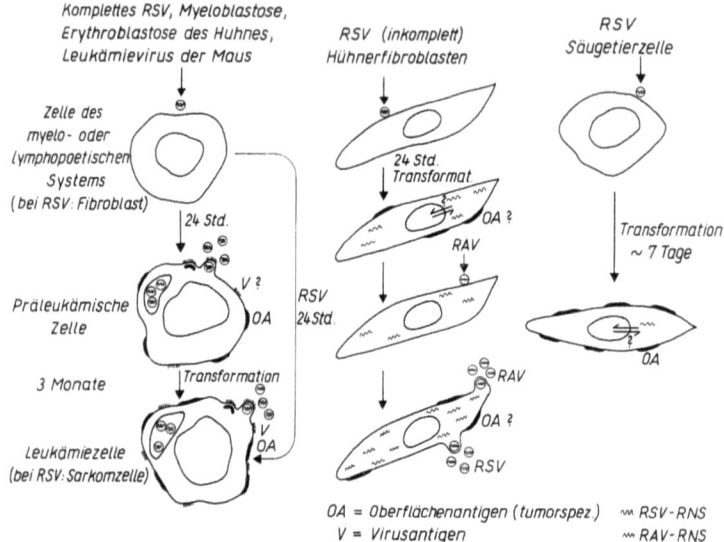

Abb. 29. Schematische Darstellung der Antigenverhältnisse nach Infektion sowie Transformation einer Zelle durch onkogene RNS-Viren: Leukämievirus der Maus und des Huhnes; Rous-Sarkomvirus (RSV) in kompletter sowie inkompletter Form und bei heterologer Wirkung auf Säugetierzellen

schaften zu bestimmten Besonderheiten der Tumorzelle (Korrelationstaubheit, Kontaktinhibition) direkte Beziehungen hat (s. später)*.

Bei den RNS-Viren wie Rous-Sarkom[95], Hühner- und Mäuseleukämien[91-93a, 97a, 83a] ist die Situation hinsichtlich der antigenen Abänderung dadurch etwas unübersichtlicher, daß hier meistens die lytische, jedoch ohne sichtbare Zellschädigung einhergehende und die transformierende Phase nicht scharf getrennt sind, sondern die Bildung von infektiösem Virus auch nach der malignen Trans-

---

* Anmerkung bei der Korrektur: Von verschiedenen Arbeitskreisen wurde in allerletzter Zeit nachgewiesen, daß virusspezifische Oberflächenantigene in verschiedenen Zelltypen auch ohne maligne Transformation bzw. noch vor deren Ausprägung nachweisbar sind.

formation der Zellen weiterläuft (Abb. 29). Es sieht im Gegenteil so aus, als ob hier eine laufende Virusvermehrung mit der Aufrechterhaltung der Malignität sogar eng gekoppelt ist. Durch das in die Zelle aufgenommene Viruspartikel werden im Zuge seiner Vermehrung zwei neue Antigene induziert: 1. Virusantigen, das den Virions, speziell ihren Proteinanteilen zukommt, und 2. ein neues tumorspezifisches, celluläres Oberflächenantigen, wie dies auch bei den DNS-Viren der Fall war. Anscheinend kann weiterhin virusspezifisches Antigen im Zuge der Abschnürung der Viruspartikel an der Oberfläche in der äußeren Zellmembran zurückbleiben und ihr zusätzlich virusspezifische Antigenqualitäten vermitteln. Für die maligne Zelltransformation dürften die neuen Oberflächenantigene, deren Bildungsmechanismus noch ungeklärt ist, von ähnlicher Bedeutung sein, wie wir dies für die DNS-Viren besprochen haben. Nach neuesten Untersuchungen von PAYNE et al.[93b] an durch Rous- und Leukosevirus der Hühner transformierten Zellen sollen auch hier zusätzlich im Zellkern auftretende Neoantigene feststellbar sein, womit die Untersuchungen von TEMIN[124a] eine weitere Untermauerung finden würden, nach denen auch bei den onkogenen RNS-Viren speziell der Hühner eine virusspezifische DNS im Zellkern in den Reproduktionsvorgang des Virusgenoms einbezogen ist.

Beim Rous-Sarkom[85] gibt es außerdem noch zwei Sonderfälle (Abb. 29), einmal den Zustand des inkompletten Rous-Virus bei den sog. defekten RSV-Stämmen, die zwar in der Lage sind, die Zelltransformation mit dem neuen Oberflächenantigen oft schon innerhalb von 24 Std zu induzieren, aber infolge eines genetischen Defekts (Fehlen der genetischen Information für die Synthese der Hüllproteine) unfähig sind, komplette infektiöse Viruspartikel zu produzieren. Letzteres ist jedoch möglich durch zusätzliche Infektion mit dem sog. RAV-Virus, einem dem Lymphomatosevirus nahestehenden Agens, das nach seiner Aufnahme in die Zelle Hüllproteine nicht nur für den eigenen Bedarf, sondern auch für die sich vermehrenden defekten RSV-Genome liefert, so daß nunmehr diese transformierte Zelle gleichzeitig infektiöses RSV-Virus, bei welchem das Rous-Virusgenom in falscher, nämlich RAV-Verpackung vorliegt, als auch RAV-Virus produziert bei gleichzeitiger Ausbildung der zur malignen Transformation wahrscheinlich in Beziehung stehenden Oberflächenantigene. Eine dritte Reaktionsweise existiert zwischen RSV-Virus und der durch dieses Virus malignisierbaren Säugetierzellen (Maus, Ratte, Hamster, Mensch), die durch komplettes RSV unter Ausbildung von Oberflächenantigenen in wenigen Tagen transformiert werden, ohne daß jedoch Virus neu gebildet wird, womit ein ähnlicher Zustand wie der bei den kleinen DNS-Viren im Zuge der Zelltransformation zu verzeichnen ist.

Es ist zu erwarten, daß die künftigen immunologischen Untersuchungen im Zuge der malignen Transformation noch weitere wesentliche Einblicke in den Vorgang der malignen Entartung ermöglichen werden. Gegenüber einer durch chemische oder physikalische Substanzen ausgelösten Cancerisierung besteht vor allem darin eine wesentliche Differenz, daß die neuen Tumorantigene, mit deren Hilfe eine Immunisierung gegen Tumorwachstum ausgelöst werden kann und die sich wahrscheinlich zum großen Teil an der Zelloberfläche befinden, bei den chemisch induzierten Tumoren individualspezifischen Charakter besitzen, d. h. selbst bei gleicher auslösender Noxe von Tumor zu Tumor unterschiedliche Antigenspezifität haben, während, wie wir sahen, bei den Virustumoren das neue Zell- speziell Oberflächenantigen für das *jeweilige* Virus charakteristisch und in allen durch das gleiche Virus ausgelösten Tumoren identisch ist. Aus beiden Tatsachen ergibt sich der Hinweis, daß die maligne Zelltransformation möglicherweise auf unterschiedlichen Wegen, d. h. durch sehr differente Änderungen des genetischen Informationsgehalts der Zelle zustande kommen kann, also im molekularen und genetischen Bereich eventuell *heterogener* Natur ist (s. GRAFFI, 1963[87]).

VI. Biochemische Charakterisierung der onkogenen Viren [98—119, 142, 146]

Wie wir aus den elektronenmikroskopischen Untersuchungen ersahen, sind am Virion zwei verschiedene morphologische Komponenten zu unterscheiden, und zwar das aus NS bestehende Nucleoid und die verschiedenen Hüllmembranen. Letztere sind nochmals unterteilbar in die aus den Capsomeren bestehende Capsidhülle, die das Nucleoid unmittelbar umgibt und mit diesem gemeinsam als Nucleocapsid bezeichnet wird, und die als Envelopes bezeichneten Sekundärhüllen des Virions, die bei großen DNS-Viren und vor allem bei den RNS-Viren vorkommen. Schon auf Grund der morphologischen Entstehungsweise kann angenommen werden, daß beide Sorten von Hüllmembranen zwar in der Hauptsache aus Proteinen, aber doch von unterschiedlicher chemischer Zusammensetzung sein dürften. Die teilweise Ableitung der äußeren Hüllmembran von der Zellmembran läßt auch chemisch eine gewisse Verwandtschaft vermuten, die auch aus immunologischen Untersuchungen, z. B. Gegenwart von Forssman-Antigen bei den Hühnerleukoseviren usw. erkennbar ist. Die biochemischen Untersuchungen der onkogenen Viren wurden 1. mit den üblichen analytischen Methoden der Biochemie, 2. mit histochemischer Methodik im elektronenmikroskopischen Bereich unter Anwendung spezifischer Fermente (DNase, RNase, Proteasen) sowie 3. speziell beim Fehlen genügend hoch gereinigten Virusmaterials durch spezifische Resistenzprüfungen gegenüber verschiedenen Fermenten, lipotropen Substanzen und anderen Chemikalien mit bekannter spezifischer Wirkung und anschließender Prüfung der Änderung der Virusinfektiosität durchgeführt.

In einer Tabelle (Tab. 3) sollen einige Befunde zur biochemischen Charakterisierung der onkogenen Viren summarisch dargestellt werden[98,111a,103,142,146,103]. Ihre Einteilung auf Grund der Natur ihrer NS (kleine und große DNS-Viren, RNS-Viren) wurde schon früher besprochen (s. Tab. 1). Die Molekulargewichte der NS sind für die kleinen DNS-Viren bekannt und liegen zwischen 3×10^6 bis 6×10^6 für die Papovagruppe und bei etwa 10^7 für die verschiedenen Adenovirustypen. Kürzlich konnte auch aus einem Leukämievirus der Maus, und zwar aus dem Rauscher-Virus intakte RNS unter sehr schonenden Bedingungen gewonnen werden (MORA et al.[115]) mit einem Molekulargewicht von 13×10^6, also ungefähr von gleicher Größenordnung wie die DNS der Adenoviren. Es han-

Tabelle 3. *Chemisches Verhalten verschiedener onkogener Viren*

Agens	Verhalten gegenüber					Chemische Zusammensetzung				Molgewicht der NS	Struktur der NS
	56 °C	Trypsin	Formalin	Desoxychols.	Äther	Proteine	Lipoide	RNS	DNS		
Rous-Virus	+	—	+	+	+	+	+	+++			
Hühnerleukosevirus	+	+	+	+	+	+	+	+++			
Mäuseleukämievirus	+	—	+	+	—	+	+	+++		$13 \cdot 10^6$	Einstrang
Bittner-Virus	+	+	+	+		+	(+)				
Polyomavirus	—	+	—	—	—	+	—	—	+	$3 \cdot 10^6$	Doppelstrang O
Shope-Papillom						+			+		
Shope-Fibrom						+		(+)	+	~ $5 \cdot 10^6$	
SV 40		+				+		—	+	$3 \cdot 10_7$	Doppelstrang
Adeno	+	+		+	—	+	—	—	+	$1 \cdot 10^7$	O

+ = empfindlich bzw. vorhanden   — = unempfindlich bzw. fehlt

delt sich dabei um eine Einzelstrang-RNS, während die aus den Papovaviren isolierte DNS in allen Fällen doppelsträngig ist und gleichzeitig Ringform besitzt. Auf eine nähere chemische Charakterisierung der NS onkogener Viren soll hier nicht eingegangen werden, da Herr BIELKA zu diesem Problem gesondert Stellung nehmen wird.

Die chemische Natur der verschiedenen onkogenen Viren konnte vor allem über die Gewinnung infektiöser NS und deren Resistenz gegenüber DNase bzw. RNase geklärt werden. Die Isolierung infek-

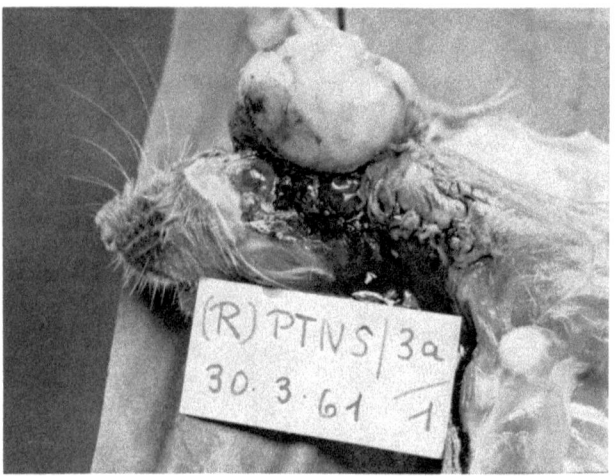

Abb. 30. Knochentumor am Schädeldach einer Ratte und kleines subcutanes Lipom (rechts unten) nach Applikation von DNS aus 2 Jahre laufend transplantierten Polyomasarkom des Goldhamsters (nach GRAFFI und GIMMY; Acta biol. med. germ. **1964**, 210)

tiöser NS gelang bereits vor einigen Jahren aus Polyomavirus (DI MAYORCA u. a.[119], GRAFFI und FRITZ[118]) (s. auch Abb. 7), wobei es sich um DNS handelt, sowie aus Virusmaterial der myeloischen Mäuseleukämie, die sich auf Grund ihrer Empfindlichkeit gegenüber RNase als RNS erwies (BIELKA und GRAFFI, 1959[111], MOLONEY und DALTON, 1961[112]). In der Folgezeit konnten dann infektiöse DNS aus Shopeschem Papillomvirus[109], SV40- und Adenovirus gewonnen werden. Der Nachweis der infektiösen Natur dieser NS wurde dadurch erbracht, daß entweder in vivo oder in vitro neue Virusbildung oder eine Geschwulsterzeugung erzielt werden konnte, letzteres beispielsweise mit RNA aus der myeloischen Mäuseleukämie[111,112] und der DNS aus Kaninchenpapillom-material (ITO, 1961)[109]. In vorläufigen Untersuchungen konnten wir auch mit DNS-Präparationen aus virusfreien, bereits 2 Jahre lang laufend transplantierten Polyomatumoren der Ratte und des Goldhamsters eine Geschwulstinduktion erzielen (7 Tumoren/29 Ratten) (Abb. 30). Dagegen ist trotz mehrfacher Bemühungen die Isolierung einer infektiösen RNS aus Rous-Virusmaterial bisher ergebnislos verlaufen (s. [98]).

Die Temperaturempfindlichkeit der verschiedenen onkogenen Viren hängt in erster Linie mit besonderen Eigenschaften ihrer Hüllproteine zusammen, und hierzu ist generell zu vermerken, daß die kleinen DNS-Viren als nackte Nucleocapside im allgemeinen eine um etwa 10 bis 15 °C höhere Temperatureinwirkung vertragen als die onkogenen RNS-Viren der Maus und der Hühner (Leukämieviren, Milchfaktor, Rous-Virus) sowie die großen DNS-Viren der Poxgruppe (Shope-Fibrom usw.), die alle durch sekundäre Hüllmembranen im Sinne eines Envelopes charakterisiert sind. Eine ähnliche Aufteilung der onkogenen Viren in resistente und empfindliche Arten finden wir gegenüber einer Behandlung mit Fettlösungsmitteln wie Äther und Aceton[99,98] oder gegen bestimmte, vor allem Lipoproteide angreifende Detergenzien (Desoxycholsäure, Digitonin)[111a,88]. Auch diesen Reagentien gegenüber sind die kleinen nackten DNS-Viren ziemlich unempfindlich, während vor allem sämtliche onkogenen RNS-Viren, aber auch die großen DNS-Viren empfindlich sind. Dies spricht für die Gegenwart von Lipoproteinen in den Envelopehüllen dieser Virions. Bei den RNS-Viren läßt sich dieser Lipoidgehalt schon aus der topographischen Bildungsweise, nämlich der Abschnürung an der Zellmembran ziemlich zwanglos ableiten, ist doch bekanntlich die äußere Zellgrenzphase, die bei der Virusabschnürung in die äußere Virushülle wenigstens teilweise mit eingeht, sehr reich an Lipoiden, speziell auch Phosphorlipoiden. Der hohe Lipoidgehalt verschiedener RNS-Viren, z. B. der Hühnermyeloblastose wie auch des Rous-Sarkoms konnte auch mit analytischer Methodik am angereicherten Virusmaterial nachgewiesen werden. So geben z. B. BEARD u. Mitarb.[103] für hoch gereinigtes Myeloblastosevirus einen Lipoidgehalt von 35% des Trockengewichtes an bei einem RNS-Gehalt von 2,17%. Die inaktivierende Wirkung der proteolytischen Fermente (Trypsin, Papain) auf RNS-Viren, z. B. der myeloischen Mäuseleukämie[114], spricht für die essentielle Bedeutung der Hüllproteine für die Infektiosität der

Virions. Gleiches gilt für die Unversehrtheit der aus Lipoproteinen bestehenden äußeren Hüllen auf Grund der leichten Virusinaktivierbarkeit durch lipotrope und lipolytische Agentien sowie Wärmeeinwirkung. Dagegen scheint die Inaktivierungswirkung von Formalin nicht nur auf einer Einwirkung auf die Proteine des Virus, sondern auch auf dessen NS-Komponente zu beruhen, da gegenüber Formalin auch die nackten Nucleocapside der kleinen DNS-Viren wie z. B. Polyoma sehr empfindlich sind[98].

Abschließend zu diesem Kapitel seien noch einige biochemische Besonderheiten einiger Viren kurz erwähnt, wie z. B. die schon genannte hohe ATPase-Aktivität des Myeloblastosevirus (BEARD[105]), die auf Grund histochemischer Reaktionen im elektronenmikroskopischen Bereich an der äußeren Hülle des Virus lokalisiert ist. Beim Rous-Sarkom konnte EPSTEIN[50] an Ultradünnschnitten des Virus durch Behandlung mit RNase, DNase und Proteasen überzeugend nachweisen, daß das Nucleoid hauptsächlich aus RNS besteht und die Hüllen vorwiegend aus Proteinen aufgebaut sind. Eine besondere Wirkungsweise des Shopeschen Kaninchenpapilloms in der Wirtszelle ist die Induktion einer starken Arginaseaktivität, über deren Bedeutung für den Cancerisierungsprozeß noch völlige Unklarheit besteht.

Eine Reihe von Arbeiten beschäftigen sich in den letzten Jahren unter Anwendung der verschiedensten Methoden der Nucleinsäurehybridisierung mit Fragen einer Homologie der Virus-NS, 1. bei verschiedenen onkogenen Viren, wie etwa den verschiedenen Adenotypen untereinander, und 2. der Homologie zwischen Virus-NS und cellulärer NS (speziell DNS oder Messenger-RNS) vor und nach der malignen Transformation. Diese noch nicht völlig klar übersehbaren Resultate möchte ich hier nur kurz erwähnen, zumal sicher auch auf diese Probleme Herr BIELKA in seinem Vortrag ausführlicher eingehen wird.

Zusammenfassend ergibt sich, daß auch im Hinblick auf die biochemische Zusammensetzung und Struktur die bereits in morphologischer und biologischer Hinsicht nachgewiesene Heterogenität bei den onkogenen Viren vorhanden ist.

Es gibt auch keine sichtbaren Hinweise dafür, daß die große heterogene Gruppe der onkogenen Viren wenigstens durch partielle Übereinstimmung ihres genetischen Informationsgehaltes in ihrem Genom als eine einheitliche Gruppe mit identischer Wirkungsweise charakterisiert ist, so daß, wie bereits kurz angedeutet wurde, mit der Möglichkeit gerechnet werden muß, daß die maligne Transformation durch die Einführung sehr unterschiedlicher, neuer genetischer Informationen in die Zelle bewerkstelligt werden kann. Schließlich ist auch in biochemischer Hinsicht keine scharfe Grenze zwischen onkogenen und nichtonkogenen, rein cytociden Viren festzustellen, bei denen die gleiche Mannigfaltigkeit des biochemischen Aufbaues zu verzeichnen ist, entsprechend ihrer morphologischen Struktur und systematischen Stellung.

VII. Wirkung der onkogenen Viren auf die Zelle[120—127,142,146]

Viren können, wie bereits kurz erwähnt wurde, zu tierischen Zellen in mehrfacher Weise in Beziehung treten:

1. In Form eines lytischen Wirts-Virus-Zellverhältnisses, wobei sich das Virus bei intensiver Neubildung infektiöser Virions gegenüber der Zelle cytocid verhält und sie oft bis zum Zelltod schädigt. Dieses Verhältnis finden wir z. B. beim Polyomavirus gegenüber Mäuseembryonalzellen in der Gewebekultur als vorrangige Reaktionsweise.

2. Ein Virus-Wirts-Zellverhältnis, bei dem sich das Virus in der Zelle ebenfalls vermehrt, ohne daß jedoch die Zelle zugrunde geht. Dieser Fall ist bei verschiedenen onkogenen RNS-Viren realisiert und wird als moderierter Zustand des Virus bezeichnet.

3. Die Integration oder Inkorporation des Virusgenoms in das Zellgenom, wonach sich das Virusgenom (oder zumindest Teile desselben) zwar ebenfalls weiter vermehrt, aber im Gegensatz zu 1. und 2. nur noch synchron mit dem Zellgenom. Es ist also ein ähnlicher Zustand wie die Lysogenie bei den Bakteriophagen. Dieses inkorporierte Virusgenom kann in der betreffenden Zellart neue genetische Effekte hervorrufen. Vermutlich spielt die Inkorporation des Virusgenoms in das Zellgenom bei der onkogenen Wirkung und der malignen Umwandlung der Zellen, speziell durch die kleinen DNS-Viren (Polyoma-, SV40-, Adenoviren), die entscheidende Rolle.

Über die Virus-Wirts-Reaktionsweise 1. und 2., d. h. die cytocide und moderierte Virus-Zell-Relation mit starker oder beträchtlicher Neubildung von Virions ist bereits in dem Abschnitt über die elektronenmikroskopischen Befunde bei der intracellulären Virusvermehrung das Wichtigste gesagt worden.

Man kann diese Phänomene besonders in der Gewebekultur verfolgen, indem man z. B. Kulturen von embryonalen Mäusezellen mit Polyomavirus beimpft. Nach 3 bis 4 Tagen treten in zunehmendem Maße cytopathische Veränderungen an den Zellen auf bei gleichzeitigem starken Anstieg des Virustiters in der Nährlösung. In 5 bis 6 Tagen ist die größte Zahl der Zellen in der Monolayerkultur infolge der starken intranucleären Virusneubildung dem Untergang verfallen und nur vereinzelte Zellen überleben die Infektion. Diese aber sind besonders disponiert, die dritte Form der Virus-Wirtszell-Relation zu erfahren, nämlich die Virus-Genom-Inkorporation in das Zellgenom, womit gleichzeitig neue Zelleigenschaften auftreten[126,124,121]. Bei anderen Zellarten, z. B. embryonalen Hamsterzellen[124,126] in der Kultur, kommt es nach einer gleichartigen Beimpfung mit Polyomavirus infolge der anders gearteten genetischen Konstitution der Zelle gar nicht zum cytociden, also mit Virusvermehrung einhergehenden Virus-Zell-Verhältnis, sondern es dominiert von vornherein die zur malignen Transformation führende Virusintegration in das Zellgenom ohne Neubildung von Virions. Bei der Beimpfung von Mäuseembryonalzellen oder HeLa-Zellen mit dem Virus der myeloischen Mäuseleukämie oder anderen Leukämieviren kann über mehrere Zellgenerationen und Kulturpassagen, also über viele Wochen *ohne* maligne Transformation der Zellen, aber auch völlig ohne Anzeichen einer cytociden Wirkung auf die Zellen eine Virusvermehrung und Freisetzung in die Nährlösung nachgewiesen werden.

Die gleiche kontinuierliche Virusvermehrung findet auch statt, wenn man in vivo malignisierte Leukämiezellen der Maus oder des Huhnes in vitro züchtet, ebenso bei der Kultivierung des Rous-Sa und des den Milchfaktor enthaltenden Mamma-Ca der Maus. Bei diesen RNS-Viren werden also auch von den bereits malignisierten Zellen Virions laufend weiter produziert (s. Kap. III).

Für die Analyse und das Verständnis der geschwulstauslösenden Wirkung der onkogenen Viren ist, wie bereits gesagt wurde, die mit Zelltransformation und bei den DNS-Viren mit Virusgenomintegration verbundene Virus-Wirtszell-Relation von entscheidender Bedeutung. Schon 1 bis 2 Wochen nach Beimpfung sind an vielen der in der Kultur überlebenden und nicht dem Zelltod verfallenden Zellen die ersten Anzeichen einer Transformation nachweisbar in Form von Wachstumsfoci mit zwar noch zweidimensionaler, aber ansonsten stark unregelmäßiger Anordnung der Zellen. 3 bis 4 Wochen nach der Infektion der Kulturen wird die Wachstumsanomalie noch deutlicher, indem sich dreidimensionales, also turmförmiges Zellwachstum in den Proliferationsfoci einstellt. Die Transformation, die sich in diesen Wachstumsanomalien sowie in weiteren Eigenschafts- und Formänderungen der Zellen kundgibt, hat also fortschreitenden Charakter. Folgende Eigenschaftsänderungen können bei der malignen Zelltransformation im Zuge der Integration onkogener DNS-Viren, wie z. B. des Polyomavirus, beobachtet werden[126,124,121]:

1. Gesteigerte Wachstums- und Vermehrungstätigkeit der Zellen, was vor allem einer gesteigerten DNS-Synthese gleichkommt.

2. Das bereits erwähnte dreidimensionale sogenannte Criss-Cross-Wachstum, das darauf zurückzuführen ist, daß den Zellen die normale Kontakthemmung (contact inhibition) verlorengeht. Diese besteht darin, daß Zellen in der Monolayerkultur ihre Wachstumstätigkeit einstellen, sobald sie sich mit ihren Oberflächen berühren, wodurch verhindert wird, daß dreidimensionales Wachstum in der normalen Kultur zustande kommt (ABERCROMBIE[130]).

3. Erhöhte Clonierbarkeit der Zellen als Zeichen gesteigerter Autonomie.

4. Vermehrte Säurebildung in der Kultur durch gesteigerte Glykolyse.

5. Ausbildung neuer Zellantigene, die sich von den viralen Antigenen unterscheiden (s. Kap. V).

Wichtig sind dabei vor allem, wie bereits ausgeführt wurde, die im Bereich der Zellmembran sich ansammelnden neuen Antigene, weil sie wahrscheinlich in erster Linie für den Verlust der normalen Kontaktinhibition verantwortlich sind, indem sie die normalen Zellreceptoren an der Oberflächenmembran einfach verdrängen. Die Ausbildung neuer Zellantigene bei gleichzeitigem Fehlen viraler Antigene ist ein wichtiger Hinweis für die Persistenz des Virusgenoms in der transformierten Zelle, das in allen folgenden Zellgenerationen nachweisbar ist, etwa mit der Immunofluorescenzmethode oder mit dem Cytotoxtest. Diese Antigene könnten entweder auf eine genetische Wirkung des integrierten Virusgenoms oder aber auch auf einen Effekt des Virusgenoms auf das Zellgenom etwa im Sinne einer Derepression bestimmter Zellgene zurückgeführt werden.

6. Stark vermehrtes Auftreten von Chromosomenabnormitäten, speziell Chromosomenbrüchen in den transformierten Zellen, die gleichfalls über viele Generationen nachweisbar sind und damit ebenfalls für die Persistenz des Virusgenoms in der transformierten Zelle sprechen. Es ist möglich, daß diese durch die Chromosomen-

brüche ausgelösten Chromosomenumbauten in der Zelle für die volle Ausprägung der Malignität von Bedeutung sind.

7. Überimpfbarkeit der transformierten Zellen auf isologe Tiere und Auslösung maligner Tumoren. Dieser letzte und eindeutigste Beweis der malignen Zelltransformation ist jedoch meistens erst mehrere Wochen nach der Virusinfektion gegeben, woraus sich besonders deutlich ergibt, daß die Transformation im Laufe mehrerer Wochen fortschreitenden Charakter aufweist.

Bei den onkogenen RNS-Viren (Leukämie des Huhnes und der Maus, Rous-Sarkom), die entsprechend dem moderierten Virus-Wirtszell-Verhältnis sich dauernd und oft unvermindert auch in den durch sie maligne transformierten Zellen vermehren können, findet die maligne Entartung der Target-Zellen ebenfalls allmählich innerhalb weniger Tage (Rous-Sarkom, Hühnerleukämie) oder zuweilen auch erst in Wochen oder Monaten (Leukämien und Mamma-Ca der Maus) statt. Auch bei dieser Transformation scheint die Bildung eines neuen Zellantigens im Bereich der Zellmembran für die Malignisierung von großer Bedeutung zu sein (s. auch Abb. 29). Es ist vorstellbar, daß bei den RNS-Viren die RNS nicht nur als Matrize der eigenen Reduplikation dient, sondern ein Teil der in ihr enthaltenen genetischen Information gleichzeitig direkt als Messenger fungiert, durch den unter Benutzung der Ribosomen sowie des Protein-synthetisierenden Apparates (s-RNS, pH 5-Enzyme usw.) der Zelle außer den Hüllproteinen für die Virions noch die neuen mit der malignen Transformation eventuell direkt kausal verbundenen Zellantigene synthetisiert werden*. Es würde damit ein Unterschied zu den integrierten onkogenen DNS-Viren darin bestehen, daß bei letzteren die inkorporierte DNS (entweder direkt oder indirekt über das Zellgenom) erst über eine neugebildete Messenger-RNS diese neuen Zellantigene ausbilden würde. Es sei jedoch andererseits an die bereits erwähnten Ergebnisse von TEMIN[125] erinnert, der auf Grund der von ihm festgestellten Hemmwirkung von Cytomycin auf die Rous-Virusbildung auch für das Genom bestimmter RNS-Viren eine Neusynthese über eine im Zellkern verankerte, durch die Virus-RNS induzierte Virus-DNS annimmt.

Diese von DULBECCO und VOGT, STOKER, SACHS und MEDINA u. a. mit dem Polyomavirus und von RUBIN und HANAFUSA, s. [91]

---

* s. jedoch Fußnote S. 220.

sowie TEMIN[125], BEARD u. a. mit dem Rous- und Hühnerleukosevirus durchgeführten Transformationsversuche in vitro betreffen das zentrale Problem virologischer Cancerogenese, von deren Fortsetzung man weitere entscheidende Fortschritte bei der Aufklärung des Mechanismus der malignen Transformation erwarten kann.

VIII. Mögliche Bedeutung von Viren für die Entstehung menschlicher Tumoren[128—141]

Die außerordentlich weite Verbreitung von Virustumoren im Tierreich drängt die Frage auf, ob nicht auch menschliche Tumoren virusbedingte Erkrankungen sein könnten, zumal einige gutartige Geschwülste des Menschen wie Papillome und Warzen sicher viraler Ätiologie sind und außerdem durch Versuche in der Gewebekultur mit Rous-Virus und SV40-Virus gezeigt werden konnte, daß auch menschliche Zellen einer malignen Transformation durch bereits bekannte Viren unterliegen. Schließlich ist noch an die weite Verbreitung von Adenoviren beim Menschen in latenter Form, vor allem im Respirationstrakt zu erinnern, für die eine onkogene Wirkung im Tierversuch eindeutig nachgewiesen wurde.

Für den Nachweis einer Beziehung zwischen Viren und menschlichen Tumoren sind, da der homologe Infektionsversuch entfällt, vor allem folgende Methoden anwendbar:

1. Der heterologe Versuch zellfreier Tumorübertragung menschlicher Geschwülste unter Verwendung neugeborener Tiere verschiedenster Species und unter eventueller Einbeziehung der Gewebekultur mit der Absicht einer vorausgehenden Virusanreicherung.

2. Der elektronenmikroskopische Nachweis von Viruspartikeln in menschlichen Tumoren in vivo und nach Kultivierung in vitro.

3. Der Nachweis von spezifischen neuen Antigenen in bestimmten Tumortypen möglichst unter Verwendung von homologem Serum in erster Linie von Tumorträgern.

Die Wahrscheinlichkeit einer Virusätiologie erscheint vor allem für die Leukämien und andere vom blutbildenden Gewebe ausgehende Tumoren wie Reticulosen, Lymphogranulomatosen relativ hoch, weil derartige Erkrankungen bei sämtlichen Tierarten zumindest in einem sehr beträchtlichen Maße sicher viraler Herkunft sind (Kaltblüter, Hühner, Nagetiere, Katzen, wahrscheinlich auch Rind). Es ist daher nicht verwunderlich, daß eine Reihe von Autoren sich speziell der Frage der Virusätiologie dieser Krankheits-

gruppe beim Menschen zugewandt hat. Es wurden bisher folgende Ergebnisse erzielt:
Versuche einer heterologen zellfreien Übertragung mit menschlichem Tumormaterial bei Verwendung verschiedener Tierarten und unter Einbeziehung von Anreicherungsverfahren in der Kultur oder durch fraktionierte Zentrifugation sind bis jetzt praktisch negativ verlaufen sowohl mit Leukämien und ähnlichen Erkrankungen als auch sonstigen Tumortypen. Dies gilt für umfangreiche Untersuchungen von GRACE et al.[134], MOORE et al.[136] gleichermaßen wie

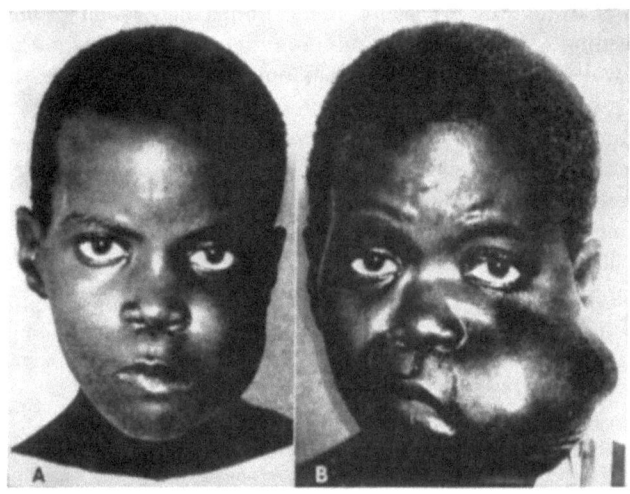

Abb. 31. Burkitt-Lymphom bei einem 10jährigen Negerkind im Bereich der linken Maxilla. Rechts Zustand 8 Wochen nach der Aufnahme links (nach BURKITT)

für unsere eigenen diesbezüglichen Versuche mit rund 10000 Tieren (Mäuse, Ratten, Goldhamster). Es ist hervorzuheben, daß für den heterologen Versuchsansatz die Erfolgsaussichten von vornherein nicht sehr groß sind, da auch bei tierischen Leukämien die heterologe Übertragung, wenn überhaupt, erst nach starker Virulenzsteigerung des Virus durch häufiges Passagieren (z. B. Übertragung des Virus der myeloischen Mäuseleukämie auf Ratten[25b]) möglich war.

Einige wichtige Hinweise zur möglichen Virusätiologie menschlicher Leukämien kamen aus der geographischen Krebsforschung. Vor einigen Jahren wurde von BURKITT in Äquatorialafrika bei Kindern ein stark gehäuftes Vorkommen eines malignen Lymphoms

(Abb. 31) im Bereich des Halses und der Unterkiefergegend festgestellt, dessen Verbreitung in Abhängigkeit von der geographischen Lage und der klimatischen Verhältnisse (Fehlen in höheren Gebirgslagen) an eine durch Insekten, eventuell Stechmücken, verbreitete Krankheit denken ließ. Von verschiedenen Autoren wurden mit Hilfe des Elektronenmikroskops zum Teil nach Kultivierung in vitro Viruspartikel nachgewiesen, die teils der Herpesgruppe ange-

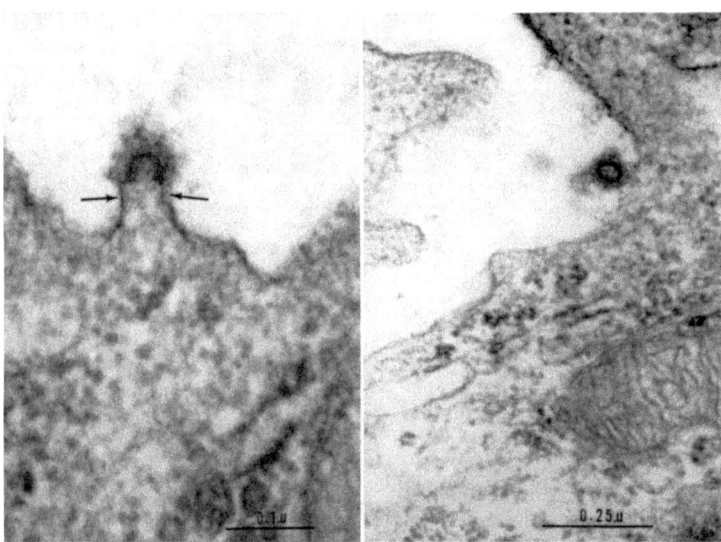

Abb. 32. Viruspartikel im Stadium der Abschnürung von der Zellmembran einer menschlichen Leukämiezelle (nach L. DMOCHOWSKI: Cancer Res. **25**, 1654 (1965)]

hörten (EPSTEIN[133], STEWART et al.), teils sich als Reoviren erwiesen (SIMONS und ROSS[139]) oder auch als Mycoplasmen angesprochen wurden (DALLDORF et al.[129b]). Die ätiologische Bedeutung dieser Viren für das Burkitt-Lymphom ist jedoch völlig offen, da bei der starken viralen Durchseuchung der Gesamtbevölkerung durchaus damit gerechnet werden muß, daß es sich dabei um Passengerviren handelt, die sich sekundär in dem Tumor eingenistet haben. Nach neuesten Ergebnissen soll es KLEIN und unabhängig auch G. u. W. HENLE gelungen sein, in einer großen Zahl von Burkitt-Tumoren ein neues, jeweils identisches Zellantigen zu finden, was in Analogie zu tierischen Virustumoren mit der Virusätiologie gut übereinstimmen würde.

In menschlichen Leukämien, die zum Teil unter Einbeziehung einer Züchtung der Leukämiezellen in vitro vor allem auf embryonaler menschlicher Grundlage durchgeführt wurden, sind elektronenmikroskopisch folgende für Virus oder auch Mycoplasma sprechende Befunde erhoben worden:

Von DMOCHOWSKI[130-132] wurden bei sehr verschiedenen, vor allem lymphatischen Leukämieformen in dem Leukämiegewebe und den Leukämiezellen Partikel festgestellt, die einerseits den Viren der

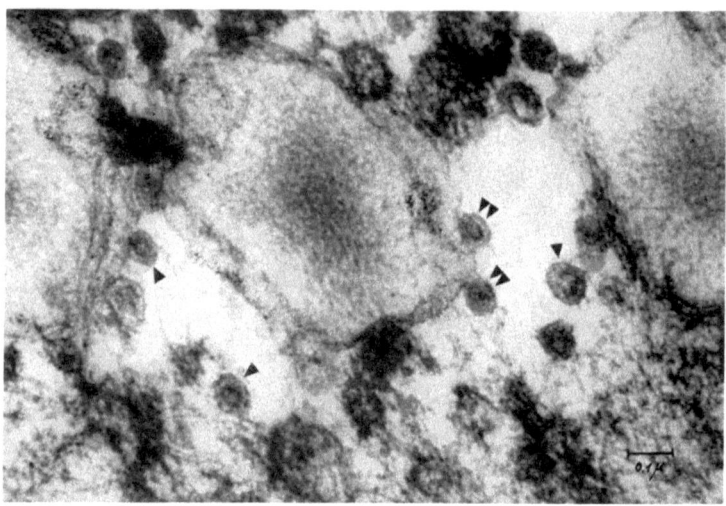

Abb. 33. Viruspartikel (Pfeile) im Cytoplasma einer myeloischen Leukämie des Menschen im Stadium eines akuten Schubs nach viertägiger Züchtung der Leukämiezellen in vitro. Ultradünner Schnitt im elektronenmikroskopischen Bild; 100000fach. Doppelpfeile: Abschnürung der Viruspartikel an cytoplasmatischen Membranen (nach GRAFFI, BIERWOLF, BAUMBACH, BLANKENHAGEL, WIDMAIER und RANDT, 1964)

Mäuseleukämien sehr ähnlich sind (Abb. 32), wobei sogar einzelne Abschnürungen an Zellmembranen dargestellt werden konnten. Eine andere, etwas größere Partikelform wird eher als ein zu den Mycoplasmen (PPLO) gehöriges Formelement angesehen. Von PORTER et al.[138] wurden im Blut bei akuten Leukämien gleichfalls Partikel von 80 bis 90 mμ Durchmesser mit ein bis zwei Membranen gesehen. In eigenen Untersuchungen[135] fanden wir bei 8 von 30 erwachsenen Patienten mit akuter myeloischer Leukämie nach kurzer Züchtung der Leukämiezellen in der Gewebekultur virusverdächtige Partikel, und zwar 1. kleinere, kompakte Formen von etwa 40 mμ Durchmesser im Zellkern, 2. 60 bis 80 mμ große, mit einer Doppelmembran und einem Nucleoid ausgestattete Elemente im Cytoplasma (Abb. 33), wobei Abschnürungen der Partikel von glatten Membranen des endoplasmatischen Reticulums wahrscheinlich gemacht werden konnten. Weitere virusverdächtige Befunde an menschlichem Leukämiematerial liegen noch von BERNHARD et al.[141], BRAUNSTEINER[128a], BERGOLZ[128] u. a. vor. Von NEGRONI[137] wurden aus einer größeren Zahl menschlicher Leukämien *Mycoplasmen* in der Gewebekultur isoliert, die auch elektronenmikroskopisch nachgewiesen werden konnten.

Zusammenfassend ist jedoch zu sagen, daß die ätiologische Bedeutung von Viren für menschliche Tumoren ein noch völlig offenes Problem darstellt und die dargelegten Befunde bestenfalls als eine Verdachtsverdichtung in dieser Richtung für bestimmte Leukämieformen betrachtet werden können. Vielleicht sind immunologische Untersuchungsreihen und Fahndungen nach neuen, durch Viruswirkung induzierten, spezifischen Zellantigenen, wie sie von SABIN, MELNICK u. a. vorgeschlagen wurden, für die Zukunft ein brauchbarer Weg bei der Fährtensuche nach Virus auch in menschlichen Tumoren.

Nichtsdestoweniger hat sich die breite experimentelle Bearbeitung des Tumorvirusproblems für die Krebsforschung schon jetzt gelohnt, da speziell durch die Analyse der Virus-Zell-Relation tiefe Einblicke in das fundamentale Problem der Carcinogenese, nämlich in den Mechanismus der malignen Transformation auf cellulärer Ebene möglich wurden.

Literatur

zu Abschnitt I: Historische Entwicklung

[1] BITTNER, J., J.: Science **84**, 162 (1936).
[2] BORREL, A.: Ann. Inst. Pasteur **17**, 81 (1903).
[3] DURAN-REYNALS, F.: Virus-induced tumors and the virus theory of cancer. In The Physiopathology of Cancer, p. 298. New York: 1953.
[4] ELLERMANN, V., u. O. BANG: Zbl. Bakt. Abt. I, **46**, 595 (1908).
[5] GROSS, L.: Oncogenic Viruses. Band 11. Pergamon Press 1961.
[6] — Proc. Soc. exp. Biol. (N. Y.) **76**, 27 (1951).
[7] — Proc. Soc. exp. Biol. (N. Y.) **83**, 414 (1953).
[9] GYE, W. E.: Lancet **209**, 109 (1925).
[9] OBERLING, CH.: Le Cancer. Paris: Gallimard 1954.
[10] —, and M. GUERIN: Advanc. Cancer Res. **2**, 354 (1954).

[11] ROUS, P.: J. Amer. med. Ass. **56**, 198 (1911).
—, and J. B. MURPHY: J. exp. Med. **19**, 52 (1914).
[12] SANARELLI, G.: Zbl. Bakt. Abt. I, **13**, 865 (1898).
[13] SHOPE, R. E.: J. exp. Med. **56**, 803 (1932).
[14] — J. exp. Med. **58**, 607 (1933).
[15] STANLEY, W. M.: Ann. N. Y. Acad. Sci. **71**, 1100 (1958).
[16] ZILBER, L. A.: Progr. exp. Tumor Res. **1**, 2 (1960); **7**, 1 (1965).

zu Abschnitt II: Onkogene Viren und die durch sie erzeugten Tumoren
zusätzlich zu den in Abschnitt I angeführten Literaturzitaten [1,4,5,7,9,10,12,13,14,16]

[17] AHLSTRÖM, C. G., and N. FORSBY: J. exp. Med. **115**, 839 (1962).
[18] ANDERVONT, H. B.: Ann. N. Y. Acad. Sci. **54**, 1004 (1952).
[19] BIERWOLF, D., H. BIELKA und A. GRAFFI: Geschwulsterzeugung durch Viren. In Handbuch der experimentellen Pharmakologie, Band 16, Teil 12, S. 243, Berlin-Göttingen-New York: Springer 1966.
[20] BRYAN, W. R.: J. nat. Cancer Inst. **6**, 225 (1946).
[21] — J. nat. Cancer Inst. **16**, 285 (1955); **16**, 843 (1956).
[22] BURMESTER, B. R., C. O. PRICKETT, and T. C. BELDING: Cancer Res. **6**, 189 (1946).
[23] DMOCHOWSKI, L.: In Cancer. Progress Volume, p. 214. London: 1957.
[23a] EDDY, B. E., G. S. BORMAN, W. H. BERKELEY, and R. D. YOUNG: Proc. Soc. exp. Biol. (N. Y.) **107**, 191 (1961).
[24] FRIEND, CH.: J. exp. Med. **105**, 307 (1957).
[24a] FURTH, J.: J. exp. Med. **58**, 254 (1933).
[25] GRAFFI, A., H. BIELKA, F. FEY, F. SCHARSACH und R. WEISS: Naturwissenschaften **41**, 503 (1954).
[25a] —, F. FEY und H. BIELKA: Acta haemat. (Basel) **15**, 145 (1956).
[25b] —, u. J. GIMMY: Naturwissenschaften **44**, 518 (1957).
[26] — Progr. exp. Tumor Res. (Basel) **1**, 112 (1960).
[27] —, u. H. BIELKA: Probleme der experimentellen Krebsforschung. Leipzig: Akad. Verlagsges. Geest & Portig K. G. 1959.
[28] HAGUENAU, F., and J. W. BEARD: The avian sarcoma leukosis complex, its biology and ultrastructure. In Tumors induced by Viruses. Herausgeber: Dalton und Haguenau. New York/London: Academic Press 1962.
[29] HARRIS, R. J. C.: In Viruses, Nucleic Acids and Cancer, p. 331. Baltimore: 1963.
[30] — In Cancer. Progress Volume, p. 1. London: 1953.
[31] HUEBNER, R. J., W. P. ROWE, and W. T. LANE: Proc. nat. Acad. Sci. (Wash.) **48**, 2051 (1962).
[31a] LUCKÉ, B. J.: J. exp. Med. **68**, 457 (1938).
[32] MAZURENKO, N. P.: Die Rolle der Viren in der Ätiologie der Leukosen (russ.). Kiew: Verlag Gosmediskat 1962.
[32a] MELNICK, J. L.: Science **135**, 1128 (1962).
[33] MOLONEY, J. B.: J. nat. Cancer Inst. Monograph No. 4 (1960).
— Fed. Proc. **21**, 19 (1962).
[34] RAUSCHER, F. J.: J. nat. Cancer Inst. **29**, 515 (1962).
[35] SVOBODA, J.: Folia biol. (Kraków) **7**, 46 (1961).

35a SCHMIDT, F.: Naturwissenschaften **41**, 504 (1954).
35b SCHMIDT-RUPPIN, K. H.: Oncologia (Basel) **17**, 242 (1964).
36 STANSLY, P. G.: Progr. exp. Tumor Res. (Basel) **3**, 216 (1963).
37 STEWART, S. E., B. E. EDDY, and M. F. STANTON: Progr. exp. Tumor Res. **1**, 67 (1960).
38 TRENTIN, J. J., Y. YABE, and G. TAYLOR: In Viruses, Nucleic Acids and Cancer, p. 559. Baltimore: 1963.

zu Abschnitt III: Ultrastruktur und Bildungsweise der onkogenen Viren Übersichten:

39 BERNHARD, W.: Cancer Res. **18**, 491 (1958).
40 — Cancer Res. **20**, 712 (1960).
41 DALTON, A. J., and F. HAGUENAU: Tumors induced by Viruses. New York/London: Academic Press 1962.
42 DMOCHOWSKI, L.: Progr. med. Virol. **3**, 363 (1960).
43 — Cancer Res. **20**, 977 (1960).
44 OBERLING, CH.: Die Morphologie der onkogenen Viren. In Berl. Symp. über Fragen der Carcinogenese. Abh. Dtsch. Akad. Wiss. Berlin **3**, 211 (1960).

Untertitel: Onkogene Viren der Hühner

45 BEARD, J. W., D. G. SHARP, and E. A. ECKERT: Advanc. Virus Res. **3**, 149 (1955).
46 BENEDETTI, E. L., and W. BERNHARD: J. Ultrastruct. Res. **1**, 309 (1958).
47 BONAR, R. A., D. F. PARSON, G. S. BEAUDREAU, C. BECKER, and J. W. BEARD: J. nat. Cancer Inst. **23**, 199 (1959).
48 CLAUDE, A., K. R. PORTER, and E. G. PICKELS: Cancer Res. **7**, 421 (1947).
49 DMOCHOWSKI, L., C. E. GREY et B. R. BURMESTER: Acta Un. int. Cancr. **15**, 780 (1959).
50 EPSTEIN, M. A., and S. D. HOLT: Brit. J. Cancer **12**, 363 (1958).
51 HEINE, U., G. DE THÉ, H. ISHIGURO, and J. W. BEARD: J. nat. Cancer Inst. **29**, 211 (1962).
52 — —, D. BEARD, and J. W. BEARD: J. nat. Cancer Inst. **30**, 817 (1963).
53 OBERLING, CH., W. BERNHARD, and P. VIGIER: Nature (Lond.) **180**, 386 (1957).

Untertitel: Shope-Viren

54 BERNHARD, W., A. BAUER, J. HAREL, and CH. OBERLING: Bull. Cancer **61**, 423 (1954).
55 FEBVRE, H.: In Tumors induced by Viruses, p. 79. Hrsg. Dalton, A., and Haguenau. New York/London: Academic Press 1962.
56 SHARP, D. G., A. R. TAYLOR, D. BEARD, and J. W. BEARD: Proc. Soc. exp. Biol. (N. Y.) **50**, 205 (1942).
57 STONE, R. S., R. E. SHOPE, and D. H. MOORE: J. exp. Med. **110**, 543 (1959).
58 WILLIAMS, R. C., S. J. KASS, and C. A. KNIGHT: Virology **12**, 48 (1960).

Untertitel: Bittnerscher Milchfaktor

[59] BERNHARD, W., M. GUERIN et CH. OBERLING: Acta Un. int. Cancr. **12**, 544 (1956).
[60] DMOCHOWSKI, L.: J. nat. Cancer Inst. **15**, 785 (1954).
[61] MOORE, D.: In Tumors induced by Viruses, p. 113. Hrsg. Dalton, A., and F. Haguenau. New York/London: Academic Press 1962.

Untertitel: Mäuseleukämieviren

[62] BERNHARD, W., et L. GROSS: C. R. Acad. Sci. (Paris) **248**, 160 (1959).
[63] BIERWOLF, D., and A. GRAFFI: 3rd Europ. Reg. Conf. Electron microscopy, p. 375. Prag 1964.
[64] DALTON, A. J., L. W. LAW, J. B. MOLONEY, and R. A. MANAKER: J. nat. Cancer Inst. **27**, 747 (1961).
[65] DMOCHOWSKI, L., and C. E. GREY: Ann. N. Y. Acad. Sci. **68**, 559 (1957).
[66] GRAFFI, A. U. HEINE, J.-G. HELMCKE, D. BIERWOLF und A. RANDT: Klin. Wschr. **38**, 254 (1960).
[67] DE HARVEN, E., and CH. FRIEND: J. biophys. biochem. Cytol. **7**, 747 (1960).
[68] HEINE, U., A. GRAFFI, J.-G. HELMCKE, D. BIERWOLF und A. RANDT: Naturwissenschaften **44**, 449 (1957).

Untertitel: Polyomavirus

[69] BANFIELD, W. G., C. J. DAWE, and D. C. BRINDLEY: J. nat. Cancer Inst. **23**, 1123 (1959).
[70] BERNHARD, W., H. L. FEBVRE et R. CRAMER: C. R. Acad. Sci. (Paris) **249**, 483 (1959).
[71] BIERWOLF, D., A. GRAFFI, and L. KRAUSE: Proc. Europ. Conf. Electron Microscopy, Band II, S. 982. Delft 1960.
— —, and L. BAUMBACH: Acta morph. Acad. Sci. hung. **10**, 357 (1961).
[72] DOURMASHKIN, R., and G. NEGRONI: Exp. Cell Res. **18**, 573 (1959).
[73] HOWATSON, A. F., and J. D. ALMEIDA: J. biophys. biochem. Cytol. **7**, 753 (1960).
[74] WILDY, P., M. G. P. STOKER, I. A. MACPHERSON, and R. W. HORNE: Virology **11**, 444 (1960).

zu Abschnitt IV: Epidemiologie

[75] ANDERVONT, H. B.: J. nat. Cancer Inst. **2**, 307 (1942).
[76] — J. nat. Cancer Inst. **3**, 309 (1942).
[77] BIERWOLF, D.: s. Nr. 19.
[78] BITTNER, J. J.: s. Nr. 1.
[79] BURMESTER, B. R.: Ann. N. Y. Acad. Sci. **68**, 245 (1957).
[80] GRAFFI, A., u. W. KRISCHKE: Biol. Z. **81**, 277 (1962).
[80a] — — Naturwissenschaften **43**, 333 (1956).
[81] GROSS, L.: Acta haemat. (Basel) **13**, 23 (1955).
[82] LAW, L. W., and J. B. MOLONEY: Proc. Soc. exp. Biol. (N. Y.) **108**, 715 (1961).
[83] MÜHLBOCK, O.: J. nat. Cancer Inst. **10**, 861 (1950).

[83a] PASTERNAK, G.: Immunologische Eigenschaften virusinduzierter Leukämien. Habilitationsarbeit, Berlin 1966.
[84] SACHS, L., and E. HELLER: Brit. J. Cancer **13**, 452 (1959).
[84a] HALLAUER, C.: Die Virusaetiologie der Tumoren. Rektoratsrede Universität Bern 1960. Dies Academicus. Buchdruckerei P. Haupt Bern. 1961. S. 1—37.

zu Abschnitt V: Immunologische Eigenschaften der onkogenen Viren

[85] BEARD, J. W.: Ann. N. Y. Acad. Sci. **68**, 473 (1957).
[86] GRAFFI, A., u. F. FEY: Naturwissenschaften **42**, 652 (1955).
[87] — Arch. Geschwulstforsch. **22**, 13 (1963).
[88] HARRIS, R. J. C.: UICC Cancer **3**, Nr. 4, 6 (1965).
[89] KLEIN, G., H. O. SJÖGREN, and E. KLEIN: Cancer Res. **22**, 955 (1962).
[90] MELLORS, R. C.: Cancer Res. **20**, 744 (1960).
[91] OLD, L. J., and E. A. BOYSE: Ann. Rev. Med. **15**, 167 (1964).
[92] — — Fed. Proc. **24**, 1009 (1965).
[93] PASTERNAK, G., and A. GRAFFI: Brit. J. CANCER **17**, 532 (1963).
[93a] — J. nat. Cancer Inst. **34**, 71 (1965).
[93b] PAYNE, F. E., J. J. SOLOMON, and H. G. PURCHASE: Proc. nat. Acad. Sci. (Wash.) **55**, 341 (1966).
[94] RAPP, F., and J. L. MELNICK: Scient. Amer. **214**, 34 (1966).
[95] RUBIN, H.: Scient. Amer. **210**, 46 (1964).
[96] SJÖGREN, O.: Progr. exp. Tumor Res. **6**, 289 (1965).
[97] SLETTENMARK, B., and E. KLEIN: Cancer Res. **22**, 947 (1962).
[97a] WAHREN, B.: Immunologic aspects of virus-induced mouse leukemias, p. 1. Karolinska Institutet, Stockholm: 1966.

zu Abschnitt VI: Biochemische Charakterisierung der onkogenen Viren

[98] BIERWOLF, D.: s. Nr. 19.
[99] GROSS, L.: s. Nr. 5.
[100] OBERLING, CH.: s. Nr. 10.

Untertitel: Hühnerviren

[101] BATHER, R.: Brit. J. CANCER **11**, 611 (1957).
[102] — Brit. J. Cancer **12**, 256 (1958).
[103] BEARD, J. W., A. R. BONAR, G. S. BEAUDREAU, C. BECKER und D. BEARD: IV. Internat. Kongr. Biochemie Wien, 1958, Vordruck Nr. 3.
[104] BONAR, R., and J. W. BEARD: J. nat. Cancer Inst. **23**, 183 (1959).
[105] MOMMAERTS, E. B., E. A. ECKERT, D. BEARD, D. S. SHARP, and J. W. BEARD: Proc. Soc. exp. Biol. (N. Y.) **79**, 450 (1952).
[106] RIMAN, J., and R. THORELL: Biochim. biophys. Acta (Amst.) **40**, 565 (1960).

Untertitel: Shope-Viren

[107] BEARD, J. W., and R. W. G. WYCKOFF: J. biol. Chem. **123**, 461 (1938).
[108] GREENSTEIN, J. P.: Advanc. Protein Chem. **1944** 1.

[109] ITO, Y.: Virology **12**, 596 (1960).
[110] WATSON, J. D., and J. W. LITTLEFIELD: J. molec. Biol. **2**, 161 (1960)

Untertitel: Mäuseleukämieviren

[111] BIELKA, H. u. A. GRAFFI: Acta biol. med. germ. **3**, 515 (1959).
[111a] — — Acta biol. med. germ. **9**, 386 (1962).
[112] DALTON, A. J., and J. B. MOLONEY: 5th Internat. Congr. for Electron Microscopy. Academic Press New York, 1962, MM-7; MOLONEY Lit.-Zit.[33].
[113] GRAFFI, A., u. H. BIELKA: Acta biol. med. germ. **3**, 511, 513 (1959)
[114] GRAFFI, A., u. W. KRISCHKE: Acta biol. med. germ. **3**, 402 (1959).
[115] MORA, P. T., V. W. MCFARLAND, and S. W. LUMORSKY: Proc. nat. Acad. Sci. **55**, 438 (1966).

Untertitel: Polyomavirus

[116] BRODSKI, J. W., P. ROWE, J. W. HARTLEY, and W. T. LANE: J. exp. Med. **109**, 439 (1959).
[117] EDDY, B. E., S. E. STEWART, and G. E. GRUBBS: Proc. Soc. exp. Biol. (N. Y.) **99**, 289 (1958).
[118] GRAFFI, A., et D. FRITZ: Rev. franç. Etud. clin. biol. **5**, 388 (1960).
[119] DI MAYORCA, G. A., B. E. EDDY, S. E. STEWART, W. HUNTER, CH. FRIEND, and A. BENDICH: Proc. nat. Acad. Sci. **45**, 1805 (1959).

zu Abschnitt VII: Wirkung der onkogenen Viren auf die Zelle

[120] ABERCROMBIE, M., and E. J. AMBROSE: Cancer Res. **22**, 525 (1962).
[121] MEDINA, D., and L. SACHS: Brit. J. Cancer **15**, 885 (1961).
[122] MUNK, K.: Ergebn. Mikrobiol. **38**, 223 (1964).
[123] MCPHERSON, J.: J. nat. Cancer Inst. **30**, 795 (1963).
[124] STOKER, M.: In Viruses, Nucleic Acids and Cancer, p. 487. Baltimore: 1963.
[124a] Virology **18**, 649 (1962).
[125] TEMIN, H. M.: J. nat. Cancer Inst. **34**, 441 (1964); **35**, 679 (1965).
[126] VOGT, M., and R. DULBECCO: Proc. nat. Acad. Sci. (Wash.) **46**, 365 (1960).
[127] — — Virology **16**, 41 (1962).

zu Abschnitt VIII: Viren und menschliche Tumoren

[128] BERGOLZ, V. M.: Progr. exp. Tumor Res. **1960**, 86.
[128a] BRAUNSTEINER, H., K. FELLINGER und F. PAKESCH: Wien. Z. inn. Med. **40**, 384 (1959).
[129] BURKITT, D.: Ann. Coll. Surg. England **30**, 211 (1962).
[129a] DALLDORF, G., and F. BERGAMINI: Proc. nat. Acad. Sci. (Wash:) **51**, 263 (1964).
[130] DMOCHOWSKI, L., H. G. TAYLOR, C. E. GREY, E. DESIGNER, D. A. DREYER, J. A. SYKES, PH. L. LANGFORD, T. ROGERS, C. C. SHULLENBERGER, and C. D. HOWE: Cancer (Philad.) **18**, 1345 (1965).
[131] — Cancer Res. **25**, 1654 (1965).

Onkogene Viren 241

[132] —, C. E. GREY, D. A. DREYER, J. A. SYKES, PH. L. LANGFORD, and H. G. TAYLOR: Med. Rec. (Houston) **57**, 563 (1964).
[133] EPSTEIN, M. A., B. G. ACHONG, and Y. M. BARR: Lancet **1964**, 702.
[134] GRACE, J. T., E. A. MIRAND, and D. T. MOUNT: Arch. intern. Med. **105**, 482 (1960).
[135] GRAFFI, A., D. BIERWOLF, L. BAUMBACH, H. BLANKENHAGEL, R. WIDMAIER und A. RANDT: Dtsch. Gesundh.-Wes. **19**, 1576 (1964).
[136] MOORE, A. E.: Proc. Amer. Ass. Cancer Res. **3**, 135 (1960).
[137] NEGRONI, G.: Brit. med. J. **1**, 927 (1964).
[138] PORTER, G. H., A. J. DALTON, J. B. MOLONEY, and E. Z. MITCHELL: J. nat. Cancer Inst. **33**, 547 (1964).
[139] SIMONS, D. J., and M. G. R. ROSS: Nature (Lond.) **205**, 371 (1965).
[140] STEWART, S. H., E. LOVELACE, J. LAUDAN e J. MCBRIDE: Lav. Ist. Anat. Univ. Perugia **23**, 153 (1963).
— J. nat. Cancer Inst. **33**, 557 (1965).
[141] VASQUEZ, C., A. PAVLOVSKY et W. BERNHARD: C. R. Acad. Sci. (Paris) **256**, 2261 (1963).

Allgemeine Übersichtsarbeiten

s. lfd. Nr. 5, 9, 16, 19, 23, 26, 27, 30, 41, 42 sowie

[142] Viruses, Nucleic Acids and Cancer. — A collection of papers presented at the seventeenth Annual Symposium on Fundamental Cancer Research, 1963. Baltimore: The Williams and Wolkins Company 1963.
[143] Subcellular particles in the neoplastic process. Amer. N. Y. Acad. Sci. **68/2**, 245 (1957).
[144] Symposia Tumor Viruses. National Cancer Institute Monograph No. 4 (1960).
[145] A Ciba Foundation Symposium on Tumour Viruses of Murine Origin. London: J. A. Churchill Ltd. 1962.
[146] World Health Organization Techn. Report Series, 295, Viruses and Cancer. Report of the WHO Scientific Group, Genf 1965.
[147] Berliner Symposium über Fragen der Carcinogenese. Abh. Dtsch. Akad. Wiss. Berlin, Kl. f. Med. 1930, Nr. 3.

An Analysis of the Mechanism of Carcinogenesis by Polyoma Virus, Hydrocarbons, and X-Irradiation

By L. SACHS

Section of Genetics, Weizmann Institute of Science, Rehovoth, Israel*

With 4 Figures

Tumor cells can replicate under conditions when normal cells can not, because to varying degrees depending on the tumor, their replication is not inhibited by controls that inhibit the replication of normal cells. The basic process responsible for the change of a normal cell into a tumor cell is thus a change in the control mechanism for cell replication. The mechanisms that control the replication of normal cells can be divided into 2 types; those in which circulating substances produced by one cell control the growth of another cell, and those that appear to be mediated by cell surface interactions. This second type of control will be referred to as contact inhibition[1]. The experiments to be presented are concerned with our *in vitro* studies on the mechanism of the change in the contact inhibition control for cell replication that is induced by carcinogenic agents. Carcinogenesis can be induced by viruses, non-viral chemicals, and physical agents. The present experiments will therefore be concerned with the change in this control induced by small DNA tumor viruses with particular reference to polyoma virus, carcinogenic hydrocarbons, and x-irradiation. The change in control resulting in a change of a normal cell into a tumor cell will be referred to as transformation.

The development of a visible tumor *in vivo* requires 3 events. *A.* Induction of the change in the control mechanism for cell replication. *B.* Fixation of this change in the cell so that it can be transmitted as a hereditary property of the tumor cell, and *C.* Growth of a single tumor cell into a visible tumor. The experiments to be presented will be primarily concerned with *A.* and *B.* Five questions will be examined in relation to these 2 events.

---

* Supported by Research Grant CA 05266 from the National Cancer Institute, US. Public Health Service.

1. Can the viral, non-viral chemical, and physical carcinogens, directly induce cell transformation?
2. What is the frequency of transformation, and are all cells equally competent to be transformed?
3. Does fixation of the transformed state as a hereditary property of the cell require a process associated with cell replication and, if so, is it essential for fixation that this process occurs soon after treatment with the carcinogen?
4. If fixation requires a process associated with cell replication, can a carcinogen also induce the replication of cellular constituents?, and
5. Can the different types of carcinogens induce the same change in control mechanism by inducing the same type of change in the cell surface?

Direct induction of cell transformation in vitro by polyoma virus, carcinogenic hydrocarbons, and X-irradiation

The demonstration that normal hamster cells can be transformed in tissue culture by polyoma virus[2,3], carcinogenic hydrocarbons[1,5], and x-irradiation[6], has made it possible to show the direct induction of cell transformation by these 3 types of carcinogens. *In vitro* transformed cells can be distinguished from normal cells by a change in the cell surface control for cell replication, in that transformation results in a decrease in contact inhibition. The normal cells grow as a flat layer, whereas the transformed cells are piled up on one another. This difference in growth pattern can be found in colonies derived from single cells (Fig. 1), so that it is possible to distinguish single normal from single transformed cells by the growth pattern of the colonies derived from these cells[7,8].

The quantitation of the percent of transformed colonies after treatment of a mixed cell population of normal hamster embryo cells, has shown about 2% transformed colonies after infection with polyoma[8-11], about 3 to 20% after treatment with carcinogenic hydrocarbons[4,5], and about 0.5% after x-irradiation with 300 r[6]. No such transformed colonies have been observed in control cultures of untreated hamster embryo cells. The lack of transformed colonies in the controls, the small or no decrease in cloning efficiency with the dose of carcinogen used, and the high frequency of transformation in the treated cells, have indicated for the 3 types of carcinogens that the transformed colonies were directly induced by the carcinogen and not merely the result of a selection of spontaneously occuring transformed cells[5,6,11]. The frequency of transformation observed with all 3 carcinogens, is also much higher than that expected for a randomly occurring mutation.

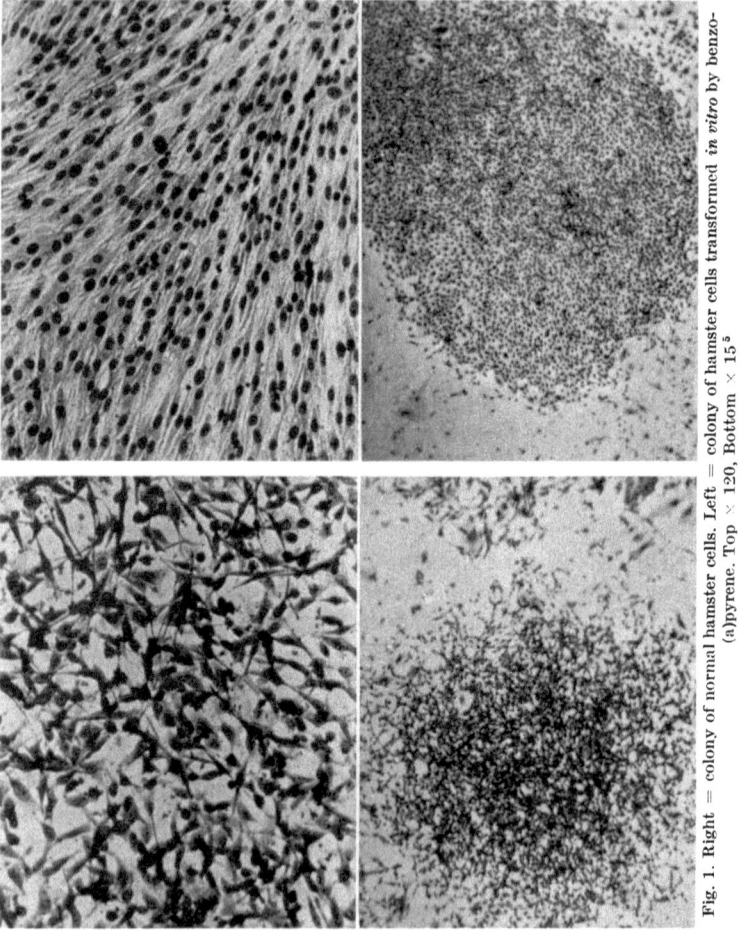

Fig. 1. Right = colony of normal hamster cells. Left = colony of hamster cells transformed *in vitro* by benzo(a)pyrene. Top × 120, Bottom × 15[5]

The relationship between the percent of transformed colonies and the dose of carcinogens has been determined for benzo(a)pyrene[12] and with polyoma virus[13], and a "one hit" dose response curve for transformation has been obtained has been obtained with both carcinogens. Experiments with 7 hydrocarbons[5], have indicated a specificity for cell transformation *in vitro* to the compounds known to be carcinogenic *in vivo* (Tab. 1).

Table 1. *Specificity of in vitro transformation to compounds carcinogenic in vivo*[5]

Compound	Tumor induction in vivo	Cell transformation in vitro
Benzo(a)pyrene	+	+
3-Methylcholantrene	+	+
7,12-Dimethylbenz(a)antracene	+	+
10-Methylbenz(a)anthracene	+	+
8-Methylbenz(a)anthracene	—	—
Pyrene	—	—
Chrysene	—	—

Cell competence for transformation

The finding of transformed and normal colonies after treatment of a mixed population of normal hamster embryo cells, raises the question whether there is a hereditary and/or physiological state of competence for transformation, and whether the same situation applies to the different carcinogenic agents. It has been shown for polyoma virus, that with rare exceptions the frequency of cell transformation after infection of hamster embryo clones derived from single cells was about the same as after infection of mixed cell populations[8,10] (Tab. 2). This indicates that for this virus, the ability of a cell to be transformed is primarily due to a physiological state of competence, and there is only a small degree of hereditary heterogeneity for competence to transformation. It would be of value for further studies to obtain virus mutants that give an increased frequency of cell transformation without an equivalent increase in the frequency of the lytic interaction induced by polyoma[9].

The percent of transformed colonies obtained after treatment of single cell clones with benzo(a)pyrene has indicated that there is also in this case a physiological state of competence, but that there is considerably more hereditary heterogeneity for competence to transformation with this agent than with polyoma[12]. Preliminary results with transformation induced by x-irradiation[14] also suggest a higher degree of hereditary heterogeneity for competence to transformation than with polyoma. The treatment of single cell clones has further shown that the cytotoxic and transforming

Table 2. *In vitro transformation of hamster enbryo cells infected with polyoma virus*

Celle type	Total no. colonies	% colonies transformed
Mixed cell cultures	449	1.8
	2216	1.8
	3322	2.0
Single cell clones	916	0
	611	1.0
	1332	1.2
	546	1.6
	966	1.8
	670	1.8
	1162	2.1
	681	2.2
	1203	2.2
	473	2.3

Colonies stained at 9 days after cell plating[8,10].

activity of a carcinogenic hydrocarbon are 2 different events. Normal hamster cell clones can be either susceptible or resistant to the cytotoxic action of benzo(a)pyrene, and in clones susceptible to transformation, transformed colonies can be obtained after treatment of either type of clone[12].

The development of a transformed colony requires a change in the control mechanism for cell replication, and the fixation of this change so that it can be transmitted as a hereditary property of the transformed cell. Competence can be expressed at either of these stages, so that in order to further analyse the nature of competence, it is necessary to examine the requirements for fixation of the transformed state.

Cell replication and fixation of the transformed state

In the standard transformation assay with benzo(a)pyrene, the colonies are usually stained for scoring at 10 days after the cells are plated for cloning. The addition of this hydrocarbon to cells at different times after they were plated for cloning, gave the usual percent of transformed colonies when benzo(a)pyrene was added up to 8 days after cell plating, but almost no transformed colonies when it was added at 9 days[5]. These results indicate that after the addition of this carcinogen, between 1 to 2 days are required for expression of the transformed state. Since this time corresponds to about the time required for a cell generation, it was originally suggested from these results that the expression of transformation requires a process associated with cell replication[5].

Further studies were carried out with x-irradiation induced transformation, to determine whether a correlation between cell replication and the expression of transformation could be made with another type of carcinogen, and whether it was essential for expression of the transformed state that the suggested process associated with cell replication occurs soon after treatment. It was found that about the same frequency of transformed colonies was obtained after irradiating cells one day after they were plated for cloning, sparse cultures, or confluent cultures cloned immediately after irradiation, i.e. irradiating cells that were able to undergo cell replication soon after treatment with this carcinogen. But no transformed colonies were obtained when confluent cultures were irradiated and the cells only allowed to replicate 3 to 5 days later[6] (Tab. 3). Inhibition of cell replication for 3 to 5 days after irradiation thus prevented expression of the transformed state.

Table 3. *In vitro transformation of hamster embryo cells after x-irradiation with 300r*

Condition of cultures at time of irradiation	Days after irradiation when cells plated for cloning	No. of experiments	No of experiments with transformation	Total No. of colonies	% cloning efficiency	% colonies transformed
Cells plated for cloning	—	5	5	6875	0.5—2.1	0.5—0.7
Sparse	3	1	1	1136	1.1	0.5
Confluent	3—5	6	0	10647	0.7—1.0	0
Confluent	0	3	3	7231	0.7—1.0	0.7—0.8

Colonies stained at 10 days after cell plating[6]

It can therefore also be assumed for x-irradiation induced transformation that a process associated with cell replication is required for expression of the transformed state. The results further indicate that the fixation of the transformed state as a hereditary property of the cell requires that this process has to occur soon after treatment with carcinogen. A requirement for cell replication soon after infection in order to fix the transformed state, has also recently been noted for transformation by simian virus 40[15]. The increased frequency of polyoma induced transformation that has been obtained by treatments such as exposure of the cells to 24 °C[2,9], may also be due to the treatment inducing an increased frequency of the process associated with cell replication that is required for fixation. It will be of interest in future studies to determine whether the event required for fixation is the same for the 3 types of carcinogens; whether fixation requires duplication of some cellular constituent, e.g. cellular DNA to allow fixation of a change in the DNA before repair, or in the case of a virus the incorporation of viral genes into the cell genome; or whether fixation of the new cell surface control mechanism in the transformed cell requires the rapid dilution of the old control and this is achieved by cell replication. Neither of these alternatives requires that fixation is necessarily an irreversible event.

Induction of the replication of cellular constituents

The evidence given above indicates that fixation of the transformed state requires some process associated with cell replication soon after carcinogen treatment. An efficient carcinogen would thus be one that could induce both the change in control mechanism in the cell, and the replication of the cellular constituents required for fixation. Since the process required for fixation could be the replication of cellular DNA, the ability of a carcinogen to induce the replication of cellular DNA was first examined with polyoma virus.

Cellular DNA can be distinguished from viral DNA, in that cellular DNA contains 5-methylcytosine whereas the virus DNA contains little or no 5-methylcytosine[16,17], by fractionation on a methylated albumin kieselguhr column[18-21] (Fig. 2), or by sedimentation velocity analysis[22]. The synthesis of cellular DNA can thus be identified, and it was found that polyoma virus can induce the synthesis of cellular DNA after DNA synthesis in normal cells has been repressed by contact inhibition [16, 18, 19, 22] or by a large dose, 4000r, of x-irradiation[19]. Studies on the number of DNA synthesizing cells and the increase in the amount of DNA have indicated that the induced cells approximately double their

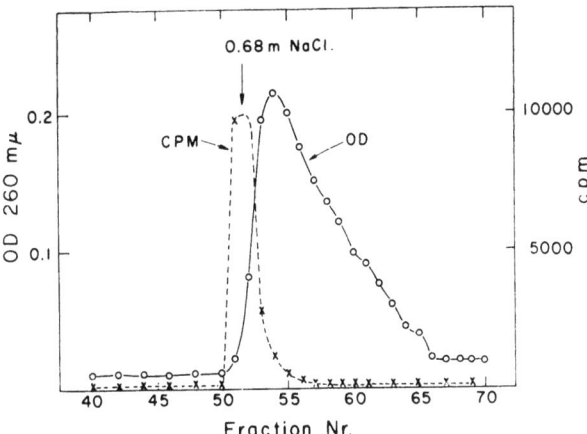

Fig. 2. Separation of cellular and polyoma virus DNA on the methylated albumin kieselguhr column. Nonradioactive cellular DNA and H³-thymidine labelled polyoma DNA were mixed and fractionated[19]

DNA content[19], like the doubling that occurs after induction of the normal process of cell replication. Rat embryo cells are sus-

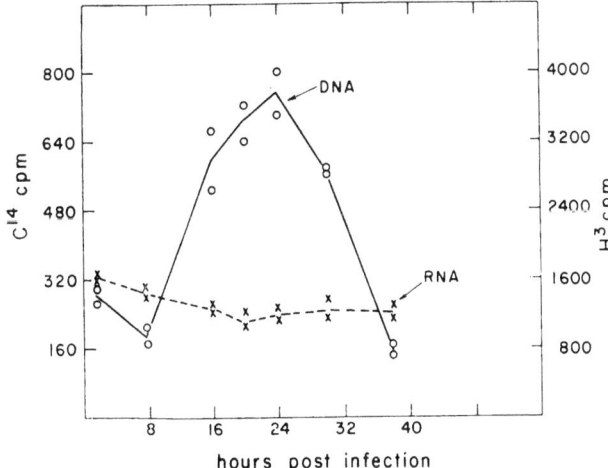

Fig. 3. Rate of DNA and RNA synthesis in polyoma virus infected x-irradiated rat embryo cell cultures. Cultures were infected 2 days after x-irradiation with 4000r, and then labelled at different times for one hour with C¹⁴-thymidine and H³-uridine. The rate of DNA synthesis in the uninfected controls was comparable to that of infected cultures ar 2 hours post infection[19]

ceptible to polyoma induced transformation, but show no detectable synthesis of infectious virus, viral coat antigen, or virus DNA[2,19,23]. The finding of an induction of cellular DNA synthesis after polyoma infection of rat embryo cells (Fig. 3) indicated that the induction of cellular DNA synthesis by polyoma does not require virus DNA

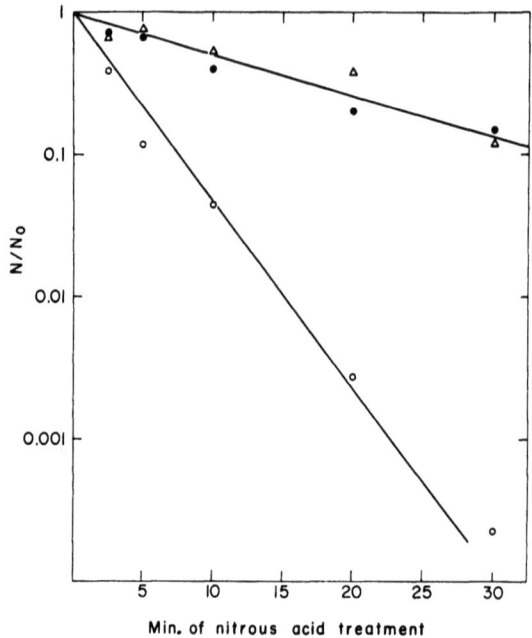

Fig. 4. Inactivation by nitrous acid of plaque forming ability, transforming ability, and DNA inducing capacity, of polyoma virus. Virus was treated with nitrous acid for different times, and the plaque forming ability (○), cell transforming (●), and DNA inducing capacity (△) of the different samples was determined[19]

synthesis[19]. The induction of cellular DNA synthesis in the absence of detectable virus DNA synthesis, after normal cellular DNA synthesis has been repressed by contact inhibition or by x-irradiation with 5000r, has also been found after infection with simian virus 40[21].

Nitrous acid inactivation experiments with polyoma have shown that the induction of cellular DNA synthesis depends on a function of the viral genome, and that the rate of inactivation of the DNA inducing capacity of the virus was similar to that found for the cell transforming capacity[19] (Fig. 4). This result supports the assumption that cell transformation and induction of cellular DNA synthesis are expressions of the same function of the viral genome. These data thus indicate that there is a messenger coded by the viral DNA that can initiate both the change in contact inhibition control in the transformed cells, and the replication of cellular DNA synthesis that may be required for fixation of the transformed state.

It will be of interest in further studies to determine whether non-viral chemical carcinogens, and x-irradiation, can also directly induce the replication of cellular constituents that may be required for fixation. In addition to a direct induction of replication of cellular constituents, as found for cellular DNA synthesis with polyoma and simian virus 40, carcinogenic agents may also indirectly stimulate the replication required for fixation by producing a cytotoxic effect on neighbouring cells.

The change in control mechanism by a change in the cell surface

The small DNA tumor viruses such as polyoma and simian virus 40, contain an amount of DNA that can code for only a small number of proteins. Thus the DNA of polyoma, with a molecular weight of about 3×10^6 [24,25] has about 4500 nucleotide pairs and these can code for about 1500 amino acids. Studies on the viral coat protein suggest that about a third of the DNA codes for this protein[26]. In addition, the finding that a small plaque mutant of polyoma[27-29], but not the large plaque virus, has a temperature sensitive block for the synthesis of virus DNA[30,31] suggests that there is a virus coded enzyme required for virus DNA synthesis, and there may be other virus coded functions required for virus development. The nitrous acid inactivation curves for the cell transforming capacity, and for the capacity for synthesis of new virus as tested by plaque forming ability[32,33], suggest that about 20% of the viral genome can be sufficient for transformation[19].

The virus DNA that codes for transformation may thus be enough to code for only one or possibly two proteins, and it is therefore of interest to examine whether one or two virus coded functions are sufficient to explain carcinogenesis. The data with polyoma indicating that the change in contact inhibition control in the transformed cells and the induction of cellular DNA synthesis are expression of the same function of the viral genome, also raise the question whether both processes may be the consequence of the same change in the cell surface.

The experimental results on cell virus interactions with the small DNA tumor viruses[31] are summarized in Tab. 4. The data indicate

Table 4. *Experimental results on cell-virus interactions with small DNA tumor viruses*

Early RNA and protein is synthesized early during virus development, before the synthesis of viral DNA.

A nuclear tumor antigen is synthesized early during virus development.

All tumors synthesize a transplantation, presumably cell surface, tumor antigen. For both the nuclear and the transplantation tumor antigen, different tumors induced by the same virus have the same antigen, and different viruses induce the synthesis of different antigens.

Virus infection can lead to an induction of cellular DNA synthesis.

The cell-virus interaction can result either in the development of virions and cell lysis, or in the transformation of normal cells to tumor cells.

Tumor cells generally do not produce detectable virus.

References in [34].

that the change in the cell surface associated with the change in contact inhibition control in the transformed cells is also expressed in the synthesis of a surface tumor antigen of the transplantation type (References in [34]). It has been suggested[34], that all the findings associated with carcinogenesis by these viruses can be explained by 2 virus coded functions in the case of cells that are able to synthesize infectious virus particles (virions) (Tab. 5). One postulated function of the virus is to code for an enzyme, recognised as a nuclear tumor antigen, that produces the change in the cell surface that is recognised as the transplantation tumor antigen and the change in contact inhibition control. It is assumed that the control mechanism for cell replication is determined by the structure of the cell surface, so that the change in regulation as a result of the change in cell surface can also explain the induction of cellular DNA synthesis[31].

Virus yielding cells lyse[35]. The maintenance of the transformed state in a cell that is able to synthesize infectious virions thus requires a second function, which may be virus coded, that represses further virus development after synthesis of the RNA that codes for the nuclear antigen (Tab. 5). However in the case of cells that are not capable of synthesizing infectious virions, e.g. rat cells infected with polyoma virus, the one virus coded function for the nuclear tumor antigen, that is assumed to induce the change in cell surface and with it the induction of cellular DNA synthesis that may be

Table 5. *Sequence of events that result either in the development of virions and cell lysis, or in carcinogenesis. Two virus coded functions are postulated for carcinogenesis*[34]

First function coded by virus DNA. For development of virions and cell lysis, and for carcinogenesis.

Synthesis of early RNA
↓
Synthesis of early protein i.e. nuclear tumor antigen
↓
Nuclear tumor antigen produces change in cell surface i.e. transplantation tumor antigen
↓
Change in cell surface results in change in cell regulatory mechanism
↓
Change in regulation results in induction of cellular DNA synthesis
↓
Further virus development and cell lysis
or
second function coded by virus DNA. For carcinogenesis
Repression of virus development after synthesis of early RNA. This can result in the formation of tumor cells

required for fixation of the transformed state, would be sufficient. It will be of interest to determine in these cells whether the initial block for further virus development is at the level of transcription or translation. According to the present hypothesis the only part of the viral DNA required to maintain the transformed state is the nucleotide sequence that codes for the nuclear tumor antigen. The incorporation of only this part of the viral genome into the cell genome can explain the finding that transformed cells generally do not produce detectable virus.

The results thus suggest that one virus coded function can induce the change in the contact inhibition control and the replication of cellular constituents that may be required for fixation of the transformed state, and that the transformed state can be maintained by the incorporation of viral genes into the cell genome. It can also be suggested that non-viral chemical and physical carcinogens can induce the change in cell surface control by interacting with cellular DNA. However, it remains to be established whether the change in control by non-viral carcinogens can be induced only by interaction with cellular DNA, or whether it can also be induced by interaction with some other intercellular constituent, or directly with the cell surface. It can already be concluded from the results showing that polyoma and simian virus 40 induce different transplantation antigens (Tab. 5, references in [34]), that there is more than one way of changing the cell surface even with transformation induced by two DNA viruses. It has also been reported that different transplantation antigens can be induced by different strains of polyma[36]. The observation of a number of tumor transplantation antigens in tumors induced by the same non-viral carcinogen[37], need thus not necessarily be interpreted as a basic difference in the mechanism of transformation by the different types of carcinogens[34].

As shown above, transformation by the 3 types of carcinogens induces a change in the cell surface resulting in a change in contact inhibition control. But the transformed cells do not show a complete loss of all ability to respond to contact inhibition. Thus although the transformed cells do not show contact inhibition with other transformed cells of the same type, cells transformed by polyoma[9,38], hydrocarbons[5], x-irradiation[14], and "spontaneously" transformed cells[39], can still be contact inhibited by normal cells. It will be of interest to determine whether cells transformed by one carcinogen can be contact inhibited by cells transformed by another carcinogen, and whether clones independently transformed by e.g. benzo(a)pyrene, can contact inhibit one another. Contact inhibition of transformed by normal cells, and the immune response against the tumor antigen[37], can explain why together with the induction of a change in the control mechanism for cell replication and fixation of this change in the cell, a third event is required so that a single tumor cell can grow into a large enough colony *in vivo* to permit it to develop into a visible tumor. Carcinogens, and promoters of tumor development[40–42], may be able to produce a sufficiently large colony of transformed cells that can develop into a visible tumor by a general induction of cell replication. The may also be able to remove contact inhibition by a cytotoxic effect on neighbouring normal cells[43]. In addition, visible tumors may arise from cell variants with a more complete loss[1] of response to contact inhibition.

References

[1] ABERCROMBIE, M., and E. J. AMBROSE: Cancer Res. **22**, 52 (1962).
[2] SACHS, L., and D. MEDINA: Nature (Lond.) **189**, 457 (1961).
[3] VOGT, M., and R. DULBECCO: Proc. nat. Acad. Sci. (Wash.) **46**, 365 (1960).
[4] BERWALD, Y., and L. SACHS: Nature (Lond.) **200**, 1182 (1963).
[5] — — J. nat. Cancer Inst. **35**, 641 (1965).
[6] BOREK, C., and L. SACHS: Nature (Lond.) **210**, 276 (1966).
[7] STOKER, M., and I. MACPHERSON: Virology **14**, 359 (1961).
[8] SACHS, L., D. MEDINA, and Y. BERWALD: Virology **17**, 491 (1962).
[9] MEDINA, D., and L. SACHS: Virology **19**, 127 (1963).
[10] — — Virology **27**, 398 (1965).
[11] SACHS, L.: In New Perspectives in Biology, p. 246. Elsevier Co. 1964.
[12] HUBERMAN, E., and L. SACHS: To be published.
[13] MACPHERSON, I., and L. MONTAGNIER: Virology **23**, 291 (1964).
[14] BOREK, C., and L. SACHS: To be published.
[15] TODARO, G. J., and H. GREEN: Proc. nat. Acad. Sci. (Wash.) **55**, 302 (1966).
[16] WINOCOUR, E., A. M. KAYE, and V. STOLLAR: Virology **27**, 156 (1965).
[17] — To be published.
[18] DULBECCO, R., L. H. HARTWELL, and M. VOGT: Proc. nat. Acad. Sci. (Wash.) **53**, 403 (1965).
[19] GERSHON. D., P. HAUSEN, L. SACHS, and E. WINOCOUR: Proc. nat. Acad. Sci. (Wash.) **54**, 1584 (1965).
[20] SHEININ, R.: Virology **28**, 621 (1966).
[21] GERSHON, D., L. SACHS, and E. WINOCOUR: Proc. nat. Acad. Sci. (Wash). In press.
[22] WEIL, R., M. R. MICHEL, and G. K. RUSCHMAN: Proc. nat. Acad. Sci. (Wash.) **53**, 1468 (1965).
[23] MEDINA, D., and L. SACHS: Brit. J. Cancer **15**, 885 (1961).
[24] WEIL, R., and J. VINOGRAD: Proc. nat. Acad. Sci. (Wash.) **50**, 730 (1963).
[25] CRAWFORD, L. V.: Virology **22**, 149 (1964).
[26] KASS, S.: To be published.
[27] MEDINA, D., and L. SACHS: Virology **10**, 387 (1960).
[28] SACHS, L., and D. MEDINA: Nature (Lond.) **187**, 715 (1960).
[29] WINOCOUR, E., and L. SACHS: J. nat. Cancer Inst. **26**, 737 (1961).
[30] GERSHON, D., and L. SACHS: Virology **27**, 120 (1965).
[31] OSSOWSKI, L., and L. SACHS: To be published.
[32] SACHS, L., M. FOGEL, and E. WINOCOUR: Nature (Lond.) **183**, 663 (1959).
[33] WINOCOUR, E., and L. SACHS: Virology **8**, 397 (1959).
[34] SACHS, L.: Nature (Lond.) **207**, 1272 (1965).
[35] WINOCOUR, E., and L. SACHS: Virology **11**, 699 (1960).
[36] HARE. J. D., and T. GODAL: Proc. Soc. exp. Biol. (N.Y.) **118**, 632 (1965).
[37] KLEIN, G.: In New Perspectives in Biology, p. 261. Elsevier Co. 1964.
[38] STOKER, M.: Virology **24**, 165 (1964).
[39] BARSKI, G., and J. BELEHRADEK: Exp. Cell Res. **37**, 464 (1965).
[40] FRIEDEWALD, W. F., and P. ROUS: J. exp. Med. **80**, 101 (1944).
[41] BERENBLUM, I., and P. SHUBIK: Brit. J. Cancer **1**, 383 (1947).
[42] — Cancer Res. **14**, 471 (1954).
[43] HADDOW, A.: J. Path. Bact. **47**, 581 (1938).

Struktur der Nucleinsäuren onkogener Viren

Von H. BIELKA

Institut für Zellphysiologie der Deutschen Akademie der Wissenschaften zu
Berlin, Berlin-Buch

Mit 2 Abbildungen

Die Analyse physikalischer und chemischer Eigenschaften der Nucleinsäuren von Viren ist von besonderer Bedeutung a) für die Kenntnis der Struktur hochmolekularer Nucleinsäuren allgemein; b) für die Virusarchitektur und damit die systematische Einordnung der Viren; c) für den Mechanismus der Virusreplikation und Viruswirkung.

Die verschiedenen onkogenen Viren gehören zum Teil den R-Viren (RNS-haltige Viren), zum Teil den D-Viren (DNS-haltige Viren) an.

Die onkogenen *R-Virionen* umfassen die verschiedenen Leukoseviren der Hühner und Nagetiere sowie das Rous-Sarkomvirus. Sie sind auf Grund ihrer chemischen Zusammensetzung und ihrer Sensitivität, insbesondere gegenüber lipolytisch wirkenden Agentien, der Nomenklatur und Systematik der Viren nach LWOFF et al.[33] folgend, den RHE-Viren (RNS-haltig; Helix-Struktur der Kapsiden; „Enveloped" Kapsiden) zuzuordnen und damit wahrscheinlich der Gruppe der Myxoviren.

Die RNS des Rous-Sarkomvirus (RSV) ist auf Grund der starken Abhängigkeit der Sedimentationskoeffizienten von der Ionenstärke und der Kinetik des Abbaues durch RNase als flexibles Einstrangpolynucleotid charakterisiert. Für $S_{20,w}$ ergibt sich in 0,2 M NaCl-Lösung ein Wert von 79, in 0,1 M NaCl-Lösung von 71 und in einer 0,001 M Trislösung von 27[37]. Die Veränderungen der S-Werte bei abnehmender Ionenstärke sind nicht vollständig reversibel. Aus $S_{20,w} = 71$ läßt sich nach $MG = 1550 \cdot S_{20,w}^{2,1}$ ein mittleres Molekulargewicht von $12 \cdot 10^6$ berechnen (Tab. 1).

In Tab. 1 sind weiter die Werte für die Nucleotidzusammensetzung der RSV-RNS dargestellt. Nach vergleichenden Untersuchungen von ROBINSON und BALUDA[37] weisen die RSV-RNS und die RNS des Hühnermyeloblastosevirus (AMV), Stamm BAI A, die gleiche Nucleotidzusammensetzung auf. Dieses Ergebnis ist insofern von Bedeutung, als daß das Hühnermyeloblastosevirus als Helfervirus für den im Hinblick auf die Ausbildung des Hüllproteins defekten Bryan-Stamm des RSV wirken kann und damit offenbar genetische Komponenten besitzt, die dem Bryan-RSV fehlen. Einer besonderer Erwähnung bedarf auch noch der Befund

Tabelle 1. *Nucleotidzusammensetzung, S-Werte und Molekulargewichte der RNS einiger onkogener R-Viren*

	Mol.-%				$S_{20,w}$	MG
	A	G	C	U		
Rous-Sarkom............	25	28	24	23	71	12×10^6
Myeloblastose						
Stamm BAI A.........	25	29	23	23	71	12×10^6
Stamm R	21	36	23	20		
Rauscher-Leukämie	24	26	26	24	73	13×10^6

von TEMIN[41,42], daß es nach Infektionen von Zellen mit dem RSV zur Bildung eines DNS-haltigen Provirus kommt, das als integrierter Bestandteil des cellulären Genoms angesehen wird, in seinem Status lysogenen Prophagen vergleichbar, an dem gerichtet die RSV-RNS-Synthese stattfindet. Dieses Resultat konnte mit Hilfe der Hybridtechnik erhoben werden, indem aus RSV-infizierten Zellen eine DNS isoliert werden kann, die sich mit der RSV-RNS hybridisieren läßt, also Bereiche mit Nucleotidsequenzen enthalten muß, die der RSV-RNS basenkomplementär sind.

Versuche, aus dem RSV eine infektöse RNS zu isolieren, sind bisher negativ verlaufen[2,46].

Die wenigen, bisher an der RNS des Hühnermyeloblastosevirus durchgeführten Untersuchungen haben für die RNS einen Wert von $9{,}7 \cdot 10^6$ Molekulargewichtseinheiten pro Partikel ergeben[9]. Nach ROBINSON und BALUDA[37] ergibt sich aus der Sedimentationskonstante $S_{20,w} = 71$ ein mittleres Molekulargewicht von $12 \cdot 10^6$. Auch die AMV-RNS weist in vitro die Struktur eines einfachen, flexiblen Polynucleotidstranges mit etwa gleichen Werten für die Sedimentationskoeffizienten und deren Abhängigkeit von der Ionenstärke des Lösungsmittels auf[37,38], wie vorausgehend für die RSV-RNS dargestellt.

Die Ribonucleinsäuren verschiedener Stämme des Myeloblastosevirus sind zum Teil durch signifikante Unterschiede in der Nucleotidzusammensetzung ausgezeichnet[7,37,43].

Von den murinen Leukämieviren wurde bislang nur die RNS des Rauscher-Virus etwas näher untersucht[17,20,34a]. In bezug auf die Nucleotidzusammensetzung der RNS[20] (Tab. 1) ergeben sich zum Teil signifikante Abweichungen im Vergleich zur RNS der Hühner-Leukämieviren. Der Sedimentationskoeffizient in 0,2 M Na$^+$ beträgt für die RNS dieses Virus 70[20] bzw. 73[34a], woraus sich ein Molekulargewicht von etwa $1,3 \cdot 10^7$ errechnen läßt[34a]. Aus der Sensitivität gegenüber RNase[20] und dem UV-Temperaturprofil ($T_m = 59\,°C$ in 0,2 M Na$^+$)[34a] ergibt sich, daß es sich um ein Einstrangmolekül handelt.

Aus verschiedenen murinen Leukämieviren konnten leukämogen wirksame RNS-Präparationen gewonnen werden, so aus dem der myeloischen Leukämie[3,4], der AKR-Leukämie[25] und der Moloney-Leukämie[34].

Die onkogenen *D-Viren* sind auf Grund ihrer architektonischen Struktur und chemischen Zusammensetzung sämtlich den DCN-Viren (*D*NS-haltig; *c*ubische Symmetrie der Kapsiden; „*n*aked" Kapsiden) zuzuordnen, und zwar der Gruppe der Adenoviren, von denen die Papillom-, Polyoma- und SV-40-Viren (PaPoVa-Viren) dem 42c-Typ (Anzahl der Capsomeren = 42), die onkogenen Adenoviren dem 252c-Typ zugehören.

Im Vergleich zu den R-Viren sind die Nucleinsäuren der D-Viren insbesondere in bezug auf ihr physikochemisches Verhalten bei weitem intensiver untersucht worden. Die Versuchsergebnisse sollen für die PaPoVa- und Adenoviren nachfolgend vergleichend dargestellt werden.

Die Werte für die Nucleotidzusammensetzung sind in Tab. 2 zusammengestellt. Der Mol-%-Anteil für G+C ist für die Desoxyribonucleinsäuren der verschiedenen onkogenen D-Viren verschieden. Den höchsten G+C-Gehalt weisen die Desoxyribonucleinsäuren des Polyomavirus und der onkogenen Adenoviren Typ 12 und 18 auf, den geringsten die DNS des SV-40-Virus sowie der menschlichen Warzenviren. Bei den verschiedenen Papillomviren unterschiedlicher Speziesspezifität treten zum Teil beträchtliche Differenzen im G+C-Gehalt der DNS auf. Einer besonderen Erwähnung bedarf noch der ohne Zweifel interessante Befund, daß sich die DNS der onkogenen (Typen 12 und 18) und der nichtonkogenen (Typen 2 und 4) Adenoviren beträchtlich in ihrer chemischen Zusammensetzung unterscheiden, wobei insbesondere darauf hinzuweisen ist, daß der G+C-Gehalt der DNS der onkogenen Adenoviren dem G+C-Gehalt von Säugetier-DNS näher kommt als der der nichtonkogenen. Insgesamt zeigen die Desoxyribonucleinsäuren der onkogenen Viren eine gewisse Ähnlichkeit in der Basenzusammensetzung untereinander und mit der cellulären DNS von Säugetieren, ein Befund, der möglicherweise einige Rückschlüsse auf Gemeinsamkeiten in der Evolution und auf den Mechanismus der onkogenen Wirkung in der Zelle zuläßt.

Tabelle 2. *Nucleotidzusammensetzung der DNS onkogener Viren*

	Mol.-%					Autor
	A	T	G	C	G+C	
					a | b	
Polyomavirus					49 | 49	WEIL[47]
					47 | 48	CRAWFORD[10,11]
Papillomviren						
Kaninchen	26,6	24,7	24,4	24,2	47	WATSON[15], CRAWFORD[11]
Rind..........					45,5	CRAWFORD[11]
Hund..........					43	CRAWFORD[11]
Mensch					41	CRAWFORD[11]
Adenoviren						
Typ 12[+]			48			GREEN[23]
Typ 18[+]			49			GREEN[23]
Typ 2[++]			56			GREEN[23]
Typ 4[++]			57			GREEN[23]
SV 40-Virus					41	CRAWFORD[13]

a) nach T_m; b) nach Dichte; [+] onkogene Typen; [++] nichtonkogene Typen

Die DNS der onkogenen und nichtonkogenen Adenoviren unterscheiden sich weiterhin auch im Molekulargewicht[24]. Der $S^0_{20,w}$-Wert der DNS von Typ 2 beträgt in 1 M NaCl-Lösung 32, der der DNS des onkogen wirksamen Typs 12 = 30,7[24]. Damit erklärt sich auch der um etwa 10% geringere DNS-Gehalt des Adenovirus Typ 12 im Vergleich zu Typ 2.

Auch hinsichtlich der physikochemischen Eigenschaften und der daraus ableitbaren molekularen Konfigurationen der DNS der verschiedenen onkogenen Viren ergeben sich prinzipielle Gemeinsamkeiten (Abb. 1). Am umfangreichsten wurde bisher die DNS des Polyomavirus untersucht; die nachfolgenden Beschreibungen werden sich daher in erster Linie auf die mit der Polyomavirus-DNS erhaltenen Versuchsergebnisse beziehen.

260   H. BIELKA:

Die Polyomavirus-DNS kommt in Abhängigkeit von den Präparations- und Untersuchungsbedingungen in verschiedenen molekularen Konfigurationen vor[10,11,19,44,47,48]. Bei schonender Darstellung erhält man eine schnell sedimentierende DNS, die als „twisted circular structure", verdrillte zirkulare Form (= III in Abb. 1) charakterisiert ist und einen $S^0_{20,w}$-Wert von 20 besitzt. Diese „twisted circular structure" ergibt sich aus den hydro-

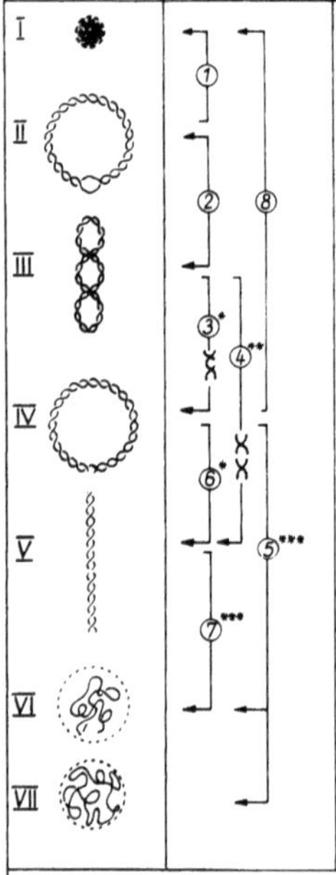

Konfiguration	Umwandlungen		Polyoma	Papillom		SV40	Adeno	
				SHOPE	Mensch		onkogen	nicht onkogen
I			$S_{20,W}$ 53					
II			16					
III			20	28	28			
IV			16		20	21		
V			14	21	18	16		
VI			16					
VII			18					
		T_m [°C]	89	89,5			89	92,5
		d [g/ml]	1,709	1,711			1,708	1,717
		MG (×10⁶)	7,5	4,2 - 5,5	4 - 5,3	2,7 - 3,2	5	

③ red. Agenzien, DNase (Schell); ④ DNase, E.coli - Endonuklease; ⑤ DNase, Alkali (pH 12,5), Hitze;
⑥ DNase (langsam); * Einstrang - Spaltung; ** Zweistrang - Spaltung; *** Strangtrennung
⑦ Alkali, Hitze; ⑧ Alkali

Abb. 1. Molekulare Konfigurationen und physikochemische Eigenschaften der DNS verschiedener onkogener Viren

dynamischen Eigenschaften, speziell aus dem Verhalten der S-Werte als Funktion des pH und der Temperatur, aus der Kinetik des enzymatischen Abbaues und ist auch durch elektronenmikroskopische Untersuchungen[44] belegt (Abb. 2). Durch Spaltung eines der beiden Polynucleotidstränge kommt es zur Aufhebung dieser verdrillt zirkularen Struktur unter Ausbildung einer einfachen, doppelhelikalen Ringstruktur (= IV in Abb. 1; Übergang [3]) mit einem Wert für den Sedimentationskoeffizienten von etwa 16 und einer Länge des DNS-Moleküls von etwa 1,6 μ. Eine solche Strukturumwandlung von Konfiguration III und IV durch Spaltung eines

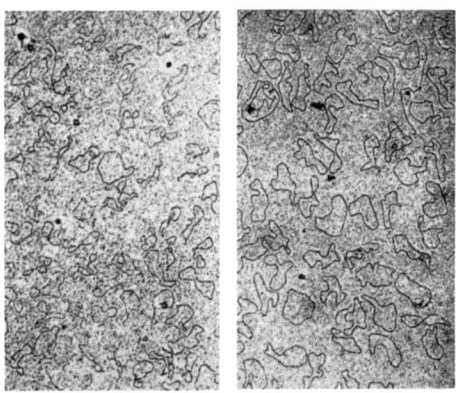

Abb. 2. Elektronenmikroskopische Darstellung der Polyomavirus-DNS [nach J. VINOGRAD et al., aus Proc. nat. Acad. Sci. (Wash.) **53**, 1104 (1965)]

Polynucleotidstranges erfolgt durch Behandlung mit reduzierenden Agentien (Hydrochinon, $FeCl_2$) und durch DNase. Die enzymatische Umwandlung von III in IV verläuft mit großer Geschwindigkeit nach einer Reaktion erster Ordnung. Eine milde Behandlung mit DNase wandelt sowohl die Konfiguration III als auch IV in ein einsträngig zirkulares (VII) und ein einsträngig lineares (VI) Polynucleotid um (Übergänge [3] und [5]), mit S-Werten von 18 bzw. 16. Bei längerer Einwirkung von DNase erfolgt eine Umwandlung von Konfiguration III über IV in Struktur V, wobei der Übergang [6] (IV → V), im Gegensatz zu III → IV, langsam verläuft und im statistischen Mittel bei etwa 50 Spaltungen des noch ringförmig geschlossenen Stranges der Konfiguration IV erfolgt. Der Übergang [6] von IV in Konfiguration V findet dann statt, wenn eine Spaltung des noch ringförmig geschlossenen Stranges in der Struktur IV benachbart zu dem bereits geöffneten zweiten Strang dieser Konfiguration IV eintritt. Eine direkte Überführung (= [4]) von Konfiguration III in V ist auch möglich durch Einwirkung von Endonuclease aus E. coli. Die Konfiguration V läßt sich in VI durch

Schmelzen, d. h. thermische Denaturierung und Alkalibehandlung (pH > 11,8) überführen. Während die thermisch bedingte Umwandlung von Konfiguration V in VI (Übergang [7]) schnell abläuft, ist die DNS in den Konfigurationen III und IV beträchtlich thermostabil; erst bei längerem Erhitzen erfolgt eine Umwandlung IV → V → VI (Übergänge [6] und [7]).

Die Herausbildung der Konfigurationen VI und VII beruht auf Strangtrennung der DNS in Struktur IV und kann auch durch Hitzebehandlung und Alkali (pH > 12,5) induziert werden. Neben der Alkalidenaturierung der DNS in der Konfiguration IV unter Herausbildung der Strukturen VI und VII (Übergang [5]) kommt es, ebenfalls als Ausdruck einer Denaturierung, auch zur Bildung eines doppelsträngigen, sehr kompakten cyclischen Knäuels mit einem S-Wert von 53 (Konfiguration I; Übergang [8]).

Die Herausbildung der DNS-Konfiguration III mit tertiären Windungen, wie sie offenbar im intakten Virus vorliegt, erfolgt nach beendeter DNS-Replikation und vor der Virusreifung über die Konfiguration V → II → III. Die Replikation der Virus-DNS führt also zunächst zur Konfiguration V. Bevor die beiden Polynucleotidstränge in dieser Anordnung vollständig rechtswindend doppelhelikal konfiguriert sind, kommt es zum Ringschluß unter Ausbildung der Konfiguration II. Danach erfolgt die weitere Organisation der beiden Polynucleotidstränge im Sinne der Herausbildung einer vollständig strukturierten Doppelhelix, wodurch es zur Organisation tertiärer Windungen des ringgeschlossenen Moleküls kommt. Umgekehrt führt, wie vorausgehend dargestellt, die Spaltung eines Polynucleotidstranges der DNS in Konfiguration III zu einer Aufhebung der linksdrehenden Tertiärstruktur unter Ausbildung der Strukturform IV.

Vergleichende Untersuchungen[19,44,47] über Beziehungen zwischen den verschiedenen Strukturformen und infektiöser Wirkung, getestet an der Plaque-Bildung, haben ergeben, daß die DNS des Polyomavirus in den Konfigurationen III, IV und V onkogen wirksam ist. Die biologische Wirkung der DNS in Konfiguration V geht bei pH > 11,8 verloren, nicht dagegen die der Komponente III. Wird die DNS in Konfiguration III auf pH 12,5 gebracht, d. h. denaturiert, und anschließend neutralisiert, so bilden sich die Strukturformen VI und VII unter Anstieg des infektiösen Titers aus. Der Anstieg der Infektiosität beruht wahrscheinlich darauf, daß denaturierte DNS etwa 3 bis 5mal stärker an die Zellmembran adsorbiert wird[47]. Ein Anstieg des infektiösen Titers kann auch durch thermische Denaturierung der DNS, d. h. Erhitzen und schnelles Abkühlen, erreicht werden[47]. Ähnliche Beziehungen zwischen molekularer Konfiguration und biologischer Aktivität der DNS, wie für die Plaque-Bildung dargestellt, ergeben sich auch für die transformierende Wirkung. Transformationen lassen sich mit der DNS in den Konfigurationen III, IV und V erzielen[12].

Nahezu analoge Ergebnisse, wie vorausgehend für die Polyomavirus DNS beschrieben, wurden in bezug auf die verschiedenen molekularen Konfigurationen, ihre Umwandlungen und biologischen Aktivitäten auch für die DNS des Shope-Papillomvirus[14] und des menschlichen Warzenvirus[15] erhalten. Different sind lediglich die Werte für die Sedimentationskoeffizienten der verschiedenen Strukturformen und die Molekulargewichte (Abb. 1). Unterschiede in der Stabilität der Konfiguration III der DNS des Polyomavirus einerseits und der Papillomviren andererseits sind insofern vorhanden, als daß die den Ringschluß bedingenden Bindungen in der Polyomavirus-DNS stabil gegen niedriges pH und Phenol sind, während sie in der DNS der Papillomviren instabil sind und durch Phenol und im sauren pH-Bereich gespalten werden[14]. Daher erhält man bei Extraktion der DNS aus dem Shope-Papillomvirus mit Phenol nur die 21-S-Komponente, bei Präparation mit Dodecylsulfat auch die 28-S-Komponente.

Die cyclische Struktur der Shope-Papillomvirus-DNS ist ebenfalls durch elektronenmikroskopische Bilder belegt[32].

Eine cyclische Struktur haben auch die Desoxyribonucleinsäuren des SV 40-[13] sowie der verschiedenen onkogenen und nichtonkogenen Adenoviren[24,40]. In DNS-Präparationen aus dem SV-40-Virus konnte in Dichtegradienten eine ringförmige Komponente (Konfiguration IV) mit einem S-Wert von 21,2 und eine linear konfigurierte Komponente (= V) mit S = 16,1 nachgewiesen werden.

Die DNS des menschlichen Warzenvirus sowie des Shope-Papillomvirus weisen demnach im Sedimentationsverhalten und der molekularen Konfiguration gewisse Gemeinsamkeiten auf, sind jedoch in der Nucleotidzusammensetzung verschieden (Tab. 2) und dürften damit auch im genetischen Informationsgehalt different sein. Das ergibt sich überzeugend aus Experimenten von CRAWFORD[15], der zeigen konnte, daß sich mit Shope-Papillom-DNS bzw.

mit DNS aus dem menschlichen Warzenvirus als Templates und mit Hilfe von RNS-Polymerase ^{3}H-Uridin-markiert synthetisierte RNS nur mit der jeweilig homologen DNS hybridisieren lassen, nicht dagegen im heterologen Ansatz. Damit in Übereinstimmung steht der Nachweis einer fehlenden immunologischen Cross-Reaktion der beiden Viren.

Eine engere strukturelle und möglicherweise auch genetische Beziehung besteht dagegen zwischen den Desoxyribonucleinsäuren des SV-40-Virus und des Adenovirus Typ 7, und zwar insofern, als daß genetisches Material des SV 40-Virus in das Adenovirus eingebaut werden kann[35,36,39]. Eine solche genetische Hybridisation der Desoxyribonucleinsäuren dieser beiden Viren ergibt sich aus dem experimentellen Befund, daß eine aus einer Mischinfektion von SV-40-Virus und Adenovirus Typ 7 herausgezüchtete SV-40-freie Linie des Adenovirusstammes Typ 7 in infizierten Zellen die Bildung eines intranucleär lokalisierten SV-40- Antigens induziert.

Versuche zur Darstellung infektiös wirksamer Nucleinsäuren aus onkogenen D-Viren sind sehr erfolgreich verlaufen. Bislang konnten aus folgenden Viren onkogen wirksame DNS-Präparationen dargestellt werden: Shope-Papillomvirus[28,29,30,31], Polyomavirus[16,22,26,27], Rinderpapillomvirus[5], SV-40-Virus[6,21].

Den Nucleinsäuren der geschwulsterzeugenden Viren kommt neben der onkogenen Wirkung ganz offenbar auch eine spezifische morphogenetische Wirkung zu. In dieser Hinsicht unterscheiden sich RNS und DNS der onkogenen Viren. Durch onkogene R-Viren transformierte Zellen weisen meist einen differenzierten Status auf, während durch onkogene D-Viren infizierte Zellen meist entdifferenziert sind[18]. Auch im Antigenbestand transformierter Zellen sind Unterschiede vorhanden. Nach Transformation von Zellen durch D-Viren treten Tumorantigene und neue Oberflächenantigene auf, in der Regel jedoch keine virusspezifischen Antigene (diese werden nur bei der lytischen Reaktion der Zellen gebildet). Im Gegensatz dazu kommt es in Zellen nach Transformation durch R-Viren auch zur Synthese von Virusantigenen. Eine Ausnahme machen die bei Säugetieren durch RSV indizierten Tumoren. Diese Unterschiede lassen sich möglicherweise mit dem Zustand der Nucleinsäuren in der Zelle in Zusammenhang bringen. Während offenbar die RNS der onkogenen Viren ständig in einer reproduktiven Phase in der Zelle vorliegt, kann sich die DNS der geschwulsterzeugenden Viren in ihrer ringförmigen Konfiguration nicht reproduzieren und wird in das zelleigene Genom integriert. Die zirkulare Struktur schützt die DNS gleichzeitig weitgehend gegen den Angriff zelleigener Nucleasen. Für eine solche Integration der DNS onkogener Viren in das zelleigene Genom sprechen Versuche von AXELROD et al.[1] mit Polyomavirus-DNS. Diesen Autoren gelang es, mit Hilfe der Hybridisierungstechnik in virusfreien Tumoren und transformierten Zellen eine DNS nachzuweisen, die sich mit ^{3}H-markierter DNS aus gereinigtem Polyomavirus hybridisieren läßt, während in nicht infizierten Zellen eine solche DNS nicht nachweisbar ist. Die Ergebnisse dieser Experimente lassen den Schluß zu, daß mindestens ein Virusgenom pro Zelle integriert ist und somit einen Status besitzen dürfte, der dem von Prophagen in lysogenen Bakterien analog sein könnte.

Literatur

[1] AXELROD, D., E. T. BOLTON, and K. HABEL: Fed. Proc. **23**, 401 (1964).
[2] BIELKA, H., u. A. GRAFFI: Naturwissenschaften **45**, 320 (1958).
[3] — — Acta biol. med. german. **3**, 515 (1959).
[4] — — und CHANG YÜ YEN: Acta biol. med. german. **10**, 63 (1964).
[5] BOIRON, M., M. THOMAS, and PH. CHENAILLE: Virology **26**, 150 (1965).
[6] —, J. P. LÉVY et M. THOMAS: Ann. Inst. Pasteur **108**, 298 (1965).
[7] BONAR, R. A., R. H. PURCELL, D. BEARD, and J. W. BEARD: J. nat. Cancer Inst. **31**, 705 (1963).
[8] BONHOEFER, F., and H. K. SCHACHMAN: Biochem. biophys. Res. Commun. **2**, 366 (1960).
[9] CRAWFORD, L. V., and E. M. CRAWFORD: Virology **13**, 227 (1961).
[10] — Virology **19**, 279 (1963).
[11] —, and E. M. CRAWFORD: Virology **21**, 258 (1963).
[12] —, R. DULBECCO, M. FRIED, L. MONTAGNIER, and M. STOKER: Proc. nat. Acad. Sci. (Wash.) **52**, 148 (1964).
[13] —, and P. H. BLACK: Virology **24**, 388 (1964).
[14] — J. molec. Biol. **8**, 489 (1964).
[15] — J. molec. Biol. **13**, 362 (1965).
[16] DI MAYORCA, G. A., B. E. EDDY, S. E. STEWART, W. S. HUNTER, C. FRIEND, and A. BENDICH: Proc. nat. Acad. Sci. (Wash.) **45**, 1805 (1959).
[17] DUESBERG, P. H., and W. S. ROBINSON: Proc. nat. Acad. Sci. (Wash.) **55**, 219 (1966).
[18] DULBECCO, R.: Science **142**, 932 (1963).
[19] —, and M. VOGT: Proc. nat. Acad. Sci. (Wash.) **50**, 236 (1963).
[20] GALIBERT, F., C. BERNARD, PH. CHENAILLE, and M. BOIRON: Nature (Lond.) **209**, 680 (1966).
[21] GERBER, P.: Virology **16**, 96 (1962).

[22] GRAFFI, A., et D. FRITZ: Rev. franç. Etud. clin. biol. **5**, 388 (1960).
[23] GREEN, M., and M. PINA: Proc. nat. Acad. Sci. (Wash.) **50**, 44 (1963).
[24] — — Proc. nat. Acad. Sci. (Wash.) **51**, 1251 (1964).
[25] HAYS, E. F., and H. SIMMONS: Proc. Amer. Ass. Cancer Res. **3**, 26 (1959).
[26] —, and J. A. CARR: Cancer Res. **22**, 1319 (1962).
[27] — Cancer Res. **24**, 1741 (1964).
[28] HODES, M. E., C. G. PALMER, L. E. BEATY, M. K. SWENSON, and J. DONALD: J. nat. Cancer Inst. **30**, 1 (1963).
[29] ITO, Y.: Virology **12**, 596 (1960).
[30] — Proc. nat. Acad. Sci. (Wash.) **47**, 1897 (1961).
[31] —, and CH. A. EVANS: J. nat. Cancer Inst. **34**, 431 (1965).
[32] KLEINSCHMIDT, A. K., S. J. KASS, R. C. WILLIAMS, and C. A. KNIGHT: J. molec. Biol. **13**, 749 (1965).
[33] LWOFF, A., R. HOME, and P. TOURNIER: Quant. Biol. **27**, 51 (1962).
[34] MOLONEY, J. B.: Acta Un. intern. Cancer. **19**, 250 (1963).
[34a] MORA, P. T., V. W. McFARLAN, and S. W. LUBORSKY: Proc. nat. Acad. Sci. (Wash.) **55**, 438 (1966).
[35] RAPP, F., J. L. MELNICK, J. S. BUTEL, and T. KITAHARA: Proc. nat. Acad. Sci. (Wash.) **52**, 1348 (1964).
[36] REICH, P. R.: Proc. nat. Acad. Sci. (Wash.) **55**, 336 (1966).
[37] ROBINSON, W. S., and M. A. BALUDA: Proc. nat. Acad. Sci. (Wash.) **54**, 1686 (1965).
[38] —, A. PITKANEN, and H. RUBIN: Proc. nat. Acad. Sci. (Wash.) **54**, 137 (1965).
[39] ROWE, W. P., and S. G. BAUM: Proc. nat. Acad. Sci. (Wash.) **52**, 1340 (1964).
[40] SMITH, K. O.: Science **148**, 100 (1965).
[41] TEMIN, H. M.: Proc. nat. Acad. Sci. (Wash.) **52**, 323 (1964).
[42] — Virology **23**, 486 (1964).
[43] TRÁVNIČEK, M., L. BURIČ, J. ŘIMAN, and F. ŠORM: Neoplasam (Bratisl.) **11**, 571 (1964).
[44] VINOGRAD, J., J. LEBOWITZ, and R. RADLOFF: Proc. nat. Acad. Sci. (Wash.) **53**, 1104 (1965).
[45] WATSON, J. D., and J. W. LITTLEFIELD: J. molec. Biol. **2**, 161 (1960).
[46] WEIL, R., A. BENDICH, and CH. M. SOUTHAM: Proc. Soc. exp. Biol. (N. Y.) **104**, 670 (1960).
[47] — Proc. nat. Acad Sci. (Wash.). **49**, 480 (1963).
[48] —, and J. VINOGRAD: Proc. nat. Acad. Sci. (Wash.) **50**, 730 (1963).

Untersuchungen über die cytocide und onkogene Wirkung des Simian-Virus (SV-)40

Von R. HAAS

Hygiene-Institut der Universität Freiburg

Mit 8 Abbildungen

Erlauben Sie mir folgende Vorbemerkungen: Soweit sich meine Ausführungen auf eigene Befunde beziehen, stammen sie aus Untersuchungen mit meinen Mitarbeitern Dr. MAASS, MÜLLER, SEEMAYER, WERCHAU und WESTPHAL.

Wegen der Kürze der Zeit werde ich auf die Fragen der Hybridisierung mit Adenoviren nicht eingehen.

SV-40 ist das erste aus Primaten isolierte onkogene Virus. SWEET und HILLEMAN[1,2] entdeckten es in den Überständen von Kulturen der Nieren von Rhesus- und Cynomolgusaffen, die sie auf Gewebekulturen der Nieren der afrikanischen Affenart Cercopithecus aethiops übertrugen. Da für den sich dabei einstellenden cytopathologischen Effekt die Bildung intracytoplasmatischer Vacuolen charakteristisch ist, nannte man dieses Virus anfänglich auch vacuolating agent. Kurze Zeit danach beschrieben EDDY und M.[3], daß die Beimpfung neugeborener Hamster mit Zellextrakten aus Nierengewebekulturen von Macacus rhesus das Auftreten maligner Tumoren vom Typ der Fibrosarkome zur Folge hatte. Wenig später wurde die Identität beider Agentien gesichert. Eine pikante Note erhielt die Geschichte dieses Virus, als man es in vermehrungsfähiger Form in Poliomyelitisimpfstoffen sowohl vom Salktyp als auch vom Lebendtyp nachwies.

SV-40 bildet mit dem Polyomavirus und den Papillomviren nach einem Vorschlag von MELNICK die Gruppe der Papovaviren. Ihre Nucleinsäure ist DNS, deren Anteil bei SV-40 ungefähr 9% des Partikels beträgt. Die DNS scheint in zwei Formen vorliegen zu können, einer mit 21,2 S sedimentierenden Ringform und einer offenen Form mit 16,1 S. Nach CRAWFORD[4] geht beim Lagern die schnell sedimentierende Komponente teilweise in die langsamere über. Das Capsid ist nach den Untersuchungen von MELNICK wahrscheinlich aus 42 Capsomeren aufgebaut, die auf der Oberfläche eines regelmäßigen Ikosaeders um die Virus-DNS angeordnet sind. Mit der Dichtegradientenzentrifugation läßt sich zeigen, daß die in infizierten Zellen gebildeten SV-40-Viren zwei Arten von Partikeln enthalten: nucleinsäurehaltige Teilchen, die sich bei einer Dichte von 1,32 einstellen und leere Capside, die eine Bande bei 1,29 bilden.

Von den im Verlauf des cytociden Cyclus gebildeten Virusteilchen ist nur ein sehr kleiner Anteil infektiös. Nach BLACK und Mitarb.[5] liegt der Quotient physikalische Partikel zu infektiösen Partikeln in der Größenordnung von 100:1 bis 200:1.

Eine Basenanalyse der Virus-DNS liegt bei SV-40 noch nicht vor. Auf Grund der Gleichgewichtsdichtegradientenzentrifugation nach MESELSON[6] wurde der Anteil von G+C in der SV-40-DNS auf 41% geschätzt. Ist dieser Wert richtig, so würde der Anteil G+C in der SV-40-DNS ziemlich gut mit dem G+C-Anteil der Wirtszellen übereinstimmen und sich deutlich von dem entsprechenden Wert für Polyomavirus unterscheiden.

SV-40 kann als onkogenes Virus Zellen transformieren. Die Transformation gelingt in vivo und in vitro. Als Folge der in-vivo-Transformation können sich Tumoren entwickeln. Sie entstehen meist am Applikationsort des Virus. Außer bei neugeborenen Hamstern kommt es nach SV-40-Injektion noch bei dem in Afrika lebenden Nagetier Rattus (Mastomys) natalensis zur Tumorbildung. Beim neugeborenen Hamster beträgt die durchschnittliche Latenzperiode bis zum Auftreten sicht- und tastbarer Tumoren etwa 200 Tage. Die Streuung dieses Wertes ist sehr groß. Die beiden wichtigsten die Tumorrate beeinflussenden Faktoren sind die Virusdosis und das Alter der Tiere zum Zeitpunkt der Injektion. Je größer die Virusdosis, umso höher die Tumorrate. Sie sinkt unter gleichzeitiger Verlängerung der Latenzperiode mit zunehmendem Alter. Bei 3 Monate alten Tieren konnte durch SV-40-Injektion keine Tumorbildung mehr ausgelöst werden.

Das Spektrum der in vitro durch SV-40 transformierbaren Zellarten ist wesentlich breiter als das Spektrum jener Tierspecies, bei denen SV-40-Infektion zu Tumoren führt. Unter den in vitro transformierten Zellen befinden sich neben Hamsterzellen, Affenzellen und anderen Säugerzellen auch menschliche Zellen. Die Übertragung in vitro transformierter Hamsterzellen auf Hamster läßt

Tumoren des gleichen histologischen Typs entstehen wie die Infektion mit SV-40-Virus. Analoge Beobachtungen liegen vom Menschen vor.

Wie jede virusbedingte Transformation wirft auch die durch SV-40-Virus eine Reihe von Fragen auf. In erster Linie sind es folgende: bilden die transformierten Zellen das transformierende infektiöse Virus und kann man an den transformierten Zellen irgendwelche in den aufeinanderfolgenden Zellgenerationen vorhandene, d. h. vererbbare Eigenschaften nachweisen, die mit der transformierenden Ursache zusammenhängen, die es mit anderen Worten gestatten, aus den Ergebnissen der Untersuchung transformierter Zellen auf die transformierende Ursache zu schließen. Sodann ist es das Problem, von welcher Komponente des Virusteilchens die transformierende Wirkung ausgeht und welche Bedingungen von Seiten einer Zelle erfüllt sein müssen, damit sie der Transformation durch SV-40 zugänglich ist. Endlich interessiert zu wissen, wie die SV-40-Infektion auf den Stoffwechsel sowohl der cytocid infizierten als auch der lediglich transformierbaren Zelle wirkt. Dabei käme eventuellen Unterschieden zwischen DNS-haltigen onkogenen und nichtonkogenen Viren besondere Bedeutung zu. Schließlich stellt jeder Versuch, die SV-40 induzierte Zelltransformation zu verstehen, auch vor folgende Alternativen: Integration des ganzen Virusgenoms in das Zellgenom oder nur eines Teiles oder keine Integration. Die erste Alternative würde den Gedanken einer gewissen Analogie zur Lysogenie nahe legen. Wir haben versucht, SV-40-transformierte Zellen, die spontan kein Virus freisetzen, mit Mitteln, durch die in lysogenen Bakterien der Ablauf des vegetativen Cyclus induziert werden kann, zur Bildung infektiösen Virus anzuregen. Versuche mit Mitomycin C, Proflavin und H_2O_2 hatten ein völlig negatives Resultat.

SV-40-transformierte Zellen unterscheiden sich von ihren Ursprungszellen durch einige Eigenschaften, die auch auf andere Weise transformierte Zellen erwerben. Sie gehorchen in monolayer-Kultur nicht mehr der Kontaktinhibition und sie weisen cytologische Veränderungen auf einschließlich solcher der Chromosomenzahl und Chromosomenmorphologie. Während die normale Hamsterzelle 44 Chromosomen besitzt, findet man in SV-40-transformierten Zellen sehr viel höhere und außerdem stark streuende Chromosomenzahlen. Gelegentlich sind es weit über 100 Chromosomen. Durch SV-40-transformierte Zellen besitzen in ihrem Kern ein neues Antigen, das in ihren Ursprungszellen nicht nachweisbar ist. Es wird mit Neoantigen, T-Antigen oder auch ICFA bezeichnet. Hamster, die SV-40 induzierte Tumoren tragen, bilden Antikörper gegen dieses Antigen.

Es ist schon länger bekannt, daß man bei Hamstern durch Injektion von SV-40-Virus Transplantationsimmunität gegen SV-40-Tumoren erzeugen kann. Nichts spricht dafür, daß ihre immunologische Grundlage eine Determinante des Capsids ist. Seit einiger Zeit kennt man ein zweites für SV-40 transformierte Zellen charakteristisches Antigen[7,8], das vielleicht enge Beziehung zur Trans-

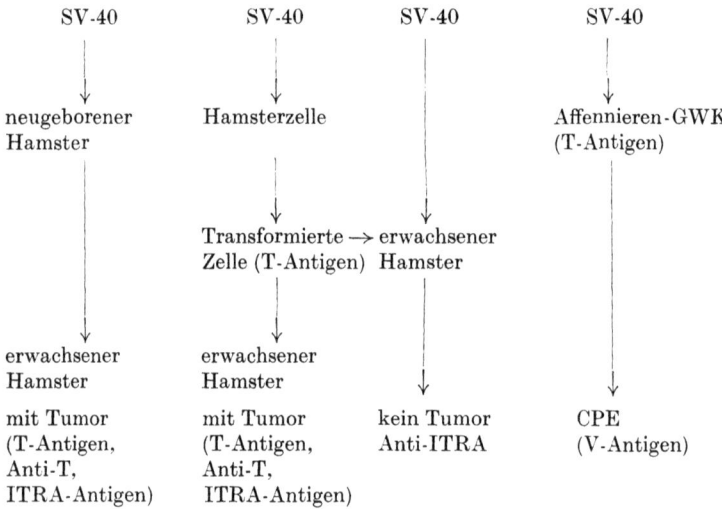

Abb. 1. Wirkungen des SV-40 in verschiedenen Zellsystemen

plantationsimmunität besitzt. Es liegt im Gegensatz zum T-Antigen nicht intranucleär, sondern an der Oberfläche SV-40-transformierter Hamsterzellen. Für die Virusspezifität dieses Antigens spricht, daß es serologisch mit dem von GIRARDI[9] in SV-40-transformierten Zellen nachgewiesenen „induced tumor resistance antigen" (ITRA) nahe verwandt zu sein scheint. Antikörper gegen dieses Antigen lassen sich im Serum tumortragender Hamster nicht nachweisen. Sie müssen durch eine spezielle Immunisation gewonnen werden. Während über dieses zweite Antigen noch relativ wenige Befunde vorliegen, wurde das erste von verschiedenen Arbeitskreisen eingehend untersucht.

Folgende Tatsachen können für das T-Antigen als gesichert angesehen werden: das T-Antigen ist virusspezifisch, aber kein Bestandteil der Viruspartikel. Zellen verschiedener Tierspecies, welche durch SV-40 transformiert sind, bilden serologisch identische T-Antigene.

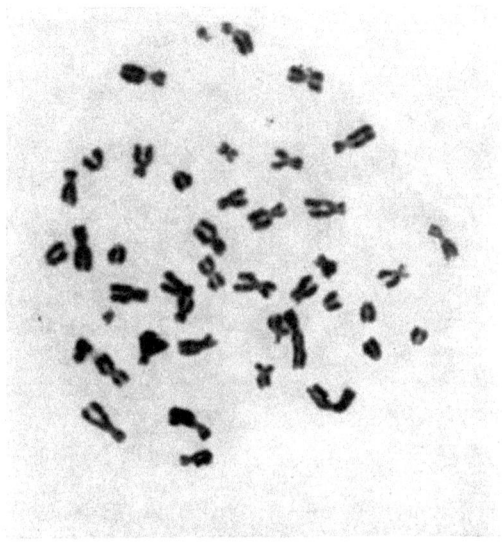

Abb. 2. Darstellung von Chromosomen aus normalen Hamsterfibroblasten. GIEMSA, 1000:1 (Ölimmersion)

Das T-Antigen dürfte infolgedessen, wofür auch andere Gründe sprechen, im Virusgenom codiert sein.

Ich möchte mir erlauben, Ihnen in der Abb. 1 in schematischer Form die verschiedenen Konsequenzen vor Augen führen, zu denen eine Infektion mit SV-40-Virus je nach Art des Wirtszellsystems führen kann. Die Abb. 2 und 3 zeigen den Chromosomensatz einer normalen und einer SV-40-transformierten Hamsterzelle.

Da das T-Antigen auch in der frühen Phase der cytolytischen Virusvermehrung auftritt, stellten wir uns die Frage, ob seine Synthese im cytociden Ablauf an eine Replikation des Virus gebunden ist. Wir haben zu diesem Zweck Versuche an SV-40-infizierten monolayer-Kulturen von Nieren von Cercopithecus aethiops vorgenommen. Die Infektionen erfolgten mit dem Virusstamm Vac 111, nachdem sich ein fast geschlossener Zellrasen gebildet hatte. Unmittelbar nach der Virusadsorption wurden die Kulturen mit Joddesoxyuridin, mit Fluordesoxyuridin, mit Actinomycin D, mit

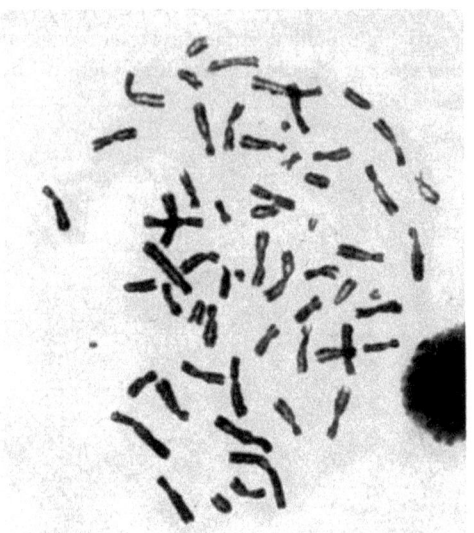

Abb. 3. Darstellung von Chromosomen aus einem durch SV-40 induzierten, in vitro gehaltenen (9. Pass.) entdifferenzierten Sarkom des Hamsters. GIEMSA, 1000:1 (Ölimmersion)

p-Fluorphenylalanin und mit Cytosinarabinosid beschickt. Ergebnisse entsprechender Untersuchungen liegen auch von RAPP u. Mitarb.[10] aus dem Laboratorium von MELNICK vor. Unsere Befunde,

Antimetabolit	Konzentration	Antigen		infektiöses Virus
		T	V	
keiner		+	+	+
JUDR	100 µg/ml	+	+	—
	50 µg/ml	+	+	(+)
FUDR	50 µg/ml	+	+	—
Cytosinarabinosid	100 µg/ml	+	—	—
p-FPA	50 µg/ml	—	—	—
Actinomycin D	0,5 µg/ml	—	—	—

Abb. 4. Bildung von T- und V-Antigen in Affennierengewebekulturen (Cerc. aeth.) nach Infektion mit SV-40 in Gegenwart verschiedener Antimetabolite

die mit denen von RAPP im großen ganzen übereinstimmen, hält Abb. 4 fest. Sie besagen, daß durch 0,5 γ/ml Actinomycin D und 50 γ/ml p-Fluorphenylalanin nicht nur die Synthese infektiösen

Virus, sondern auch von T-Antigen und der Capsomerenproteine, des sog. V-Antigens, inhibitiert wird. Mit p-Fluorphenylalanin haben wir das allerdings nur für V-Antigen geprüft. Der Nachweis der T- und V-Antigene erfolgte fluorescenzserologisch auf indirektem Wege. Das Anti-T-Serum stammte von tumortragenden Hamstern, während das Anti-V-Serum durch Immunisierung von Kaninchen mit SV-40-Virus gewonnen war. Im Gegensatz zur Wirkung von Actinomycin D und p-Fluorphenylalanin findet in Gegenwart von JUDR- und FUDR-Konzentrationen, die die Bildung infektiösen SV-40-Virus völlig blockieren, die Synthese sowohl von T-Antigen als auch von V-Antigen statt. In Übereinstimmung mit RAPP fanden wir, daß CA die Synthese infektiösen Virus und von V-Antigen blockiert, jedoch nicht von T-Antigen.

Die Zahl der in Gegenwart von 100 γ/ml JUDR T- und V-Antigen bildenden Zellen ist gegenüber einer Kultur ohne JUDR auf etwa 30 bis 50% herabgesetzt. Indessen kann diese Senkung den völlig negativen Nachweis der Bildung infektiösen Virus nicht erklären. Der Ausfall dieser Versuche ermöglicht keine Entscheidung zwischen den Alternativen der völligen Arretierung der Virus-DNS-Replikation und der Bildung einer funktionell insuffizienten Virus-DNS.

Ohne den Anspruch strenger Beweisführung sprechen für die erste Alternative folgende Beobachtungen: Die Blockade des infektiösen Cyclus durch JUDR läßt sich noch Tage nach der Infektion und nach der Zugabe des JUDR wieder aufheben. Bis zu 4 Tagen nach der Infektion gelingt das durch Ersatz des JUDR-haltigen Mediums durch eine davon freie Nährflüssigkeit. Später bis zum 8. Tage p.i. muß man dem Nährmedium zweckmäßigerweise 200 γ/ml Thymidin zugeben. Die Virussynthese kommt dann mit fast dem gleichen Tempo in Gang und führt annähernd zu den gleichen Ausbeuten, wie wenn keine Blockierung des Ablaufs des cytociden Cyclus vorgelegen hätte. Unseres Erachtens ist ein derartiger Ablauf des deblockierten Infektionscyclus wenig wahrscheinlich, wenn in Gegenwart von JUDR funktionell insuffiziente SV-40-DNS gebildet wird. Außerdem konnten wir in den JUDR-blockierten infizierten Kulturen autoradiographisch keinen Einbau von Thymidin feststellen und keine infektiöse DNS extrahieren.

Wir glauben, daß unsere und die Versuche des MELNICKschen Arbeitskreises am plausibelsten so zu interpretieren sind, daß auch in Zellen, in denen SV-40 einen cytociden Infektionscyclus durchlaufen kann, die Bildung von T-Antigen und V-Antigen ohne die bei der cytociden Infektion übliche Replikation der Virus-DNS möglich ist. Der Vollständigkeit halber möchte ich in diesem Zusammenhang erwähnen, daß wir elektronenoptisch in den Kernen unter JUDR stehender, SV-40 infizierter Cercopithecusnierenzellen bisher keine leeren Capside nachweisen konnten. Infolgedessen muß man annehmen, daß in Gegenwart von JUDR gebildetes V-Antigen in den Kernen zwar in serologisch reaktionsfähiger Form vorliegt, jedoch nicht in dem gewohnten morphologischen Arrangement von Capsomeren zu einem Capsid.

Beim Versuch, den Wirkungsmechanismus onkogener Viren zu verstehen, wurde für die DNS-haltigen Viren der Papovagruppe vermutet, ihre Wirkung auf den DNS-Stoffwechsel infizierter Zellen könne anders sein als bei DNS-haltigen nichtonkogenen Viren, wie etwa den Viren der Poxvirusgruppe oder gewisser Herpesviren. Während in letzterem Falle früher oder später eine Depression des Zell-DNS-Stoffwechsels erfolgt, könnte es durch Papovaviren zu einer Stimulierung kommen. Verschiedene Beobachtungen legen den Gedanken besonderer Beziehung DNS-haltiger onkogener Viren zum Zell-DNS-Stoffwechsel nahe: 1. die Transformationsrate ist bei logarithmisch wachsenden Zellen höher als bei stationären Kulturen; 2. die onkogenen Viruseffekte manifestieren sich in vivo umso stärker, je jünger die infizierten Tiere sind. Auch die Tatsache, daß offensichtlich der Informationsträger des T-Antigens des Virusgenoms wie ein celluläres Gen in die Zelle übernommen und von ihr gelesen und realisiert wird, führt zu solchen Überlegungen.

Ich darf hierzu bemerken, daß wir autoradiographisch mit ^{3}H-Thymidin an SV-40 transformierten Hamsterzellen die DNS-Synthesezeit und die Generationszeit bestimmt haben. Die DNS-Synthesezeit betrug 8 bis 9 Std und die Generationszeit 20 bis 21 Std. Ähnliche Zeiten fanden wir in der Literatur für die DNS-Synthesezeit und Generationszeit normaler vergleichbarer Hamsterzellen angegeben. Die Integration bestimmter Abschnitte des Virusgenoms in die Hamsterzelle kann also offensichtlich ohne Veränderung der chronologischen Rhythmik der Zelle erfolgen. Wir haben begonnen, den Effekt der SV-40-Infektion auf den DNS-Stoffwechsel zu studieren. Die Versuche wurden teilweise ebenfalls autoradiographisch vorgenommen. Daneben führten wir Untersuchungen an mit 2-C¹⁴-Thymidin markierter DNS durch, die wir zu bestimmten Zeiten aus SV-40-infizierten Kulturen extrahierten. Bei den autoradiographischen Versuchen benutzten wir infizierte und nichtinfizierte nahezu stationäre primäre Deckglaskulturen von Cercopithecusnieren und Hamsterembryonalgewebe. Zu unterschiedlichen Zeiten nach der Infektion wurde den Kulturen 40 min

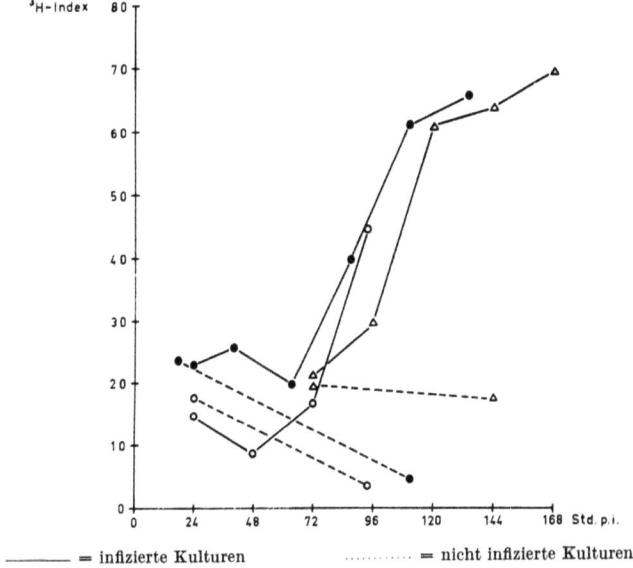

Abb. 5. ³H-Thymidineinbaurate in Gewebekulturen nach Infektion mit SV-40

lang 0,5 μC/ml Tritium-markiertes Thymidin angeboten. Die stationäre Phase der Gewebekulturen wurde gewählt, weil wir eventuelle Wirkungen auf den DNS-Stoffwechsel gegen einen möglichst niedrigen Hintergrund beobachten wollten.

Folgendes sind unsere wichtigsten bisherigen Befunde: 1. der H³-Index steigt, wie Abb. 5 erkennen läßt, nach einer Latenzperiode von 60 bis 70 Std steil an und erreicht nach 140 Std Werte zwischen 60 bis 80%. Zu dieser Zeit liegt der H³-Index der Kontrollkulturen unter 10%. Das bedeutet, daß als Folge der Virusinfektion in sehr viel mehr Zellen DNS synthetisiert wird als in den Kontrollen. Das Absinken des H³-Index in den Kontrollen dürfte Folge von Kontaktinhibition sein.

Die Latenzphase des Infektiositätsanstieges beträgt etwa einen und die des H^3-Indexanstieges etwa 3 Tage. Diese Differenz gilt es zu klären. Wir vermuten, daß der Unterschied mit der größeren Empfindlichkeit des Infektiositätsnachweises zusammenhängt. Bei gleichartigen Versuchen mit primären Hamsterembryonalkulturen, in denen SV-40 transformierend wirkt, fanden wir in 4 Wochen keine Steigerung des H^3-Index.

2. Vergleicht man die Verteilung der Silberkörner über dem Kernbereich vor und nach der Infektion markierter Zellen, in denen

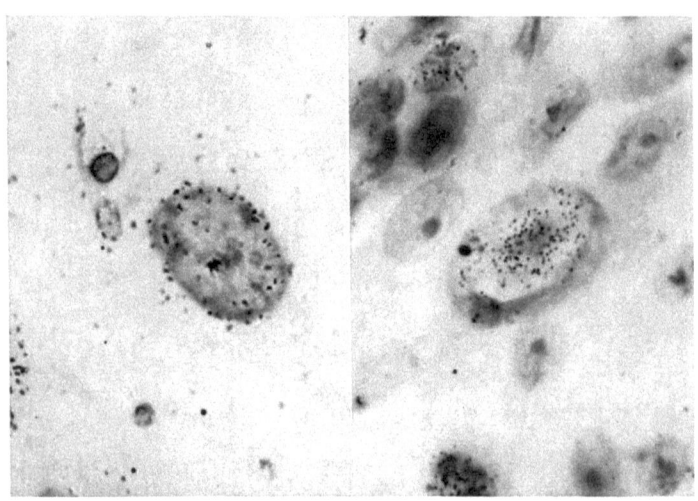

Abb. 6a. Autoradiogramm nach Einbau von ^{3}H-Thymidin in eine nichtinfizierte Nierengewebekultur, die anschließend mit SV-40 infiziert wurde. 1150:1

Abb. 6b. Autoradiogramm nach Einbau von ^{3}H-Thymidin in eine infizierte Nierengewebekultur. 1150:1

sich als Folge der Infektion gut abgegrenzte Einschlüsse gebildet haben, so stellt man bei vielen Zellen eine unterschiedliche Verteilung der Silberkörner fest. Bei den vor der Infektion markierten Zellen liegen die Silberkörner häufig vorwiegend in den Resten des randständig gewordenen Chromatins. Dagegen sieht man bei den nach der Infektion markierten Zellen die Silberkörner hauptsächlich über den Einschlüssen, d. h. über jenen Bezirken, in denen die neu synthetisierten Viren auftreten. Ich zeige diese unterschiedliche Verteilung der Silberkörner in Abb. 6. Damit soll nicht gesagt sein, daß die nach der Infektion gebildete DNS ausschließlich Virus-DNS ist. Ob letzteres zutrifft oder Virus-DNS plus Zell-DNS gebildet wird, ist eine damit nicht beantwortete Frage.

3. Wir haben aus infizierten stationären monolayer-Kulturen primärer Cercopithecusniere und aus BSC-1-Zellen DNS extrahiert, nachdem die Kulturen 1 bis mehrere Std lang vor der Extraktion

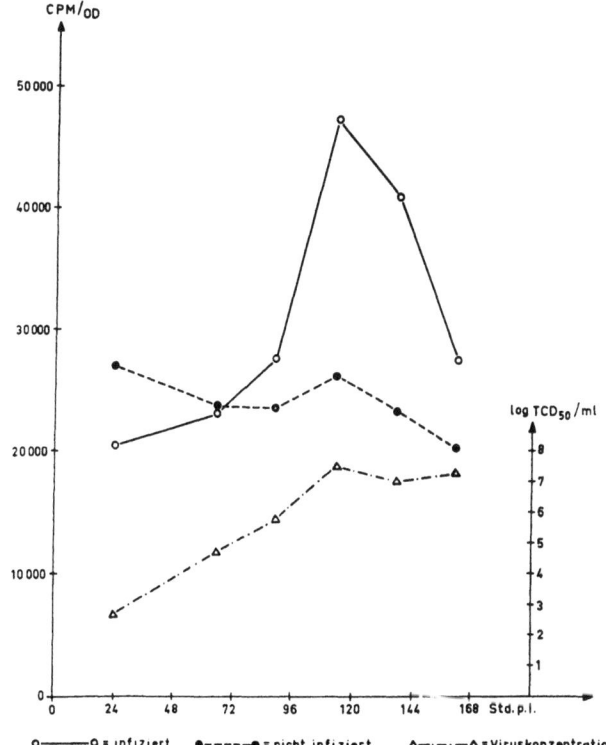

Abb. 7. Einbau von 2-^{14}C-Thymidin in die DNS von infizierten und nicht infizierten Nierengewebezellen

Pulse von 0,04 μC/ml 2-C^{14}-Thymidin erhalten hatten. Die Präparation der DNS erfolgte mit Phenol-SDS in Gegenwart von ÄDTA in Anlehnung an das Verfahren von COLTER[11].

Nach Umfällung der DNS mit Isopropanol wurde die spezifische Radioaktivität bestimmt. Die Abb. 7 hält ihren Verlauf in den Tagen nach der SV-40-Infektion fest, und zwar im Vergleich mit DNS aus nichtinfizierten Kontrollen. Wie man sieht, kommt es etwa 4 Tage nach Infektion zum deutlichen Anstieg der spezifischen Aktivität. Die erhöhte Aktivität ist mehrere Tage nachweisbar.

Eine Zunahme der absoluten Menge der aus SV-40-infizierten monolayer-Kulturen extrahierbaren DNS (vgl.[12]) konnten wir im Zeitraum bis zu 6 Tagen nach der Infektion nicht nachweisen, siehe Abb. 8. Berechnet man die aus der Virussynthese zu erwartende DNS-Menge, so kommt man auf einen Wert in der Größenordnung

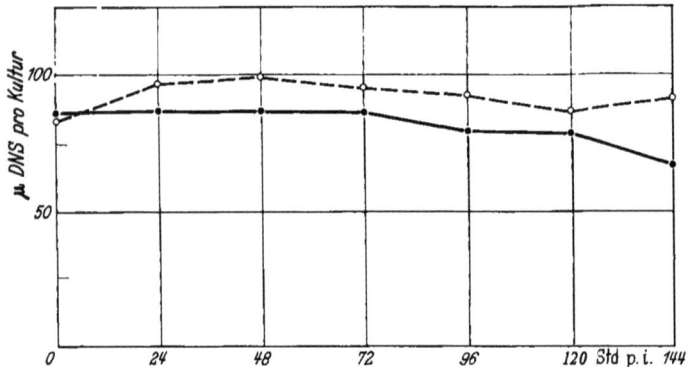

Abb. 8. DNS-Menge in SV-40-infizierten Gewebezellen (o ····· o) im Vergleich zur Kontolle (· ——— ·)

von 4 bis 5 μg pro Kultur. Der sich daraus ergebende Anstieg der Gesamtmenge DNS wäre methodisch nicht sicher gegenüber den Kontrollen nachweisbar gewesen.

Die Frage, ob es in SV-40 infizierten Gewebekulturen neben der Bildung der Virus-DNS auch zu einem Stimulationseffekt auf die Zell-DNS-Synthese kommt, können wir auf Grund unserer Versuche zur Zeit nicht beantworten. So muß ich mit der etwas resignierenden Feststellung schließen, daß wir bei dem Virus SV-40, soweit es gesicherte Fakten betrifft, dem Verständnis der cytociden und transformierenden Mechanismen noch nicht sehr nahe gekommen sind.

Die Untersuchungen wurden mit finanzieller Unterstützung der Deutschen Forschungsgemeinschaft im Rahmen der Unit „Medizinische Virologie" durchgeführt.

Literatur

[1] SWEET, B. H., and M. R. HILLEMAN: Detection of a „nondetectable" simian virus (vacuolating agent) present in rhesus and cynomolgus monkey-kidney cell culture material. 2. Intern. Conf. Live Poliovirus Vaccines Washington, D. C. 1960.

[2] — — Proc. Soc. exp. Biol. (N. Y.) **105**, 420 (1960)
[3] EDDY, B. E., G. S. BORMAN, W. H. BERKELY, and R. D. YOUNG: Proc. Soc. exp. Biol. (N. Y.) **107**, 191 (1961).
[4] CRAWFORD, L. V., and P. H. BLACK: Virology **24**, 388 (1964).
[5] BLACK, P. H., E. M. CRAWFORD, and L. V. CRAWFORD: Virology **24**, 381 (1964).
[6] MESELSON, M., F. W. STAHL, and J. VINOGRAD: Proc. nat. Acad. Sci. (Wash.) **43**, 581 (1957).
[7] TEVETHIA, S. S., M. KATZ, and F. RAPP: Proc. Soc. exp. Biol. (N. Y.) **119**, 896 (1965).
[8] —, and F. RAPP: Proc. Soc. exp. Biol. (N. Y.) **120**, 455 (1965).
[9] GIRARDI, A. J.: Proc. nat. Acad. Sci. (Wash.) **54**, 445 (1965).
[10] RAPP, F., J. S. BUTEL, L. A. FELDMAN, T. KITAHARA, and J. L. MELNICK: J. exp. Med. **121**, 935 (1965).
[11] COLTER, J. S., R. A. BROWN, and K. A. O. ELLEM: Biochim. biophys. Acta (Amst.) **55**, 31 (1962).
[12] BURTON, K.: Biochem. J. **62**, 315 (1956).

Untersuchungen am einstufigen Vermehrungscyclus von SV-40 mit Hilfe von Aktinomycin D

Von K. MUNK, H. FISCHER und W. BRÜMMER

Institut für Virusforschung am Deutschen Krebsforschungszentrum Heidelberg

Bei unseren Untersuchungen über die Viruswirtszellbeziehungen des Virus SV-40, während eines einstufigen Vermehrungscyclus, haben wir Cercopithecusnierenzellkulturen in der exponentiellen Phase mit einer Impfdosis infiziert, bei der auf jede Zelle 1,5 bis 5 infektiöse Einheiten kamen. Den Zeitraum der Infektion begrenzten wir auf 2 Std. Der Prozentsatz der infizierten Zellen, d. h. die Zahl der Zellen, in denen Antigenbildung stattfindet, wurde 50 Std p.i., etwa am Ende eines Vermehrungscyclus, mit der Immunofluorescenz unter Verwendung von Serum SV-40-infizierter Affen ermittelt. Der Virustiter wurde nach der Methode von PHILIPSON[1] bestimmt.

Um einen Einblick in den Ablauf der Prozesse nach Infektion der Zellen zu gewinnen, haben wir die Wirkung von Actinomycin D (Ac) auf die infizierten Kulturen untersucht. Wir begrenzten die Dauer der Ac-Gabe auf eine halbe Std und wählten eine Ac-Konzentration von 1 μg/ml. Die Wirkung dieser Konzentration und Dauer der Ac-Gabe prüften wir an nicht infizierten Cercopithecusnierenzellen. Der Einbau von 3H-Uridin in die säureunlösliche RNS-Fraktion sinkt 2 Std nach der Ac-Gabe auf wenige Prozent der unbehandelten Kontrolle ab. Beginnend mit der 8. Std nach der Ac-Gabe steigt die RNS-Synthese wieder an und erreicht bis 25 Std nach der Ac-Gabe 30% der Kontrolle.

Zu den infizierten Kulturen wurde das Ac zu verschiedenen Zeiten nach der Infektion gegeben. Bei einer Ac-Gabe zwischen 2½ bis 3 Std nach Infektion zeigte sich eine Steigerung der Anzahl der fluorescierenden Zellen in der Kultur gegenüber der infizierten unbehandelten Kontrolle. Ebenso lag der Virustiter in der behandelten Kultur, gemessen 50 Std p.i., höher als in der unbehandelten infizierten Kontrollkultur. Dieser Befund hat uns überrascht. Er wird weiter untersucht. Bei einer Ac-Gabe 9½ bis 10 Std p.i. ist das Phänomen der Zunahme virussynthetisierender Zellen nicht mehr zu beobachten. Auch eine Verminderung des Prozentsatzes der fluorescierenden Zellen, wie sie bei Ac-Gaben zu späteren Zeiten nach der Infektion auftritt, findet sich zu diesem Zeitpunkt noch nicht. Eine halbstündige Ac-Gabe zwischen der 14. und 30. Std p.i. führt zu einer Reduktion der Anzahl der Zellen, in denen Antigenbildung und Virusvermehrung stattfinden auf maximal 30% der Kontrolle. Auf Grund des 3H-Uridineinbaues in die RNS nichtinfizierter Zellen wäre eine vollständige Hemmung der Virusantigensynthese und der Bildung infektiöser Partikel zu erwarten gewesen. Daß bei einer halbstündigen Gabe nur eine Hemmung auf 30% der Kontrollen beobachtet wurde, könnte auf eine Asynchronie der Prozesse, auf einen langlebigen Messenger, dessen Synthese bei dieser kurzen Ac-Gabe nicht in der gesamten Kultur blockiert wird und auf ein Nachlassen der Ac-Wirkung zurückzuführen sein. In den meisten Zellen werden aber die RNS-Synthese-abhängigen Prozesse der Antigenbildung und Virusproduktion, die während der 14. bis 30. Std p.i. ablaufen, gehemmt.

[1] PHILIPSON, L.: Virology 15, 263 (1961).

Diskussion

HEIDELBERGER (Madison/Wisconsin): My colleagues Dr. IYPE and Dr. ROELLER, and myself wish to report that we have obtained transformation of normal cells *in vitro* with carcinogenic hydrocarbons in a different system than that reported so elegantly by SACHS. We have been using the mouse prostate gland maintained in organ culture by a modification of the technique originated by LASNITZKI [Cf. J. Nat. Cancer Inst. Monograph 12, 381 (1963)], in whose laboratory I was privileged to study in 1962. When little pieces of the ventral prostate glands of C3H mice are cultivated on the surface of EAGLE's medium $+10\%$ horse serum for three weeks, they remain highly differentiated with columnar rows of epithelial cells surrounding the alveoli. However, when 5 μg/ml of methylcholanthrene are added to the medium for one week and the cultivation is continued for additional two weeks extensive histological changes are observed. These include massive hyperplasia, squamous metaplasia, anaplastic cells with pleomorphic nuclei, and occasionally invasion. Some pathologists have diagnosed these histologically as being malignant, so that on purely morphological ground we might say that we have achieved carcinogenesis *in vitro* with hydrocarbons. However, in order to prove this biologically it would be necessary to implant the pieces from organ culture into C3H mice and obtain tumors. This we attempted under a variety of conditions for almost 3 years, and failed to get a single tumor in the recipient mice.

It appeared that this failure might have been due to the fact that insufficient numbers of tumor cells were implanted. Accordingly, it was considered that by dispersing the organ culture pieces and putting them into cell culture, it might be possible to grow up a population of cells with the highest degree of proliferative capacity, and which might be transformed.

The following experiment was then carried out, which is typical. Mouse prostates were treated in organ culture as described above: the controls were cultivated for three weeks, and the other group was treated for one week with 5 µg/ml of methylcholanthrene and two weeks in carcinogen-free medium. At the end of that time, the pieces within each group were combined, dispersed with pronase, and plated in dishes with EAGLE's medium and 20% fetal calf serum. Very little happened at first, but after two weeks small colonies of epithelial cells formed in both groups, after 5 weeks the first subculture was made, and at 7 weeks the second sub-culture was done. At 9 weeks the control cells had all died, but the cells derived from the methylcholanthrene-treated cultures continued to grow, and by 12 weeks had achieved a very rapid growth rate. These epithelial cells piled up when the dish was crowded, indicating a loss of contact inhibition, and have grown rapidly and continuously for more than 5 months to the present time.

That these cells have been transformed is proved by the following criteria: 1. they are a permanent line, whereas the control cells all died; 2. they pile up and have lost contact inhibition; 3. they are much less susceptible to the toxicity of carcinogenic hydrocarbons than are the normal cells; 4. on implantation of these cells into C3H mice or into the cheek-pouch of cortisonized hamsters they produce masses that are histologically compatible with malignancy, but these masses eventually regress. It is currently too early to know whether they will reappear and grow as tumors. In karyotypic analyses, these cells exhibit no chromosomal abnormalities. About half the cells are diploid and half polyploid.

This experiment has been repeated with methylcholanthrene, and similar results have been obtained in two experiments with 9,10-dimethyl-1,2-benzanthracene. Thus we can conclude at the present time that we have achieved cell transformation and possibly carcinogenesis *in vitro* with carcinogenic hydrocarbons. Much further work is in progress.

SACHS: I would like to thank Dr. HEIDELBERGER for his kind compliment, and I was interested to hear of the cell lines that he has obtained after culture of dispersed cells from mouse prostate organ cultures treated with carcinogenic hydrocarbons *in vitro*. I was also interested in his observation that in contrast to the progressive growth of tumors *in vivo* that we obtained with cells that we had transformed *in vitro* by carcinogenic hydrocarbons, inoculation of cells from his lines into animals gave growths that regressed, as we had found with cells that we had transformed *in vitro* by x-irradiation with 300 r.

I would just like to come back to a point I mentioned in my talk, regarding the cytotoxic action of carcinogenic hydrocarbons on normal cells and on tumor cells. Our results of cloning normal cells have shown that there are

Diskussion 283

normal cells that are susceptible, and normal cells that are resistant to the cytotoxic action of benzpyrene, and that transformation can be obtained with both types of cells. This raises the possibility which we are now studying further, that the resistance to the cytotoxic action of carcinogenic hydrocarbons shown by different types of transformed cells may be due to the selection of cells that were already resistant as normal cells, rather than the acquisition of resistance as a result of cell transformation.

BAUER (Tübingen): Da die RNS-haltigen Tumorviren bei der heutigen Sitzung meiner Ansicht nach etwas zu kurz kamen, möchte ich einige ergänzende Bemerkungen zum Referat von Herrn BIELKA machen und über einige Ergebnisse berichten, die aus eigenen Untersuchungen über die Nucleinsäure des Hühnermyeloblastosevirus (AMV) resultieren. Wie Herr BIELKA gezeigt hat, haben ROBINSON und Mitarbeiter erstmals aus dem Rous-Virus (RSV) eine hochmolekulare RNS isoliert. Was Herr BIELKA — wahrscheinlich aus Zeitgründen — nicht erwähnt hat, ist die Tatsache, daß diese Autoren bei dem gleichen Virus wie auch beim Newcastle Disease Virus, das zu den Myxoviren gehört und eine ähnliche Struktur hat, eine weitere Nucleinsäure-haltige Komponente isolierten, die mit etwa 4 S sedimentiert. Aus dem Myeloblastosevirus des Huhnes, das dem RSV strukturell sehr ähnlich und serologisch mit diesem verwandt ist, isolierten wir nach Natriumlaurylsulfat (NLS)-Behandlung neben einer 62 S RNS und einer 4 bis 10 S Komponente zwei weitere RNS-haltige Komponenten, die im Zuckergradienten mit etwa 45 S und 30 S sedimentieren.

In früheren Untersuchungen hatten wir im AMV zellspezifische RNS nachgewiesen. Es interessierte uns nun, welche unserer 4 RNS-haltigen Fraktionen Zell-RNS enthält. Aus diesem Grund wurde folgendes Experiment durchgeführt: Sekundäre RIF-freie Hühnerfibroblastenkulturen wurden in Gegenwart von ^{3}H-Uridin inkubiert. Nach 2 Tagen wurden die Kulturen gewaschen, mit AMV infiziert und nach der Virusadsorption im Vergleich zum ^{3}H-Uridin eine 100fache Menge an ^{14}C-Uridin zugegeben. Nach weiteren 3 Tagen wurde das neusynthetisierte Virus aus dem Gewebekulturmedium konzentriert und gereinigt und die RNS nach NLS-Behandlung im Zuckergradienten isoliert. Nach Bestimmung der Radioaktivität in den einzelnen Fraktionen ergab sich folgendes Bild: ^{14}C-Uridin war in sämtlichen Fraktionen nachweisbar. ^{3}H-Uridin dagegen, das nach der Versuchsanordnung nur in Zell-RNS eingebaut werden sollte, erschien nur in den Fraktionen mit 45 S, 30 S und 4 bis 10 S. Wir folgern daraus, daß diese Komponenten des AMV, die wohl auch im RSV ihre Parallelen haben, zellspezifische RNS enthalten und daß aller Wahrscheinlichkeit nach die 62 S Komponente die eigentliche Virus-RNS darstellt. Es bleibt zu klären, ob der im Virus enthaltenen Zell-RNS eine biologische Funktion bei der Virusinfektion zukommt.

SCHÄFER (Tübingen): Wie uns die Vortragenden dieser Sitzung zeigen konnten, hat die in den letzten Jahren einsetzende intensivere Bearbeitung der onkogenen Virusarten beachtliche Erfolge gezeitigt. Es besteht danach wohl kaum ein Zweifel daran, daß Viren eine beachtliche Rolle bei der Krebserzeugung spielen können; ob sie eine „letzte Ursache" sind, wage ich nicht zu entscheiden.

Besonderes Interesse verdient die Feststellung, daß gewisse Tumorerzeugende Viren in der Lage sind, die Synthese der cellulären DNS anzuheizen. Andererseits wissen wir, daß cytocide Viren existieren, die die Produktion von normaler RNS, normalem Protein und darüber hinaus auch von normaler DNS inhibieren. Als Virologe frage ich mich, ob es nicht möglich sein sollte, eines gegen das andere auszuspielen und die Wirkung eines onkogenen durch gewisse Wirkungen eines cytociden Virus zu paralysieren. Voraussetzung dafür wäre natürlich, daß es gelänge, den Informationsgehalt des cytociden Virus so zu dosieren, daß es nur noch die für den speziellen Zweck erwünschte Funktion ausübt. Experimentelle Ansätze zur Lösung dieses Problems sind tatsächlich vorhanden. Ich möchte in diesem Zusammenhang auf eine Arbeit von SCHOLTISSEK und ROTT verweisen, denen es gelang, die verschiedenen Funktionen eines cytociden Virus mit chemischen Mitteln zu entkoppeln. Womöglich gelingt es auf dem angedeuteten Weg eines Tages einmal, den Teufel „Krebs" durch den Belzebub „Virus" auszutreiben.

Zusammenfassung und Ausblick

Von F. BERGEL

Chester Beatty Research Institute, Institute of Cancer Research, Royal Cancer Hospital, London, Großbritannien

Mit 2 Abbildungen

Es war vor Beginn des Colloquiums nicht leicht, eine gedrängte Übersicht über die Kurzreferate der Teilnehmer des Mosbacher Colloquiums für eine Vorveröffentlichung zu verfassen. Man hoffte, daß es nach den Vorträgen, Diapositiven und Diskussionen einfacher sein werde, ein Résumé zu verfertigen. Die Sprecher berichteten jedoch über ihre Arbeiten von solch vielfältigen Gesichtspunkten aus, fingen mit so mannigfachen Arbeitshypothesen an und kamen anscheinend zu so vielerlei verschiedenen Schlüssen, daß eine einfache *Zusammenfassung* nicht möglich ist. Weniger schwierig ist es vielleicht mit einem *Ausblick*, da es sich dabei „nur" um den Versuch handelt, um die nächste Kurve zu spähen. Wie alle Teilnehmer wissen, und wie es die Tagung selbst bestätigte, verläuft die Landstraße der Krebsforschung wie eine englische Country Road, nicht geradeaus; besonders nicht vom Standpunkt der Molekularen Biologie aus gesehen. Man sollte eigentlich „Standpunkte" sagen, da die Molekulare Biologie ein Mischfach und nach CHARGAFFS Scherz „die Praxis eines Biochemikers ohne Diplom ist". Dagegen muß man einwenden, daß sich die Tätigkeiten vieler Krebsforscher — mit Diplomen vom Nobelpreis bis zur Promotion — bestens innerhalb der weiten Felder der Molekularen Biologie einfügen (und damit ist nicht nur die Erforschung der Nucleinsäuren und ihre genetische Bedeutung gemeint):

Die Referenten dieser Tagung haben innerhalb der weiten Grenzen der Molekularen Biologie die Probleme hauptsächlich von vier verschiedenen Seiten beleuchtet (s. Tabelle). Obwohl alle Themen, besonders die der Gruppen 2 und 3, ineinander übergreifen, fällt eine Übersicht leichter, wenn man die Referate der Herren WARBURG, BUSCH und PITOT, dann die von MAGEE, DANNENBERG und HECKER, danach die von LAWLEY und HEIDELBERGER und schließ-

Tabelle

Referent	Themengruppe	Zellbestandteile oder Stoffwechselzwischenprodukte	Exogene Verbindungen
WARBURG	1. Beiträge zur fundamentalen Biochemie der neoplastischen Veränderungen	Enzyme der Glykolyse und Atmung, Sauerstoff	Sauerstoff
BUSCH		DNS und RNS der Nucleolen, Histone, sauere Proteine	
PITOT		Enzyme der Leber, m-RNS, Ribosomen	Tryptophan, Aminosäuren, Actinomycin D, Buttergelb
MAGEE	2. Die Mechanismen der chemischen Krebserzeugung (Carcinogenese-cocarcinogenese)	DNS, RNS	N-Nitrosamine, Alkylierungsmittel
DANNENBERG		DNS	Aromatische Amine, Kohlenwasserstoffe
HECKER		m-RNS	Phorbolester, DMBA, Actinomycin
LAWLEY	3. Einwirkung von Drogen auf Zellen und ihre Bestandteile	DNS, Reparaturenzyme	Alkylierungsmittel, Kohlenwasserstoffe
HEIDELBERGER		Nucleoside, Nucleotide, DNS	Fluorierte Abkömmlinge des Uracils und Thymidins
GRAFFI	4. Mechanismus der Viruseffekte in vitro und in vivo	DNS usw.	Viren
SACHS		DNS	Polyomavirus, Kohlenwasserstoffe
BIELKA		RNS und DNS der Viren	Rous-Sarkomvirus, D-Viren Simianvirus (SV-)40,
HAAS		DNS	^{3}H-Thymidin, Joddesoxyuridin

lich die Vorträge der Herren GRAFFI, SACHS, BIELKA und HAAS im großen Rahmen dieses wertvollen und äußerst anregenden Colloquiums als einigermaßen zusammengehörig betrachtet.

Es wäre Raum- und Zeitverschwendung, würde man mit schlechteren und unzutreffenderen Worten die Vorträge und Diskussionen nochmals kurz erörtern. Es ist besser, die Eindrücke wiederzugeben, die die zahlreichen Tatsachen und Spekulationen in einem der Zuhörer hinterlassen haben. Anschließend könnte ein Destillat der Resultate und eine Skizze der gegenwärtigen Situation folgen, bevor man sich auf die hohe Warte eines Ausblickes begibt.

Die Referate

Erste Gruppe. Es war eine schöne Idee der Veranstalter, den Altmeister, Herrn WARBURG, wie man im englischen Cricket-Sprachgebrauch sagt, ,,to put in as first bat". Er war der Erste, der grundlegende biochemische Eigenschaften einer großen Anzahl von Krebsgeschwülsten erforschte. Er ist der Meinung, daß der Umschlag von Atmung zu Gärung die letzte Ursache dieser biologischen Katastrophe darstellt, daß ,,Krebs" irreversibel ist und daß die Gleichung Krebs = Anaerobiose die Situation thermodynamisch, chemisch und biologisch zusammenfassend beschreibt. Eine sehr lebhafte Diskussion, man möchte nahezu sagen, eine ,,Disputatio", folgte, und betonte die Glaubensstärke der Vertreter der verschiedenen Ideen über die ersten Ereignisse bei der Krebsentstehung. Wie weiter unten ausgeführt wird, gibt es neuere Beobachtungen, die die Verknüpfung von Sauerstoff, Fermenten und Stoffwechselsystemen bestätigen, aber deren direkter Zusammenhang mit malignem Wachstum im Augenblick nur als Spekulation angesehen werden kann.

Herr BUSCH verlegte den Schwerpunkt der Betrachtungen auf die Nucleinsäuren, die basischen Histone und sauren Proteine des Zellkernes und der Nucleolen. Besonders die letzteren mit ihren granulären und fibrillären Regionen zeigen in ihrer RNS mögliche Unterschiede zwischen normalen und abnormen Geweben. Natürlich wurden solche Beobachtungen nur an einigen Tumoren gemacht, und es ist wichtig, sich im Zusammenhang mit diesen sorgfältigen und allen folgenden Arbeiten an die Gefahren von Verallgemeinerungen zu erinnern.

Der Vortrag von Herrn PITOT stellte, abgesehen von seinen sehr interessanten Resultaten, eine weitere Illustration der Komplexität der gegenwärtigen Lage der fundamentalen Biochemie des Krebses dar. Mit Untersuchungen an Tierlebern und einer Reihe von Hepatomen, einschließlich der langsam wachsenden Morris-Tumoren, versuchte er in das Geheimnis der Leberkrebse einzudringen. Dazu benutzte er das unterschiedliche Verhalten von Enzymen, z. B. der Tryptophanpyrrolase und der Threonin-Serin-Dehydrase, die sowohl durch Tryptophan, bzw. Aminosäurengemische als auch durch Cortison charakteristisch induziert werden können. Die Wirkungen von Actinomycin D auf die Fermentinduktion und andere celluläre Ereignisse, zusammen mit Stabilitätsänderungen von Polysomen in Hepatomen, führte zu Ideen über Regulationsvorgänge und deren Störungen bereits auf der präancerösen Stufe, die sich auf verschiedenen Ebenen des Lebensprozesses der Zellen, wie der ,,Translation" oder der ,,Transscription" abspielen können.

Wie schon erwähnt, fand eine lebhafte Diskussion statt über Gärung, DNS, Histone, Hormone, ein Anzahl von Enzymen und den Aminosäureeinbau in die Mitochondrien, deren DNS, wie Herr BÜCHER bemerkte, sich von der des Zellkernes unterscheidet und eine circuläre Struktur besitzt.

Zweite Gruppe. Im Rahmen der Vorträge über Carcinogenese und Cocarcinogenese, berichtete Herr MAGEE über die Wirkungen von N-Nitrosaminen, Nitrosoharnstoff und Dimethylsulfat auf tierische Gewebe und verglich deren biologische Effekte mit denen anderer alkylierender Substanzen. Die Möglichkeit der Alkylierung von DNS, RNS oder anderen Zellbestandteilen wurde erörtert und die dabei vielleicht wichtige Funktion von Alkylasen, z. B. Methylasen erwähnt. MAGEE und nach ihm Herr DANNENBERG haben die Beziehungen zwischen Mutagenese und Carcinogenese diskutiert. DANNENBERG verglich besonders mehrkernige aromatische Amine, deren Struktur einen Einbau in die NS-Doppelstränge zuläßt, miteinander und mit Kohlenwasserstoffen. Herr HECKER beschrieb anschließend die schönen Arbeiten seiner Gruppe über die nahezu vollkommene Aufklärung der aktiven Bestandteile und der Synthese einiger Phorbolester des Crotonöls, das ja vor mehr als 20 Jahren von BERENBLUM und MOTTRAM als cocarcinogenes Material der polycyclischen Kohlenwasserstoffe in der Maus erkannt worden war. HECKER und seine Mitarbeiter studierten die reinen Bestandteile dieses Naturproduktes und bestätigten frühere Beobachtungen. Bei ihren Untersuchungen konnten sie auch schwache carcinogene Effekte erzielen. Was einen dabei bekümmern könnte, ist die von SHUBIK (Chicago) publizierte Tatsache, daß sich bisher die cocarcinogenen Wirkungen nur auf die Maus erstrecken.

Die Diskussion nach diesen Vorträgen war lebhaft und voller Argumente. Von WICK wurde bemerkt, daß für den Wirkungsmechanismus des Benzpyrens möglicherweise ein durch Oxydation entstehendes schwerlösliches Dimer von Bedeutung ist.

Dritte Gruppe. Wie bereits angedeutet, waren die Referate dieser Gruppe eng mit denen der vorhergehenden verknüpft, nur daß vorwiegend die Wirkung von cytotoxischen Drogen und von Antimetaboliten an Stelle der carcinogenen Substanzen untersucht wurde. Herr LAWLEY präsentierte eine Art Zusammenfassung und Schlußfolgerung der Untersuchungen, die er mit Herrn BROOKS während der letzten Jahre gemacht hat: Der Wirkungsmechanismus von Zellgiften, carcinogenen und mutagenen Substanzen, die Degradation von Nucleinsäuren, der Mechanismus der Reparatur solcher Zellschäden oder die Möglichkeiten, das Wirksamwerden solcher Reparaturmechanismen zu verhindern: all das erarbeitet mit Hilfe von Alkylierungsmitteln von Kohlenwasserstoffen, Aminen und Hydroxylaminen.

Die Frage, ob seine Schlußfolgerung, daß die DNS den Hauptangriffspunkt dieser Substanzen darstellt, den wirklichen Tatsachen entspricht oder in der augenblicklichen Situation eine ausgezeichnete Arbeitshypothese darstellt, wurde in der nachfolgenden Diskussion eifrig debattiert. Einige der Diskussionsredner bevorzugten andere Zellbestandteile, so z. B. Fermente, oder das DPN oder vielleicht Komponenten der Mitochondrien (siehe weiter oben und unten).

Herr HEIDELBERGER, der während dieser Diskussion erklärt hatte, daß er bezüglich der ,,Treffer" in der Zelle noch zu keiner definitiven Meinung gekommen wäre, berichtete den Zuhörern die interessante Geschichte seiner fluorierten Pyrimidinderivate, die sich nicht nur als chemotherapeutisch nützlich erwiesen haben, sondern auch als Sonden verwendbar sind, um einen tieferen Einblick in die enzymatischen Geschehnisse bei der Synthese der Nucleinsäuren in der Zelle zu erhalten. Sehr eindrucksvoll war es, über die Wirkungen der neueren Thymidinverbindungen — F_3TDR

und F_3TDRP — zu hören, wobei die letztere die Thymidilatsynthese stark hemmt (UDRP–||→TDRP). Das gab wieder Gelegenheit für eine weitschweifende Diskussion, die sich besonders auf den Stoffwechsel der Pyrimidine bezog.

Vierte Gruppe. Nachdem Herr GRAFFI eine ausgezeichnete Übersicht über das gesamte Gebiet der onkogenen (tumorerzeugenden) Viren, einschließlich der harmlose Geschwülste im Menschen verursachenden, gegeben hatte, berichtete Herr SACHS ausführlich und eindrucksvoll über seine Studien zur Wirkung von Viren, polycyclischen Kohlenwasserstoffen und Röntgenstrahlen auf Zellkulturen und zog aus diesen und anderen Untersuchungen seine kampflustigen Schlüsse. Da es unmöglich ist, die Besonderheit seines Stiles und seiner Persönlichkeit in wenigen Sätzen nachzuzeichnen, soll hier nur festgehalten werden, daß Carcinogenese, hervorgerufen durch verschiedene Agentien, ein cellulärer Prozeß ist, der eine Reihe von Phasen durchläuft und anscheinend nicht, wie noch einige Mediziner glauben, nur in einem kompletten Organismus vor sich gehen kann. Ob die Unsicherheit bezüglich der Rolle der Viren bei der Entstehung des menschlichen Krebses durch diese und andere Forschungen zugunsten einer sicheren Aussage aufgehoben werden kann, wird die Zukunft lehren. Es ist nicht undenkbar, daß der menschliche Organismus im Laufe der Phylogenese die Fähigkeit erworben hat, Viren, die im Tier maligne Geschwülste oder Leukämien erzeugen (wo sind hier Mycoplasmen einzuordnen?), zu relativ harmlosen, nichtpathogenen Eindringlingen zu transformieren (siehe Verrucca oder Molluscum contagiosum). Jedenfalls sind die Arbeiten, die die Herren BIELKA und HAAS referierten, und die sich auf spezifische molekulare Eigenschaften von bestimmten Viren bezogen, besonders auf deren Nucleinsäuren, äußerst wertvolle Beiträge zur Kenntnis von komplizierten hochmolekularen Naturprodukten.

Ein Destillat der Zusammenfassung

Man könnte sich zwei Fragen vorlegen:

1. Wieweit hat dieses Colloquium klargestellt, welche biochemischen Ereignisse, durch verschiedene Eingriffe in die Zelle hervorgerufen, direkte Beziehungen zu nachfolgenden biologischen Geschehnissen haben (siehe Abb. 1).

Zusammenfassung und Ausblick

2. Was sind die Treffersubstanzen oder die Receptoren exogener Angriffe? Ist es der Zellkern mit den Nucleinsäuren — DNS und RNS —, sind es Histone, saure Proteine oder Enzyme; sind es die Nucleolen oder gar das Cytoplasma mit allen seinen partikulären bzw. chemischen Bestandteilen, einschließlich der Mitochondrien, der Polysomen, Ribosomen und Plasmagene; oder ist es vielleicht die Zellmembran, die mit ihren Antigenen, Mucoproteinen, Sterinen, Phospholipiden und der Sialinsäure die Eigenschaften des Zellinneren oder die Haftfähigkeit der Zelloberflächen zumindest mitbestimmen (s. Abb. 2).

Die Antworten auf diese Fragen, die im Verlauf des Colloquiums angeboten wurden, waren leider nicht eindeutig. Eine Gesamthypothese konnte davon nicht extrahiert werden. Da war die Grundlage der Veränderung des Energiestoffwechsels mit seinen Fermenten, Substraten und dem Sauerstoff; da waren die Nucleinsäuren, besonders die DNS mit ihren Schäden und Reparaturen, gewisse Enzyme oder Coenzyme, Hormone. Mit anderen Worten: da waren alle, ja nahezu alle Zellbestandteile und Zellvorgänge an den verschiedenen

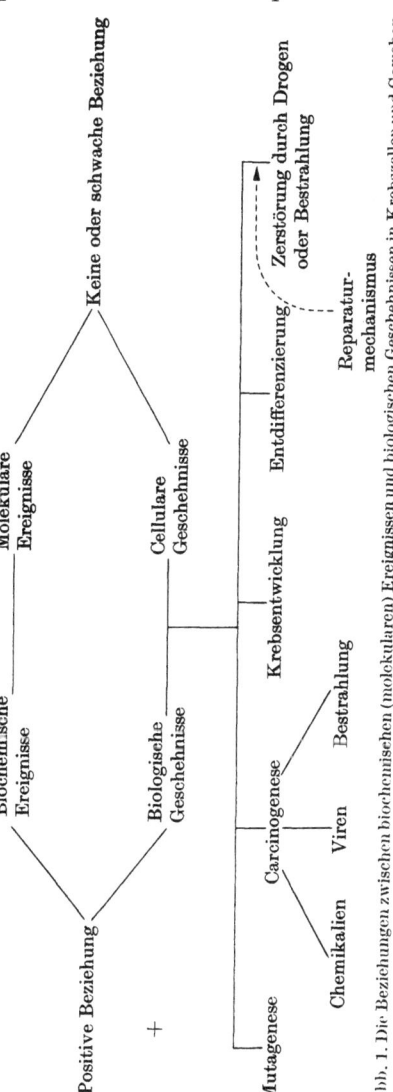

Abb. 1. Die Beziehungen zwischen biochemischen (molekularen) Ereignissen und biologischen Geschehnissen in Krebszellen und Geweben

Vorstellungen beteiligt. Herr PITOT begann seinen Vortrag mit der treffenden Bemerkung, daß nahezu jedermann eine Lieb-

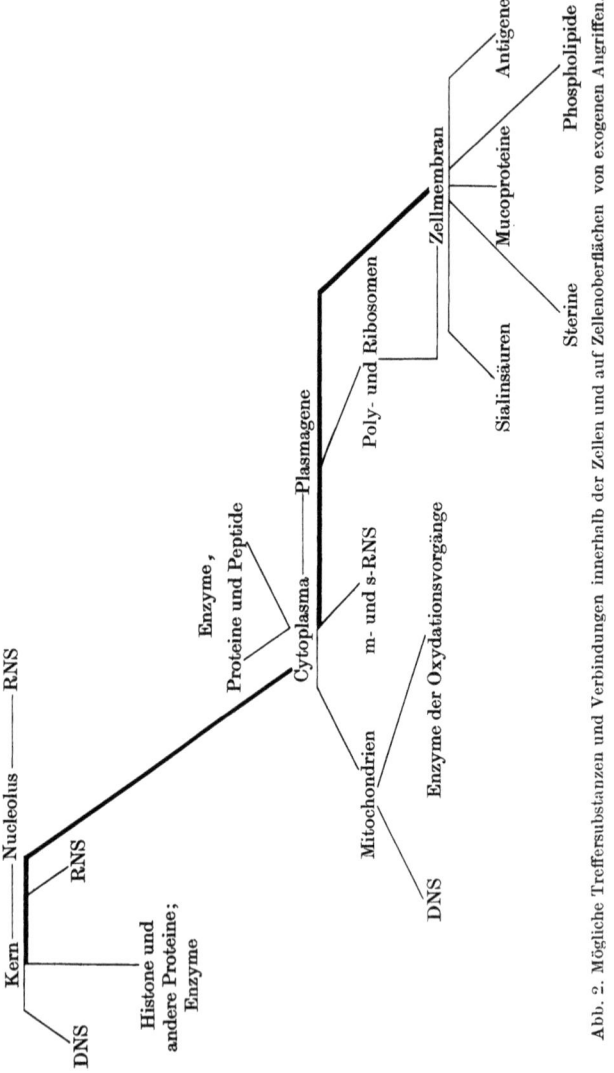

Abb. 2. Mögliche Treffersubstanzen und Verbindungen innerhalb der Zellen und auf Zellenoberflächen von exogenen Angriffen.

lingsidee über die molekularen Vorgänge der Krebsentstehung vertritt. Der Verfasser dieses Endreferates hat seine eigene

Vorzugshypothese, die den Vorteil hat, mit all den bekannten Tatsachen nicht in Widerspruch zu stehen und fast keine der vorgeschlagenen Hypothesen ablehnt: es ist nicht unwahrscheinlich, daß die Prozesse der Krebsbildung, Entwicklung und Zerstörung auf einer multimechanistischen Grundlage beruhen, d. h. daß einige oder viele der erwähnten Einzelereignisse zusammenkommen müssen, um die Verwandlung des normalen Gewebes zu verursachen, zu fixieren, zu replizieren. In dieser Hinsicht ist es ganz gleichgültig, ob die Ereignisse primär, sekundär usw. oder synchron stattfinden. Für einige muß das wie ein schwacher Kompromiß aussehen, doch ist das, so glaubt der Verfasser, kein Grund zu einer pessimistischen Auffassung.

Ausblick

Es ist viel besser und auch gerechtfertigt mit einem bescheidenen Optimismus in die Zukunft zu blicken. Einige Dinge, die während der Tagung nur leicht oder garnicht berührt wurden, sollen im Folgenden in beschränkter und persönlicher Auswahl angedeutet werden.

Der Vortrag von Herrn WARBURG endete mit der Feststellung, daß Entdifferenzierung von der Entstehung eines Gärungsstoffwechsels abhängt. Während der letzten Jahre hat die Entdeckung und Untersuchung von Isoenzymen (= Fermente verschiedener Proteinstruktur, aber mit sehr ähnlichen katalytischen Funktionen), besonders die fünf Formen der Milchsäuredehydrase (M_4 bis H_4 oder I bis V oder V bis I), eine Anzahl von Forschern mehr und mehr beschäftigt. Da eines der Tetrameren dieses Ferments, H_4, durch Brenztraubensäure stark gehemmt wird, während die Aktivität des Enzyms M_4 durch niedrige Konzentrationen dieses Substrates sogar erhöht wird (siehe u. a. WILSON, CAHN und KAPLAN, Nature 1963, *197*, 331; YASIN und BERGEL, Eur. J. Cancer, 1965, *1*, 203), ist es nicht unwahrscheinlich, daß bei Vorherrschen von M_4 — in Muskeln und in vielen Geschwülsten — ein milchsäurereiches, vielleicht gärungsgeneigtes Gewebe existiert. Diese Möglichkeit stellt eine Unterstützung der Warburg-Theorie dar und wird von neueren Untersuchungen, besonders denen von KAPLAN und seinen Mitarbeitern weiter verstärkt. Bisher besteht aber noch keine Sicherheit darüber, daß die Krebsbildung diesen Vorgängen absolut parallel verläuft. LEESE im Chester Beatty Research

Institute (Eur. J. Cancer, 1965, *1*, 209) hat unabhängig von KAPLAN et al. aufgezeigt, daß in der Schleimhaut des menschlichen Magens mit fortschreitender Verwandlung zu einer histologisch der Darmschleimhaut ähnlichen Struktur (vielleicht eine präanceröse Phase) eine Verschiebung der normalerweise dominierenden H-Tetrameren zu M-Tetrameren der Milchsäuredehydrogenase eintritt. Die Situation ist im Augenblick recht kompliziert. Studien auf diesem Gebiet sollten mit Nachdruck fortgesetzt werden und sich auch auf andere Enzyme des Kohlenhydratstoffwechsels ausdehnen. Die Untersuchungen würden es gestatten, die „Feinstruktur" der Warburg-Theorie durch Aufklärung der Beziehungen zwischen Sauerstoff, Fermentsynthese, Veränderungen der vorherrschenden Stoffwechselsysteme und den dazu parallelen morphologischen Änderungen aufzudecken.

Die der Glykolyse komplementären Vorgänge — die Atmungsprozesse — verlaufen zu einem großen Teil in den Mitochondrien. Es wurde bereits erwähnt, daß diese Zellpartikel DNS enthalten, die verschieden von der des entsprechenden Kernes ist (siehe u. a. CHEVREMONT, 1963, in Cell Growth and Cell Division, ed. R. C. J. Harris, *2*, 323, Acad. Press, New York). In den Mitochondrien finden Proteinsynthesen statt. Es ist aber noch nicht geklärt, ob außer strukturellen auch funktionelle Proteine synthetisiert werden oder ob diese Atmungsfermente und ihre Cofaktoren von anderen Zellorten einwandern. Intensive Versuche auf diesem Gebiet, erleichtert durch die sauberen Trennungsmethoden mit Hilfe der Anderson-Zentrifugen, werden vielleicht diese Teile des normalen Zellapparates und ihre Schädigung bei der Entstehung der Krebszelle aufhellen.

Es führte zu weit, würde man noch zwei weitere Ausblickspunkte erwähnen. Kurz berührt, beziehen sie sich auf das seit einiger Zeit wieder erwachte Interesse an biochemischen und biologischen Studien des *menschlichen* Krebses, wenigstens in einigen seiner Formen. Dies ist um so vernünftiger, als moderne technische Methoden diese vor einigen Jahren noch unergiebigen und wissenschaftlich nahezu unfruchtbaren Studien sehr erleichtern werden. Hinzu kommt, daß der Kliniker und Histopathologe ein steigendes Verständnis für die Tätigkeit der medizinischen Laboratoriumsforscher zeigen, die von ihrer Seite mehr und mehr bemüht sind, ihre Resultate mit den histologischen Eigenschaften des Untersuchungsmaterials zu vergleichen und in Einklang zu bringen. Das sollte zwangsläufig zu einer Verschmelzung der klinischen, pathologischen und biochemischen Sprachgebräuche führen.

Baut man in Gedanken weiter und zieht die Möglichkeit von biochemischen Unregelmäßigkeiten oder Läsionen ganzer Volksgruppen oder in einem bestimmten Gebiet lebender Individuen in Betracht, erhält man eine Vorstellung von einer Richtung, in die die Molekulare Biologie und alle anderen Fächer, die sich mit ihr an der Krebsforschung beteiligen, reisen wird. In der augenblicklichen, anscheinend verwirrenden Situation, mag das englische Sprichwort ,,It is better to travel than to arrive" als ermutigendes Leitmotiv dienen.

MIX
Papier aus verantwortungsvollen Quellen
Paper from responsible sources
FSC® C105338

If you have any concerns about our products,
you can contact us on
ProductSafety@springernature.com

In case Publisher is established outside the EU,
the EU authorized representative is:
**Springer Nature Customer Service Center GmbH
Europaplatz 3, 69115 Heidelberg, Germany**

Printed by Libri Plureos GmbH
in Hamburg, Germany